Oxidative Stress Mechanisms and their Modulation

Mohinder Bansal • Naveen Kaushal

Oxidative Stress Mechanisms and their Modulation

Mohinder Bansal
Department of Biophysics
Panjab University
Chandigarh, India

Naveen Kaushal
Department of Biophysics
Panjab University
Chandigarh, India

ISBN 978-81-322-2879-0 ISBN 978-81-322-2032-9 (eBook)
DOI 10.1007/978-81-322-2032-9
Springer New Delhi Heidelberg New York Dordrecht London

Softcover reprint of the hardcover 1st edition 2014

Printed on acid-free paper

Springer is part of Springer Science+Business Media (www.springer.com)

Preface

Both the authors have been involved in oxidative stress-related experimental research for the past many years, and therefore, an idea was generated to compile the molecular mechanisms involved in oxidative stress influencing the cell functions and further developing various pathologies.

In this collection, while drafting the contents of the book, two chapters including the basics of oxidative stress-generating systems and the regulation of cell signaling by oxidative stress have been included. In the rest of the chapters, molecular mechanism(s) linking oxidative stress to various pathologies has been taken up. In the end, future perspectives are highlighted on the subject. Limited figures have been included to highlight the important aspects of the text. At the end of each chapter, references including some reviews have also been placed for the reader's convenience for detailed reading on a particular subject matter.

Since oxidative stress continues to be a very large field of study with a great deal of new literature coming up, here the aim was to describe the main fundamental components in each section to have a generic approach. In this approach, some research aspects might be overlooked, and therefore, the authors apologize for that.

The authors feel to convey thanks to all colleagues and associated students for their direct and indirect contribution during this write-up.

This book is addressed to all scientists interested in starting to work in oxidative stress research and graduates and postdocs who want to learn quickly and refer to the molecular mechanisms involved in varying oxidative stress-related pathologies and antioxidant interactions. Also, physicians who want to understand the molecular strategy of the oxidative stress-related diseases and chemists who want to learn about cellular regulation to find agents to treat related diseases might also have interest in this book. Overall, this book is a reference aiming at young researchers and aspiring students to understand the basis and pathophysiological implication of oxidative stress.

Chandigarh, India

Mohinder Bansal
Naveen Kaushal

Contents

About the Authors

Dr. Mohinder Bansal did MSc (H) in biophysics in 1972 and PhD in 1980 on carcinogenesis from Panjab University, Chandigarh (India). He worked as a faculty member since 1978 in the Department of Biophysics, Panjab University (Chandigarh, India), and superannuated in 2008 as professor. He has also worked as a research associate and visiting faculty at Baylor College of Medicine, Houston, Texas, USA. During this period, he has worked on the isolation and characterization of selenoproteins and also published work in reputed scientific journals. His core area of the study involved molecular pathophysiology under the influence of experimental oxidative stress and its modulation in spermatogenesis and atherosclerosis. Regulation of gene expression related to the endogenous antioxidative enzymes, transcription factors, and their upstream activators (MAP kinases) was studied at transcriptional and translational level.

Dr. Naveen Kaushal is an Assistant Professor in the Department of Biophysics at Panjab University, Chandigarh. He received his bachelor's and master's degrees in biophysics(H) in 2004 and earned his PhD in biophysics from Panjab University, Chandigarh (India), in 2008. Kaushal's doctoral research demonstrated the role of selenium, an essential dietary micronutrient and antioxidant, which acts in a well-coordinated manner in maintaining the cellular redox status and male germ cell maturation during spermatogenesis. Dr Kaushal has also worked as Assistant Professor of physiology at Gian Sagar Medical College and Hospital, Punjab (India), in 2007. Thereafter, he joined The Pennsylvania State University, USA, as a postdoctoral fellow. Dr Kaushal's research has focused on scientific areas pertaining to delineate the redox-regulated molecular pathways of pathogenesis and their plausible modulation by nutritional interventions, by selenium and EPA. His research in this direction led to the discovery of a novel prostaglandin that holds potential to cure leukemia and prevent its relapse. For his contributions in the field he has been recognized as "Future Leader in Nutrigenomics-2013" by Nebraska Gateway to Nutrigenomics (NGN), USA. With his keen interest in the redox biology, he aims to translate the role of active dietary antioxidants in developing superior or adjunct therapies for various diseases.

Abbreviations

AA	Arachidonic acid
Aβ	Amyloid β
AD	Alzheimer's disease
ADPP	Advanced oxidative protein products
AGE	Advanced glycation end products
AIDS	Auto immune diseases
Akt	Protein kinase B (PKB)
ALS	Amyotrophic lateral sclerosis
AMH	Anti-Mullerian hormone
AMPK	AMP-activated protein kinase
AP-1	Activator protein 1
APE	Apurine/apyrimidinic endonuclease
Apo-A	Apolipoprotein A
APP	Amyloid precursor protein
ARE	Antioxidant responsive element
ASK1	Apoptosis signal regulated kinase 1
ATF	Activating transcription factor
ATP	Adenine triphosphate
BBB	Blood brain barrier
BCl_2	B-cell lymphoma 2 protein
BCNU	Bis-chloroethylnitoso urea
Ber	Ab1 tyrosine kinase
BH_4	Tetrahydrobiopterin
BMI	Body mass index
BMK-1	Big MAP kinase 1
Bzip	Basic region leucine zipper
cAMP	Cyclic adenosine monophosphate-Bzip
CAT	Catalase
CCCP	Carbonyl cyanide m-chlorophenyl hydrophenyl hydrazine
CCl_4	Carbon tetrachloride
CDC	Centers for disease control and prevention
Cdk5	Cyclin-dependent kinase 5
cGMP	Cyclic guanosine monophosphate
CML	Carboxy methyl lysine
cJUN	Protein coded by JUN gene
CNS	Central nervous system

Co-Q10	Coenzyme Q10
COX	Cyclooxygenase
CRE	cAMP responsive element
CRP	C-reactive protein
CSF	Cerebrospinal fluid
c-Src	Cellular tyrosine protein kinase
Cul3	Cullin-3E.3 ubiquitin ligase
CVD	Cardiovascular disease
Cys	Cysteine
DHA	Dehydroascorbic acid
DIO	Deiodinase
DNA	Deoxy nucleic acid
DNK	Deoxyribonucleotide kinase
DNPH	Dinitrophenylhydrazine
EGF	Epidermal growth factor
EGFR	Epidermal growth factor receptor
Egr-1	Early growth response protein
eNOS	Endothelial NOS
epiPGF2a	F2-isoprostanes
EpRE	Electrophile response element
ER	Endoplasmic reticulum
ERK	Extracellular signal-related kinases
ESR	Erythrocyte sedimentation rate
Ets-1	E-twenty six family transcription factor
EUK	Eukaryon
EVA	Epidemiology of vascular aging
Fas	Cell surface receptor protein
FFA	Free fatty acid
FOR	Free oxy radical
FSH	Follicle-stimulating hormone
G6PD	Glucose 6 phosphate dehydrogenase
GAPDH	Glyceraldehyde 3-phosphate dehydrogenase
GCL	Glutamyl cysteine ligase
GCS	Glutathione cysteine synthetase
GFAT	Glutamine fructose-6-phosphate amidotransferase
GLUTs	Glucose transporters
GMP	Gonadin mono phosphate
GnRH	Gonadotropin-releasing hormone
GPx	Glutathione peroxidase
GR	Glutathione reductase
GRK	G protein-coupled receptor kinase
GSH	Reduced glutathione
GSH-Px	Glutathione peroxidase
GSK-3β	Glycogen synthase kinase 3
GSSeH	Selenopersulfide
GSSG	Oxidized glutathione
GS-Se-SG	Seleno-trisulfide
GST	Glutathione S-transferase

GTP	Guanosine triphosphate
GTPase	Guanine triphosphatase
HAART	Highly active antiretroviral therapy
HAND	HIV-associated neurological diseases
HAP	Human apurine/apyrimidine endonuclease
hCG	Human chorionic gonadotropic hormone
HD	Huntington's disease
HDL	High-density lipoprotein
HFS	High-frequency stimulation
HIF	Hypoxia-inducible factor
HIV	Human immune deficiency virus
HNE	Hydroxyl nenal
HO	Heme oxygenase
H_2O_2	Hydrogen peroxide
HOCL	Hydrochlorous acid
HOO.	Hydroperoxyl radical
HRE	Hormone responsive element
H_2Se	Hydrogen selenide
H_2SeO_3	Selenite
HSF	Heat shock factor
HSP	Heat shock protein
$HtrA_2$	Serine protease coded by HTRA2 gene
IB	Inhibitory protein B
ICSH	Interstitial
ICU	Intensive care unit
IGF	Insulin-like growth factor
IHC	Immunehistochemistry
IkB	Inhibitory protein kB
IKK	IkB kinase
IL-1	Interleukin-1
iNOS	Inducible nitric oxide synthetase
IR	Insulin receptor
Iso-PGF2α	Iso-prostaglandin F2α
IRS	IR substrate
IUGR	Intrauterine growth restriction
JAK	Janus kinase
JNK	c-jun N-terminal kinases
Keap1	Kelch Ech-associated protein 1
LA	Lipoic acid
LDL	Low-density lipoprotein
LH	Luteinizing hormone
LPO	Lipid peroxidase
LPS	Lipopolysaccharide
Maf	Musculoaponeuretic fibrosarcoma
MAOA	Monoamine oxidase A
MAPK	Mitogen-activated protein kinase
MCP	Monocytic chemotactic protein
M-CSF	Monocyte colony-stimulating factor

MDA	Malondialdehyde
MI	Myocardial infarction
MKP	MAPK phosphatases
MM-LDL	Minimal oxidized LDL
MMPs	Matrix metalloproteinase
MnSOD	Manganese superoxide
MPO	Myeloperoxidase
MPP^+	1-methyl-4-phenylpyridinium
MPTP	Methyl phenyl tetrahydropyridine
Mrps	Multi-drug resistance associated proteins
MS	Multiple sclerosis
MSCs	Mesenchymal stem cells
MsrB1	Methionine-R-sulfoxide reductase
MT	Metallothionein
Myb	Myeloblastosis protein
NAC	N-acetyl cysteine
NADH	Nicotinamide adenine dinucleotide reduced
NADPH	Nicotinamide adenine di nucleotide phosphate reduced
NADPH	Nicotinamide adenine dinucleotide phosphate reduced
NES	Nuclear export signal
NFAT	Nuclear factor of activated T cells
NF-kB	Nuclear factor kB
NIK	NFkB inhibitory kinase
NO	Nitric oxide
NO_2CL	Nitryl chloride
NOS	Nitric oxide synthase
NOX	NADPH oxidase
NPSH	Non protein sulfhydryl
NQO1	NADPH quinine oxidoreductase 1
Nrf2	Nuclear factor (erythroid derived-2)-like 2
Nrf2-KO	Nrf2 knockout
$O^{\cdot -}$-	Superoxide
OFR	Oxygen free radical
O-GlcNAc	O-linked N-acetyl glucosamine
$ONOO^-$	Peroxynitrite anion
oxLDL	Oxidized LDL
oxo-dG	Oxo-2-deoxygaunosine
p21	Cyclin-dependent kinase inhibitor-1
p38	p38 kinase
p53	Protein 53
p65	Transcription factor P65
PAI-1	Plasminogen activator inhibitor-1
PCR	Polymerase chain reaction
PD	Parkinson's disease
PDGF	Platelet-derived growth factor
PDK	Phosphoinositide-dependent protein kinase
PDTC	Pyrrolidine dithiocarbamate
PECAM	Platelet endothelial cell adhesion molecule

PHGPx	Phospholipid glutathione peroxidase
PI3K	Phosphoinositide 3-kinase
PLA_2	Phospholipase A_2
PINK	PTEN-induced kinase
PIP3	Phosphatidylinositol triphosphate
PKA	cAMP-dependent kinase
PKC	Protein kinase C
PKG	cGMP-dependent protein kinase
PMA	Phorbol 12-myristrate 13-acetate
POF	Premature ovarian failure
PON	Para oxygenase
PPAR	Peroxisome proliferator-activated receptor
PTEN	Phosphatase and tensin homologue
PTK	Protein tyrosine kinase
PTP	Protein tyrosine phosphatase or permeability transition pore
PUFA	Poly unsaturated fatty acid
RA	Rheumatoid arthritis or retinoic acid
RAGE	Receptor of AGE
RARE	Retinoic acid responsive element
Ref-1	Redox factor 1
RLS	Reactive lipid species
RNS	Reactive nitrogen species
ROO.	Peroxyl radical
ROS	Reactive oxygen species
rT_3	Reverse T_3
RTK	Receptor tyrosine kinase
rTMS	Repetitive transcranial magnetic stimulation
SAPK	Stress-activated protein kinase
SBP	SECIS-binding protein
Se	Selenium
Sec	Selenocysteine
SELECT	Selenium and vitamin E cancer trial
Sep	Selenoprotein
Ser	Serine
sGC	Guanylate cyclase
SHP-1	Src homology-2-domain protein
SICIS	Selenocysteine insertion sequence
SIRS	Systemic inflammatory response syndrome
SLN	Solid lipid nanoparticles
SMC	Smooth muscle cells
SNP	Single nucleotide polymorphism
SOD	Superoxide dismutase
Sp-1	Specificity protein 1
SR-A	Scavenger receptor A
T_3	Triiodothyronine
T_4	Tetra iodothionine(thyroxine)
TBARS	Thiobarbituric acid reacting substance
tBHQ	Tert-butyl hydroquinone

tDCS	Transcranial direct current stimulation
TF	Transcription factor
TG	Triglyceride
TGF	Transforming growth factor
TGR	Thioredoxin glutathione reductase
TH	Tyrosine hydroxylase
Thr	Threonine
TNF	Tumor necrosis factor
TPA	Phorbol-12-myristate-13acetate
TPR	Tetratricopeptide
TRAF	TNFα receptor associated factor
TRAPS	TNF receptor associated periodic syndrome
TRE	TPA response element
Trx	Thioredoxin
TrxR	Thioredoxin reductase
Tyr	Tyrosine
UGT	UDP-glucuronosyl transferase
uPA	Urokinase plasminogen activator
UPR	Unfolded protein response
UV	Ultraviolet
VCAM	Vascular cell adhesion molecules
VEGF	Vascular endothelial growth factor
VLDL	Very-low-density lipoprotein
VSMC	Vascular smooth muscle cells
XO	Xanthine oxidase
XOR	Xanthine oxidoreductase

Introduction to Oxidative Stress 1

Oxygen is one of the most abundant and essential elements for all the life forms on the earth. It is critical for the energy production in both prokaryotes and eukaryotes via electron transport chain. Not only in its diatomic form (O_2) in the atmosphere but also as a triatomic molecule (O_3) such as ozone, oxygen has been beneficial for the existence of organisms on earth (Kohen and Nyska 2002: 31). Its role in survival is linked to its high redox potential, which makes it an excellent oxidizing agent capable of accepting electrons easily from reduced substrates. Often referred as the Janus gas, oxygen, however, has both positive benefits and potentially damaging side effects for biological systems. Therefore, oxygen is also often referred as "necessary evil," "essential poison," "dangerous friend," and so on. These nomenclatures may appear paradoxical due to the known functions of this element; however, research over the decades has unraveled that otherwise is also true and oxygen is toxic. This existence of concept of "oxygen paradox" spanning the two extremes of physiological spectrum – from essential to toxic as prooxidant – is an index of equilibrium between the production of "oxygen-centered radicals," and the effective physiological strategies involved in the removal of these species leading to pathological effects is called "oxidative stress" (Hauptman and Cadenas 1997). This paradoxical effect of oxygen on life forms necessitated the evolution of antioxidant systems to protect against over-oxidation and to combat reactive oxygen species (ROS).

In the recent years, the classical term "oxidative stress" indicating the imbalance between the oxidant exposure and antioxidant protections (Forman and Torres 2001: 24) has been redefined in more contemporary terms. Recently, Jones (2008) has redefined oxidative stress as "two different mechanistic outcomes, macromolecular damage and disruption of thiol redox circuits leading to the aberrant cell signaling and dysfunctional redox control" (Jones 2008: 28). However, both the traditional and contemporary concepts of oxidative stress hold the common basis suggesting this phenomenon to be central to pathogenesis, and the biomolecular damages are linked to the free radicals.

Free radicals are small, diffusible molecules that are highly reactive because of the unpaired electron (Jones 2008: 28). Though free radicals were originally considered to be oxygen-centered radicals called the reactive oxygen species (ROS), they also include a subgroup of reactive nitrogen species (RNS) which are also the products of normal cellular metabolism. These free radicals act as the secondary messengers that interfere with the normal physiological processes at multiple levels to initiate a cascade of harmful chain reactions that propagates to cause molecular damage to biological tissues and signaling mechanisms. Both ROS and RNS have well-recognized beneficial and deleterious effects depending upon the concentrations (Valko et al. 2006: 161). Under the normal physiological conditions, these molecules are produced at low/moderate concentrations and

M. Bansal and N. Kaushal, *Oxidative Stress Mechanisms and their Modulation*,
DOI 10.1007/978-81-322-2032-9_1, © Springer India 2014

have a positive role in the mitogenic responses and cellular responses to noxia, defenses against infections, etc. (Valko et al. 2007: 160). Alternatively, the overwhelming production of these molecules or failure of their combating strategies leads to oxidative stress-mediated damage to lipids, proteins, and DNA culminating in pathological conditions and aging.

It is evident that the delicate balance between the beneficial and harmful effects of free radicals is a very important aspect of the living organisms and is achieved by mechanism called "redox regulation" which offers protections against oxidative stress and maintains the "redox homeostasis" by regulating oxidative stress and the redox status in vivo. Presently we discussed the molecular mechanisms of the oxidative stress, its physiological role in pathogenesis, and strategies to modulate it. However, before addressing these subject matters, it is mandatory to be familiar with the nature, cellular sources, molecular markers, and the basic outcomes of free radical mediated oxidative stress. In the next sections, we provide a brief and compact description of these areas to familiarize with the various aspects of oxidative damage.

Sources of Oxidative Stress

Free radicals are the basic units or quanta of oxidative stress. Varied sources of free radicals have been discovered which can essentially be classified into exogenous and endogenous (Fig. 1.1). Exogenous sources primarily include ionization radiations, drugs, pollutants, xenobiotics, toxins, etc. (Fig. 1.1). However, the endogenous sources and mechanisms involved therein are more complex and extensive, spanning the lifetime of each cell in an organism.

Mitochondria

The most important and key source of ROS production and hence oxidative stress is mitochondria (Richter et al. 1995: 38). Mechanistically, the ROS production occurs during the physiological process of ATP generation, i.e., respiratory chain. During this process, the molecular ground-state oxygen can be activated to form singlet oxygen ($^1O_2$), by means of energy transfer, or by electron transfer, forming the superoxide anion radical ($\bullet O_2^-$) (Boveris 1984: 11; Chance et al. 1979: 14). Subsequently, $\bullet O_2^-$ can be converted into other ROS and RNS by a cascade of other enzyme-catalyzed and spontaneous reactions. Most of these reactions are in fact a kind of homeostatic or physiological response imposed in an effort to combat these free radicals to avoid their deleterious effects. The various components of this homeostatic response are cumulatively referred to as "antioxidant defense system."

The first-line of defensive enzymes which act on $\bullet O_2^-$ generated in mitochondria are manganese superoxide dismutase (Mn-SOD) in mitochondria and copper–zinc SOD (Cu/Zn-SOD) in the cytosol. These enzymes not only nullify the highly reactive and toxic $\bullet O_2^-$ radicals but also are the secondary

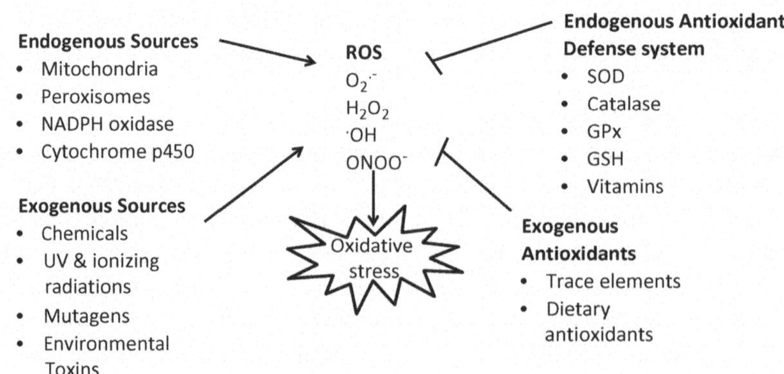

Fig. 1.1 Sources of reactive oxygen species (ROS) leading to induction of oxidative stress

sources of ROS by converting $\bullet O_2^-$ molecules to H_2O_2 (Faraci and Didion 2004; Mendez et al. 2005). H_2O_2 thus produced has various outcomes and has been discussed elsewhere in this section.

Cellular Oxidases (NOX and XOR)

Although mitochondrial respiratory chain is the major source of $\bullet O_2^-$, these molecular species can also be generated by one-electron reduction of oxygen by several different oxidases under certain conditions (Guzik et al. 2002; Mehta et al. 2006: 132). These oxidases include NAD(P)H oxidase (NOX family) and xanthine oxidase (XO). NOX enzymes are present in the lymphocytes, fibroblasts, endothelial cells, myocytes, and chondrocytes, where moderate amounts of ROS are produced and serve as a regulator of cell responses (Preiser 2012: 144). In response to infections or microbial invasion, the NOX family of enzymes is activated followed by a respiratory burst. These series of events lead to the increased oxygen consumption, glucose utilization, and increased production of reduced nicotinamide phosphate dinucleotide (NADPH) by the pentose phosphate pathway (Babior et al. 2002: 9). NADPH thus formed serves as an electron donor to an activated NADPH oxidase enzymatic complex in the plasma membrane to produce superoxide radicals ($\bullet O_2^-$) from the oxygen molecule (Kohen and Nyska 2002: 124).

Xanthine oxidase (XO) a cytosolic, nonheme enzyme is another prominent source of $\bullet O^-_2$ and H_2O_2 especially during the hypoxic conditions. Xanthine oxidase is a xanthine oxidoreductase (XOR), which exists primarily in the dehydrogenase form under normal physiological conditions. However, during hypoxia, XOR is converted to an oxidase form that can produce $\bullet O^-_2$ and hydrogen peroxide (H_2O_2) by using O_2 as an electron acceptor (Poss et al. 1996: 143).

Metal-Catalyzed Reactions

H_2O_2 produced in the above reactions can have a variety of cellular fates. It can be detoxified to H_2O and O_2 by the glutathione peroxidase (GPx) (in the mitochondria in conjunction with glutathione reductase) and catalase in peroxisomes, or it can act as a precursor for the more reactive species such as highly reactive hydroxyl radical (HO·) (Mohora et al. 2007). HO• formed during this reaction is the strongest oxidizing agent known and reacts with organic molecules at diffusion-limited rates (Kohen and Nyska 2002: 31).

Haber and Weiss illustrated that the superoxide and hydrogen peroxide will form the destructive hydroxyl radical (HO•) and initiate the oxidation of organic substrates by Haber–Weiss reaction (Haber and Weiss 1934).

$$\bullet O_2^- + H_2O_2 \rightarrow HO\bullet + OH^- + O_2$$
$$(\text{Haber} - \text{Weiss reaction})$$

However, this reaction of creating the hydroxyl radical requires a metallic catalyst (Cu^{2+} or Cu^{3+}) to proceed and is a combination of the transition metal mediated, chemical reactions called Fenton reaction (Liochev and Fridovich 2002).

$$Fe^{3+} + \bullet O_2^- \rightarrow Fe^{2+} + O_2$$

$$Fe^{2+} + H_2O_2 \rightarrow Fe^{3+} + OH^- + HO\bullet$$
$$(\text{Fenton reaction})$$

The bioavailability of ferrous ions is the rate-limiting step in this reaction, but the recycling of iron from the ferric to the ferrous form by a reducing agent such as superoxide ions can maintain an ongoing Fenton reaction, leading to the generation of hydroxyl radicals (McKersie 1996). In the presence of trace amounts of iron, the other transition metals may also participate in these electron transfer reactions by cycling between oxidized and reduced states.

Myeloperoxidase (MPO)

Similar to the above reaction, H_2O_2 can also be converted to another free radical HOCl by reacting with Cl^- ions by the enzyme myeloperoxidase (MPO)-catalyzed reaction (Hawkins et al. 2001: 25).

$$H_2O_2 + Cl^- \xrightarrow{MPO} HClO + OH\bullet$$

MPO is present mainly in the neutrophils and at lower levels in monocytes and eosinophils. Upon infections and inflammatory conditions, MPO is released from the neutrophils, and its concentration in the blood is a marker of neutrophil activation and oxidative insult (Preiser 2012: 37).

NO Synthases (NOS)

The enzyme nitric oxide synthase (NOS) produces NO• from O_2 and L-arginine, in the presence of NADPH, calcium, and/or biopterin as cofactors (Vincent et al. 2000: 46). NOS exists in three forms: endothelial NOS (eNOS) (Lamas 1992: 32), neuronal NOS (nNOS) (Bredt et al. 1991: 12), and inducible NOS (iNOS) (Xie et al. 1992: 47). eNOS and nNOS are constitutive "housekeeping" enzymes, which maintain the physiological levels of NO• needed for vascular tone and for the regulation or neurotransmission. However, iNOS is activated during the inflammatory conditions in macrophages, neutrophils, endothelial cells, smooth muscle cells, and hepatocytes and releases much higher amounts of $NO^{\bullet}$ (Preiser 2012: 37) after stimulation by the cytokines, lipopolysaccharides, and other immunologically relevant agents (Bogdan et al. 2000). Under certain circumstances, NOS can produce $^{-}O_2\bullet$ together with NO•, increasing the risk of in situ generation of $ONOO^-$ (Preiser 2012: 37).

Other Sources

In addition to the above-mentioned noteworthy ROS sources, a plethora of other "radical" enzymes present in many cell types and tissues contribute toward the oxidative stress. These include mixed-function oxidases of the endoplasmal reticulum, the cytosolic enzymes such as lipoxygenases or cyclooxygenases, the peroxisomal enzymes (glycolate oxidase, D-amino acid oxidase, urate oxidase, fatty acyl Co-A oxidase), and even DNA methylating enzymes and enzymes involved in the synthesis of hormones and neurotransmitters.

Cumulatively oxidative stress is a broad term used to define the "outcome of cascade of ROS-mediated numerous chain reactions." These ROS themselves and by oxidizing the other molecules also act a source of secondary ROS. The next section is focused on the oxidants and various types of ROS that cause changes in the redox status of the cell leading to oxidative stress.

Cellular Redox Status: Free Radicals and Oxidative Stress. What Are Oxidants? Various Oxygen Radicals

Owing to its electronic configuration, oxygen is prone to gain electrons and is thus a potent oxidant. However, kinetic considerations limit the reactivity of the dioxygen molecule O_2. During the respiration process, O_2 is progressively reduced by a controlled supply of four electrons to yield water. During the reduction of oxygen to water in the normal biological system, the electrons are transferred either from the electron transfer chain (four-electron reduction) or at random from the organic/inorganic species in their immediate vicinity (one-electron reduction). However, the incomplete reduction of O_2 is possible and leads to the formation of chemical entities that are still potent oxidants. This monovalent reduction of dioxygen involves the sequential addition of four single electrons which at each intermediate stage result in the production of potentially damaging molecular species. These molecules are known as ROS. Following a one-, two-, or three-electron reduction, O_2 may generate successively superoxide radical anion $\bullet O^-_2$, H_2O_2, or hydroxyl radicals (HO•). The modern use of the term ROS includes both oxygen radicals and nonradicals that are easily converted into free radicals (O_3, H_2O_2, $^1O_2$) (Halliwell 2006).

ROS generation is a part and parcel of all the aerobic cells involved in routine physiological processes, and the maintenance of a reduced cellular microenvironment is essential for cells' well-being. The redox status of the cell is maintained by the critical balance between the amount of oxidants and the efficient availability of antioxidant strategies to combat these oxidants. Under the physiological conditions, ROS produce subtle and/or transient changes in the cellular redox state. At low physiological concentrations, these reactive oxidants have biopositive effects and act selectively. Besides being metabolic intermediates, they are also involved in gene regulation, cellular growth, and signal transduction cascades (Saran and Bors 1989: 39; Demple and Amabile-Cuevas 1991: 18; Joseph and Cutler 1994: 29). Furthermore, they play an essential role in the microbial defense and immunological surveillance. But exposure to a changing environment routinely causes cells to face conditions that shift their redox status to a more oxidized state. This shift characterizes the cellular condition known as oxidative stress. Excess ROS react with and modify all classes of the cellular macromolecules and critical cellular targets leading to behavioral abnormalities, cytotoxicity, and mutagenic damage (Sohal and Allen 1990: 41; Floyd 1991: 88). For these reasons, the aerobic organisms wage a constant battle to maintain redox homeostasis. Maintaining this balance becomes even more when cells are exposed to the exogenous oxidants such as ultraviolet and ionizing radiation, heavy metals, and redox-active chemicals, anoxia, and hyperoxia, all of which increase ROS production (Floyd 1991: 88; Carney et al. 1991: 13).

Earlier we have discussed that ROS are generated from the molecular oxygen either by exposure to endogenous or exogenous oxidant sources such as ultraviolet radiation or chemical reduction of oxygen by the cellular oxidases, peroxidases, and mono- and dioxygenases or by the mitochondrial electron transport chain (Janssen et al. 1993: 26). These ROS include superoxide ($\bullet O_2^-$) anion, hydrogen peroxide (H_2O_2), peroxyl (ROO•) radicals, and the very reactive hydroxyl (OH•) radicals. In addition, there are nitrogen-derived free radicals, e.g., nitric oxide (NO•) and peroxynitrite anion ($ONOO^-$) also referred to as RNS, which form a class of ROS (Darley-Usmar et al. 1995: 15; Davidson et al. 1997: 16). These free radicals are the central key players regulating the molecular pathways of oxidative stress and downstream events.

Singlet Oxygen ($^1O_2$)

It is generated from the molecular ground-state oxygen by means of energy transfer (Boveris 1984: 11; Chance et al. 1979: 14). In biological systems, inflammatory processes and photosensitization are the major sources of $^1O_2$ (Klotz et al. 2000). A series of reactions involving myeloperoxidase (MPO) leads to the formation of $^1O_2$ (Steinbeck et al. 1992: 42). Alternatively, singlet oxygen may also be derived from the spontaneous dismutation of superoxide formed in NAPDH oxidase reaction (Steinbeck 1993: 43). Other reactions that can form singlet oxygen are disproportionation of the hydrogen peroxide with peroxynitrite or hypohalites and the reaction of hydroperoxides with peroxynitrite (Di Mascio et al. 1994: 79, 1997: 20; Kanofsky 1989: 119):

$$2O_2 + NADPH \xrightarrow{\text{NADPH Oxidase}} 2O_2^- + NADP^+ + H^+$$

$$2O_2^- + 2H^+ \xrightarrow{\text{Spontaneous}} H_2O_2 + {}^1O_2$$

$$2O_2^- + 2H^+ \xrightarrow{\text{SOD}} H_2O_2 + O_2$$

$$H_2O_2 + Cl^- \xrightarrow{\text{MPO}} H_2O + OCl^-$$

$$H_2O_2 + OCl^- \longrightarrow {}^1O_2 + H_2O + Cl^-$$

The singlet oxygen radicals are harmful and exert toxicity by causing DNA damage particularly oxidation of guanine residues (Piette 1991: 35) leading to G:C to T:A transversions during replication and then to mutations (Decuyper-Debergh et al. 1987: 17; Piette 1991: 35; Epe et al. 1996: 21; Jeong et al. 1998: 27).

Superoxide Radical ($\bullet O^-_2/HO\bullet_2$)

Molecular oxygen (dioxygen) possesses a unique configuration and addition of one electron to it forms the superoxide anion radical ($\bullet O^-_2$). Superoxide radicals are produced mostly in mitochondrial electron transport chain during ATP generation. During this energy transduction, some electrons "leak" prematurely and react with oxygen to form oxygen free radicals called superoxide radicals. These superoxide radicals are considered the "primary" ROS and can generate "secondary" ROS by interacting with other molecules directly or indirectly through enzymatic or metal-catalyzed reactions. Interestingly, the fate of these species varies depending on the environment and pH. Superoxide can exist in the form of either $\bullet O^{-}_2$ or, at low pH, hydroperoxyl ($HO\bullet_2$) (Bielski and Cabelli 1995; Halliwell and Gutterdgem 1999; Schafer and Buettnergr 2001: 148).

Although under physiological pH most of the superoxide is in the charged form unlike hydroperoxyl (HO_2), however, these molecules are physiologically important due to their ability to easily penetrate biological membranes than the charged form. $.O^{-}_2$ also acts as a powerful nucleophile also and is capable of attacking positively charged centers, and as an oxidizing agent, it can react with compounds capable of donating H^+ (e.g., ascorbate and tocopherol). More importantly, superoxide radicals can undergo dismutation where one superoxide radical reacts with another superoxide radical, leading to the formation of oxygen and hydrogen peroxide (Bielski et al. 1985).

Hydrogen Peroxide (H_2O_2)

It is the product of dismutation reaction of superoxide radicals. Under physiological conditions, peroxisomes are the major consumers and producers of H_2O_2. There are some enzymes that can produce H_2O_2 directly or indirectly. Since H_2O_2 contain free electrons and are not free radicals per se, they are considered reactive oxygen metabolites, because of their ability to react with the biomolecules and hence damage to the cells. Furthermore, the fact that they are freely dissolved in aqueous solution and can easily penetrate biological membranes makes them highly deleterious (Halliwell and Gutterdgem 1999; Halliwell 2000: 48). Although peroxisomes containing H_2O_2 also contain catalase, an enzyme that decomposes H_2O_2, the existence of a strikingly delicate balance is maintained to ensure no net production of ROS. If this balance is disturbed, H_2O_2 can directly cause degradation of the heme proteins, release of iron, inactivation of enzymes, and oxidation of DNA, lipids, ^{-}SH groups of proteins, and keto acids (Halliwell 2000: 48). In addition to possessing direct oxidizing properties, H_2O_2 also serves as a source for more toxic species, such as OH• or HClO.

Hydroxyl Radical (OH•)

The hydroxyl radical is a neutral form of hydroxide ion having extremely high reactivity, making it a very dangerous radical (Bielski and Cabelli 1995; Halliwell and Gutterdgem 1999; von Sonntag 1987). Unlike superoxide radicals that are relatively stable, hydroxyl radicals are short-lived species with a half-life of 10^{-9} s and therefore act with molecules at close proximity with high affinity (Pastor et al. 2000: 49; Pastor et al. 2000). OH• is produced by the two major biochemical reactions called Fenton reaction and Haber–Weiss reaction (discussed earlier). Under the conditions of stress, superoxide radicals release "free iron" from iron-containing molecules which reacts with H_2O_2 by the Fenton reaction to form hydroxyl radicals. Superoxide radicals also participate in Haber–Weiss reaction which is a combination of Fenton reaction and further reduction of Fe^{3+} by superoxide, leading to the formation of Fe^{2+} and oxygen. OH• is one of the most powerful oxidizing agent and can cause damage to almost all organic and inorganic molecules in the cell, including DNA, proteins, lipids, amino acids, sugars, and metals by hydrogen abstraction, addition, and electron transfer (Halliwell and Gutterdgem 1999).

Peroxyl Radicals (ROO•)

Hydroperoxyl radical (HOO•) is a protonated form of the superoxide radical ($\bullet O^{-}_2$) and initiates the process of fatty acid peroxidation in living systems. Though these radicals are present in minute quantities (~0.3 % of total superoxide), it exacerbates the oxidative stress by secondary reactions especially lipid peroxidation, thus leading to formation of other numerous ROS and RNS (De Grey 2002: 50; Aikens and Dix 1991: 51).

Reactive Nitrogen Species (RNS)

RNS are nitrogen-containing free radicals which possess high oxidizing potential and thus are involved in the oxidative stress. They are often classified as part of ROS, but the term RONS (reactive oxygen and nitrogen species) has also been used in the literature. The toxic effects of these molecules are often referred as "nitrosative stress" (Klatt and Lamas 2000: 122; Ridnour et al. 2004: 146). They mainly cause nitrosylation of the proteins, leading to alterations in their structure and function.

The major RNS include nitric oxide (NO•) and nitrogen dioxide ($•NO^{-}_{2}$), as well as nonradicals such as peroxynitrite ($ONOO^{-}$) besides others (Beckman and Koppenol 1996; Czapski and Goldstein 1995: 72; Halliwell and Gutterdgem 1999; Halliwell et al. 2000; Murphy et al. 1998: 137; Patel et al. 2000: 141). Out of these NO• containing an unpaired electron on antibonding, $2\pi^{*}_{y}$ orbital is the most discussed and studied. NO• is produced by oxidation of L-arginine to L-citrulline by the enzyme nitric oxide synthase (NOS) (Ghafourifar and Cadenas 2005: 91). Similar to the highly reactive hydroxyl (OH•) radicals, they are highly reactive and have a short half-life in aqueous environment. Since it is also a soluble lipid, it can diffuse through membranes and thus is critically important in neuronal signaling. NO• plays a critical role in a variety of physiological processes such as blood pressure regulation, immune regulation, and other defense strategies of the cell (Bergendi et al. 1999: 59).

In addition to the above-mentioned direct effects, NO• can also react with other radicals and molecules under certain conditions. For example, during inflammatory processes triggered by the oxidative burst, both NO• and $•O^{-}_{2}$ are produced and can react to form highly potent oxidizing agent called peroxynitrite radical ($ONOO^{-}$). Although this reaction is important for maintenance of redox status by keeping the levels of superoxide radicals and other ROS in check, the peroxynitrite radical thus formed can however cause DNA fragmentation, lipid oxidation, protein oxidation, and nitration (Carr et al. 2000: 68; Valko et al. 2007: 160):

$$NO\bullet + \bullet O_2^{-} \rightarrow ONOO^{-}$$

Additionally NO• can also react with the H_2O_2 and HClO to form nitrogen-containing molecular derivatives such as N_2O_3, NO^{-}_{3}, and NO^{-}_{2}. Protonated peroxynitrite (ONOOH) derived from NO• and $•O^{-}_{2}$ can also react with H_2O_2 or CO_2, to form adducts having numerous toxic effects on the biological systems (Kohen and Nyska 2002: 124).

Physiological Markers of Oxidative Stress

Evaluation of the oxidative stress holds great significance and has been a major challenge due to lack of sensitive and robust methods to accurately measure the levels of ROS and cellular defense systems. Another greater problem is the transient and reactive nature of ROS, which makes it even harder to monitor them in biological matrices. Over the past few years, research interest has evolved to discover new and better physiological/biochemical markers for oxidative stress. The main emphasis is to develop efficient and sensitive methods to analyze the ROS and also to make the measurement more reliable, stable, and convenient. These methods not only can delineate mechanisms by which ROS works but also help draw the correlations between the impact of ROS and their respective clinical outcomes. A number of major oxidative stress markers have been reported (Halliwell and Whiteman 2004: 99, 2009: 97), majority of which are the oxidation products of cellular and biomolecular components such as proteins (Halliwell and Gutterdgem 1999; Levine and Stadtman 2001: 127), lipid membranes (Davis 1987: 75; Halliwell and Gutterdgem 1999), and DNA (Beckman and Ames 1997: 58; Halliwell and Gutterdgem 1999).

The biomarkers for the oxidative stress can be classified in two distinct ways: first, based on the biochemical nature of the molecular targets of oxidative stress such as protein, lipids, DNA, and carbohydrates, and, second, based on the oxidation products formed as a result of enzymatic breakdown of cellular components as a result of

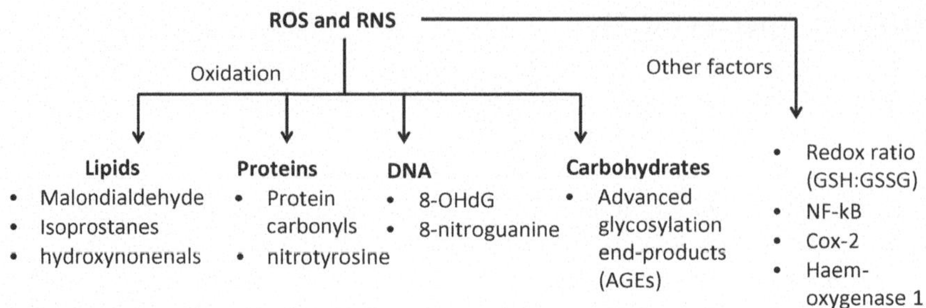

Fig. 1.2 Molecular markers of oxidative stress

modulation in the redox signaling mechanisms. The latter can be subcategorized into three classes:

1. Biomolecules modified by the free radicals, such as reactive carbonyls and 4-hydroxy-2-nonenal from protein oxidation, malondialdehyde from lipids, and 8-oxo-2-deoxyguanosine (8-oxo-dG), are derived from the nucleic acids. These are considered very sensitive and efficient markers of oxidative stress, owing to the fact that their concentrations are proportional to dose, and they are detected at the sites of free radical attack (Fig. 1.2).
2. Second, the physiological antioxidant defense system enzymes and molecules such as catalase and reduced glutathione (GSH) which are associated with the free radical metabolism (Fig. 1.2).
3. Finally, transcriptional factors are included, such as nuclear factor-κB (NF-κB) and c-myc, which are modulated by these radicals (Fig. 1.2).

Here we have discussed the most common and routinely used markers, taking both the classification criteria in consideration.

8-Hydroxydeoxyguanosine

ROS-mediated DNA strand breaks or modification to deoxyribose sugar and bases leads to the formation of several common DNA adducts such as 8-hydroxydeoxyguanosine, 8-hydroxyguanine, and DNA-MDA adducts (Bruskov et al. 2002: 65; Gedik et al. 2002: 90). 8-hydroxydeoxyguanosine (8-OHdG) is an oxidized form of the guanine, resulting in DNA damage causing mutations. This compound causes A:T to C:C or G:C to T:A transversion mutations because of its base pairing with the adenine as well as cytosine. Higher levels of 8-OHdG have been detected in cells targeted by the free radicals which have been associated with tumorigenesis indicating that oxidative stress may play a role in tumor progression (Takahashi et al. 1998: 155; Iida et al. 2001: 108). On the contrary, it has been suggested that while 8-OHdG is implicated in G-C to T-A transversions, this marker does not always correlate with the genotoxicity (Upham and Wagner 2001: 159). However, despite these contradictions linking them to tumorigenicity, 8-OHdG is a good indicator of oxidative stress along with other metabolic markers. In fact, the measurement of urinary 8-OHdG has been considered to reflect the whole body oxidative DNA damage (Halliwell and Whiteman 2004: 99).

8-Nitroguanine

Similar to the 8-OHdG, 8-nitroguanine is generated by nitration of guanine and its related nucleosides and nucleotides in the free forms or in DNA/RNA by RNS such as oxides of nitrogen (NOx) and peroxynitrite (ONOO-) (Ohshima et al. 2006: 139). Unlike normal tissues, enhanced levels of 8-nitroguanine have been detected in the nucleus of inflammatory cells and/or epithelial cells in inflamed tissues. These DNA metabolites are the most widely used approaches to quantify oxidative stress in animals and humans and form the basis for the majority of assays aimed at measuring oxidative DNA damage.

Protein Carbonyls

Studies suggest that the ROS generated during oxidative insult can potentially react with amino acid side groups as well as can cleave the polypeptide backbones (Garrison 1987). This oxidation of protein results in the formation of reactive carbonyl groups (ketones and aldehydes), which can be easily tracked experimentally and is considered as one of the best markers for oxidative stress. These carbonyl groups are mainly produced by reaction between amino acid side groups (usually Lys, Arg, Pro, or Thr) and hydroxyl radicals but can also result from the reaction between ROS and lipids (4-HNE and MDA) or carbohydrates (ketoamines and ketoaldehydes) (Berlett and Stadtman 1997: 60). Studies are suggestive of the fact that proteins are the major initial targets of ROS and the protein carbonyls thus formed have been proposed as a "sign of disease-associated dysfunction" (Shacter 2000). High levels of protein carbonyls have been detected in patients with neurodegenerative diseases, diabetes, hypercholesterolemia, arthritis, etc., suggesting carbonated proteins as biomarkers of early diagnosis of these diseases. The reason that protein carbonyl groups have been the most widely used and studied signature of protein oxidation is because they are very stable and can be readily detected by their reaction with 2–4 dinitrophenylhydrazine (DNPH).

Inducible Nitric Oxide Synthase

Nitric oxide is synthesized in a variety of tissues via the catalytic activity of nitric oxide synthase (NOS). This enzyme exists in three forms, namely, neuronal NOS, endothelial NOS (eNOS), and inducible NOS (iNOS). The inducible form, iNOS, may be induced by endotoxins and/or inflammatory cytokines and is considered as a marker of stress (Moncada and Higgs 1993: 135). Increased and extensive levels of iNOS have been associated with the human tumorigenesis and infiltrating macrophages at the sites of chronic active inflammation leading to enhanced NO production. The expression profiling of iNOS with other markers lends strong credence to the fact that oxidative stress is a key player in various pathological states including carcinogenesis (Gottschling et al. 2001: 93).

Nitrotyrosine

As discussed earlier, the NO• produced by iNOS activity further reacts with a superoxide radical to form $ONOO^-$ (Beckman and Koppenol 1996). This in itself or its secondary metabolites can cause tyrosine nitration in protein, creating 3-nitrotyrosine, a footprint of in vivo oxidation/nitration of proteins. Evidences indicate that the elevated levels of 3-nitrotyrosine occurs in diseases associated with ROS/RNS such as atherosclerosis, myocardial ischemia, inflammatory bowel disease, and amyotrophic lateral sclerosis, as well as in toxic and carcinogenic models (Beckman and Koppenol 1996; Knight et al. 2001: 123; Weinstein et al. 2000: 164). The successful detection of 3-nitrotryrosine-containing proteins in vivo using a qualitative proteomics approach offers an early diagnostic tool for diseases by defining patterns of abnormal proteins.

Malondialdehydes

Malondialdehyde (MDA) is one of the end products of lipid peroxidation in the cell membranes or in low-density lipoproteins (LDL) (Halliwell and Gutterdgem 1999). Levels of MDA are often measured by thiobarbituric acid-reacting substance (TBARS) assay. Since some aldehydes other than MDA can also be generated in peroxidizing lipids and have the same range of absorption as MDA, TBARS assay can be confounded by these chromogens (Halliwell and Gutterdgem 1999). Despite these contradictions, MDA assay remains one of the most widely and commonly used assay to establish oxidative damage.

F2-Isoprostanes

Isoprostanes are a group of bioactive prostaglandin-like compounds generated via a nonenzymatic free radical-initiated peroxidation of arachidonic acid in vivo. Out of the numerous isoprostanes formed, F-series isoprostanes especially 8-iso-PGF2α have been suggested as specific, reliable, and noninvasive markers of lipid peroxidation in vivo (Halliwell 2000: 96; Milne et al. 2005: 133). These isoprostanes can be easily measured in most biological fluids such as plasma, urine, exhaled breath condensate, and induced sputum. Furthermore, 8-iso-PGF2α is very stable in isolated samples (Griffiths et al. 2002: 95), and its levels are not influenced by lipid content in the diet unlike MDA (Gopaul et al. 2000: 92). The increased levels of 8-iso-PGF2α have been suggested to play a causative role in oxidative damage diseases like cardiovascular diseases, allergic asthma, hepatic cirrhosis, and Alzheimer's disease (Milne et al. 2005: 133; Montuschi et al. 2004: 136).

Oxidative Products of Sugars

Advanced glycation end products (AGEs) are products of the nonenzymatic glycation of proteins by reducing sugars. These AGEs accumulate in plasma and tissues as a result of aging, diabetes, renal failure, and other pathological states (Halliwell and Gutterdgem 1999; Wu 1993: 167; Miyata 1997: 134). AGEs are considered as biomarkers for the glycemic control, risk of diabetes-associated complications, mortality from cardiovascular disease, and coronary heart diseases in nondiabetics (Kilhovd et al. 2007: 121) and also an index to measure the treatment effect of diabetic patients with retinopathy, nephropathy, and neuropathy (Wu 1993: 167). Additionally, carboxymethyllysine (CML) and pentosidine which are the products of oxidation-accompanied glycation have also been regarded as representative biomarkers of AGEs (Montuschi et al. 2004: 136).

Redox Ratio (GSH/GSSG)

Glutathione (γ-glutamyl-cysteinyl-glycine; GSH) is the most abundant low-molecular-weight thiol and is the major redox couple in animal cells. Glutathione exists in two forms: the reduced tripeptide form GSH and the oxidized sulfur–sulfur-linked compounds known as glutathione disulfide or GSSG or oxidized glutathione. GSH has diverse roles ranging from signal transduction, gene expression, apoptosis, protein glutathionylation, and nitric oxide (NO) metabolism (Townsend et al. 2003: 157); (Jones 2002: 114). Most importantly, it is a major free radical scavenger (e.g., hydroxyl radical, lipid peroxyl radical, peroxynitrite, and H_2O_2) directly and indirectly through the enzymatic reactions (Fang et al. 2002: 87). During such reactions, GSH is oxidized to form GSSG, which in turn can be reduced to GSH by the NADPH-dependent glutathione reductase. The ratio of GSH/GSSG often called "the redox ratio" is considered a sensitive indicator of oxidative stress. Any shift in this ratio toward the oxidizing state (i.e., more GSSG) activates several signaling pathways (including protein kinase B, protein phosphatases 1 and 2A, nuclear factor κB, c-jun N-terminal kinase, apoptosis signal-regulated kinase 1, and mitogen-activated protein kinase), thereby affecting the cell proliferation and increasing apoptosis (Sen 2000: 151).

Nuclear Factor -κB

Nuclear factor-κB (NF-κB) is a transcriptional factor implicated in the inflammation and immune activation which is activated by oxidants and cytokines (Barnes and Karin 1997: 57). This factor normally resides in an inactive form in the cytoplasm (Hur et al. 1999: 106). Diverse stimuli, including cytokines, microbial infections, oxidants, and mitogens, lead to the activation of IκB causing the nuclear translocation of NF-κB, where it actively binds to and stimulates the transcription of target genes, including COX-2, iNOS, and several other pro-inflammatory cytokines which can be outcomes or causes of oxidative stress.

Cyclooxygenase-2

Cyclooxygenase (COX) catalyzes the formation of prostaglandins and other eicosanoids from arachidonic acid. It exists in two forms: COX-1 and COX-2. COX-1 is a homeostatic and housekeeping gene, whereas COX-2 is a target gene of NF-κB which is induced during the conditions of stress especially stress-induced inflammation. RONs, particularly H_2O_2 and $ONOO^-$, interact with various cellular molecules, to elicit pathways that lead to increased expression of inflammatory mediators such as interleukin-1 (IL-1) and tumor necrosis factor-α. These agents stimulate the mobilization of the arachidonic acid (AA) from membrane phospholipids via enzyme phospholipase A_2 (PLA_2). This AA is then acted upon by COX-2 leading to a variety of eicosanoid production. Changes in the cellular redox tone are known to impact the activation cyclooxygenases and lipoxygenases, which produce lipid mediators in the form of prostaglandins, thromboxanes, prostacyclins, and oxidized fatty acids, respectively (Lands et al. 1984: 126). As an example, PGE_2, TXA_2, LTA_4, and LTC_4 are well-known biomarkers of stress-induced inflammation.

Glutathione S-Transferase-pi

Glutathione S-transferase-pi (GST-pi) belongs to the family of phase II detoxification enzymes responsible for the intracellular detoxification reactions, including the inactivation of electrophilic carcinogens by catalyzing their conjugation with glutathione (Henderson et al. 1998: 101). Along with the antioxidant system components, such as glutathione, vitamins, catalase, and superoxide dismutase, they concertedly form two major defense systems against electrophiles and xenobiotic toxicity (Enomoto et al. 2001: 82). In addition, GSTs also act on endogenous substrates, such as lipid and nucleic acid hydroperoxides and alkenals, which result from the decomposition of lipid hydroperoxides (Coles and Ketterer 1990: 71). Under the cancerous conditions such as papillomas and squamous cell carcinomas, a wide range of GST-pi expression has been observed. The reduced expression of GST-pi in tumor cells has been seen as an indicator of altered phenotypic differentiation, and inhibition of protection incurred from oxidative or electrophilic DNA damage.

Heme Oxygenase I

Heme oxygenase (HO)-1 (HO-1), a heat shock protein, is the inducible isoform of the rate-limiting enzyme involved in heme degradation (Immenschuh and Ramadorig 2000: 109). It is induced by various stimuli, including heat shock, hyperoxia, and oxidative stress and represents a powerful endogenous protective mechanism against free radicals in a variety of pathological conditions. HO-1 has been observed in the experimental autoimmune encephalomyelitis, serving as a model for multiple sclerosis (MS) (Liu et al. 2001: 129). The co-localization of HO-1 and oxidized phospholipids in the macrophages at the site of atherosclerotic lesions indicates its potential as an oxidative insult marker. Not solely as a marker but also the HO-1 modulation affects the plasma lipid hydroperoxide and nitrite and nitrate levels by acting as a part of intrinsic antioxidant system.

Physiological Significance of Oxidative Stress

ROS-mediated oxidative stress is the root cause of almost all the pathological conditions; however, their production and release are critical in several physiological pathways. The major "redox-responsive signaling" pathways that are regulated by ROS include cell signaling, NO• production-mediated regulation of vascular tone and neurotransmission, cell adhesion, the immune response, and the sensing of hypoxia and apoptosis (Droge 2002: 81; Valko et al. 2007: 44). In this section, we briefly discussed the basic principles of the above-mentioned redox-regulated physiological functions:

1. During the inflammatory conditions, a considerable amount of ROS is produced leading to "oxidative burst." This increased ROS production activates neutrophils and macrophages producing large quantities of superoxide radical and other ROS via the phagocytic isoform of NAD(P)H oxidase (Keisari et al. 1983: 120). The activation of these key players plays an important role in defense against environmental pathogens.

2. In contrast to the neutrophils and macrophages, various nonphagocytic cells such as fibroblasts, vascular smooth muscle cells, cardiac myocytes, and endothelial cells can also produce ROS by NAD(P)H oxidase to regulate the intracellular signaling cascades (Jones et al. 1996: 116; Thannickal and Fanburg 1995: 156). These vascular cells are mainly stimulated in response to the growth factors and cytokines such as angiotensin II, thrombin, PDGF (platelet-derived growth factor) and TNF-α (tumor necrosis factor- α), and interleukin-1 (IL-1) and thus regulate the cardiac and vascular cell functioning in a ROS-dependent manner (Griendling et al. 2000: 94).
3. Hydrogen peroxide and NO• radicals activate the enzyme soluble guanylate cyclase (sGC) that catalyzes the formation of cGMP. The cGMP is used as an intracellular amplifier and second messenger in a variety of physiological responses such as the function of protein kinases, ion channels, and other physiologically important targets, the most important ones being regulation of smooth muscle tone and the inhibition of platelet adhesion (Ignarro and Kadowitz 1985: 107; Wolin et al. 1999: 166).
4. Several ROS-producing proteins involving b-type cytochrome independently act as sensor for the changes in oxygen concentration. In contrast to this, the change in the rate of mitochondrial ROS may also play a role in oxygen sensing by the carotid bodies which are sensory organs that detect changes in the arterial blood oxygen. ROS-mediated changes in the glutathione redox state have also been implicated in the control of K^+ efflux and the corresponding Ca^{2+} influx which are involved in the transduction of the sinus nerve signal in response to the changes in oxygen tension (Lopez-Barneo et al. 1999: 130).

   These changes in the oxygen tension are also sensed by changes in ROS production (Fandrey et al. 1994: 86; Huang et al. 1996: 103; Jungermann and Kietzmann 1997: 118; Neumcke et al. 1999: 138). For example, hydrogen peroxide repressed the expression of erythropoietin whose production is stimulated under hypoxia. The erythropoietin gene in turn is controlled by the transcription factor hypoxia-inducible factor 1 (HIF-1) (Wang et al. 1995: 163) existing as HIF-1α and HIF-1β. Under normoxic conditions, HIF-1α is rapidly degraded by the proteasomes in an ROS-dependent manner (Huang et al. 1998: 104), whereas hypoxia decreases the ROS-mediated degradation of HIF-1α and enhances its formation (Semenza 2000: 150; Zhu and Bunn 1999: 171). Oxygen tension does not affect the concentration of the HIF-1β subunit. This activation of HIF-1 under stress conditions then targets a variety of genes involved in the angiogenesis, energy metabolism, erythropoiesis, cell proliferation and viability, vascular remodeling, and vasomotor responses (Semenza 2000: 150) and thus influences the production of a variety of hypoxia-regulated hormones and proteins including the vascular endothelial growth factor (VEGF) that stimulates the formation of new blood vessels (Bunn 1996: 66) and the tyrosine hydroxylase (TH) that facilitates the control of ventilation by the carotid body (Semenza 2000: 150; Zhu and Bunn 1999: 171).
5. Cell adhesion is an important cellular property of great significance due to its role in embryogenesis, cell growth, differentiation, wound repair, and other processes. These adhesive properties of the cells and tissues are in turn tightly redox regulated (Albelda1994: 53; Frenette and Wagner 1996). For example, the adhesion of leukocytes to endothelial cells is induced by ROS via the induced phosphorylation of the focal adhesion kinase pp125FAK, a cytosolic tyrosine kinase that has been implicated in the oxidant-mediated adhesion process (Schaller et al. 1992: 149). Additionally, various other cell adhesion molecules can be stimulated either by the microbial invasion such as bacterial lipopolysaccharides or by various cytokines such as TNF, interleukin-1, and interleukin-1 (Albelda et al. 1994: 53). The activation of these molecules is essentially the outcome of changes in the physiological redox state.
6. The immune response is a highly regulated and complex physiological process that is

critically maneuvered by the redox status. T lymphocytes are significantly activated by the ROS or by a shift in intracellular glutathione redox state. Furthermore, it has been reported that the T-cell functions such as interleukin-2 production can be induced by the physiologically relevant concentrations of superoxide radical and hydrogen peroxide (Los et al. 1995: 131). Numerous evidences also suggest that the immunological functions of macrophages are also redox regulated (Hamuro et al. 1999). Macrophages vary strongly in their release of prostaglandins, interleukin-6, and interleukin-12, depending on the intracellular content of glutathione. This balance between "reductive" and "oxidative" macrophages regulates thereby the ratio of helper T cells of type 1 versus type 2 (TH1/TH2).

7. Apoptosis is an integral physiological process that plays an indispensable role in the development and homeostasis of multicellular organisms (Wyllie 1980: 168). Numerous apoptotic stimuli such as APO-1/Fas/CD95 ligands induce cellular ROS production as observed in apoptotic processes ROS in apoptosis (Banki et al. 1999: 56; Esteve et al. 1999: 84; Hockenbery et al. 1993: 102; Johnson et al. 1996: 113; Korsmeyer 1995: 125; Um et al. 1996: 158; Williams and Henkart 1996: 165; Zamzami et al. 1995: 170). However, opposite effects have also been reported suggesting that the prooxidative conditions are not a general prerequisite for apoptotic cell death (Hug et al. 1994: 105; Jacobson et al. 1994: 110; Castedo et al. 1996: 69). Nevertheless, high ROS concentrations induce the apoptotic cell death in various cell types (Dumont 1999: 172; Slater et al. 1995: 173), suggesting that ROS contribute to the cell death whenever they are generated in the context of the apoptotic process. However, a variety of different mechanisms have been presented for the ROS-mediated oxidative stress depending upon the cell type and the ROS involved.

   For example, relatively moderate concentrations of the hydrogen peroxide induce a CD95-independent apoptotic process in T lymphocytes that requires mitochondrial ROS production and the activation of NF-kB (Dumont, 1999: 172). Another NO-dependent apoptotic pathway is characterized by decrease in the concentration of cardiolipin, decreased activity of the mitochondrial electron transport chain, and release of mitochondrial cytochrome c into the cytosol (Brune et al. 1997: 64). However, endothelial cells are resistant to the induction of apoptosis by NO• due to high intracellular levels of glutathione (Albina and Reichner 1998: 54). From a more pathological perspective, the role of TNF-α in cell death in many types of tumor cells is noteworthy. In these transformed cell lines as well as leukocytes and fibroblasts, TNF-α induces endogenous ROS production by the mitochondria (O'donnell et al. 1995: 174; (Schulze-Osthoff 1992: 175) and by the activation of membrane-bound NADPH oxidases, respectively. The outcome of this induction can either cause proliferation or cell death depending on the condition of the ROS-producing cell (Hennet et al. 1993: 176; Klebanoff et al. 1986: 177; Meier et al. 1989: 178; Schulze-Osthoff 1992: 175; Shalaby et al. 1985: 179) and on the signaling and execution pathways that are activated (De Vos et al. 1998: 180).

8. ROS also play yet another important role in iron homeostasis. The iron–sulfur proteins, in which iron is bound simultaneously to inorganic sulfide groups and cysteine thiolate groups of the proteins, are sensitive to both ROS and RNS. Oxidation of these proteins causes dissolution of the iron–sulfur cluster and loss of function (Butler et al. 1988; Castro et al. 1994: 181; Henry et al. 1993: 182). For example, RNS inhibit the mammalian (4Fe-4S) aconitase, an enzyme involved in the citric acid cycle. RNS disrupt the Fe–S clusters and simultaneously expose an RNA-binding site with specificity for the iron-response elements of the transferrin receptor and ferritin mRNAs. In this form, the protein is called iron-regulatory protein-1 and is involved in iron homeostasis.

References

Aikens J, Dix TA (1991) Perhydroxyl radical (Hoo.) initiated lipid peroxidation. The role of fatty acid hydroperoxides. J Biol Chem 266(23):15091–15098

Albelda SM, Smith CW, Ward PA (1994) Adhesion molecules and inflammatory injury. FASEB J 8:504–512

Albina JE, Reichner JS (1998) Role of nitric oxide in mediation of macrophage cytotoxicity and apoptosis. Cancer Metastasis Rev 17:39–53

Babior B, Lambeth J, Nauseef W (2002) The neutrophil NADPH oxidase. Arch Biochem Biophys 397:342–344

Banki K, Hutter E, Gonchoroff NJ, Perl A (1999) Elevation of mitochondrial transmembrane potential and reactive oxygen intermediate levels are early events and occur independently from activation of caspases in Fas signaling. J Immunol 162:1466–1479

Barnes P, Karin M (1997) Nuclear factor J B: a pivotal transcription factor in chronic inflammatory diseases. N Engl J Med 336:1066–1071

Beckman K, Ames B (1997) Oxidative decay of DNA. J Biol Chem 272:19633–19636

Beckman J, Koppenol W (1996) Nitric oxide, superoxide, and peroxynitrite: the good, the bad and ugly. Am J Physiol 271:C1424–C1437

Bergendi L, Benes L, Durackova Z, Ferencik M (1999) Chemistry, physiology and pathology of free radicals. Life Sci 65:1865–1874

Berlett B, Stadtman E (1997) Protein oxidation in aging, disease, and oxidative stress. J Biol Chem 272:20313–20316

Bielski B, Cabelli D (1995) Superoxide and hydroxyl radical chemistry in aqueous solution. In: Foote C, Valentine J, Greenberg A, Liebman J (eds) Active oxygen in chemistry. Chapman & Hall, London, pp 66–104

Bielski B, Cabelli B, Arudi R, Ross A (1985) Reactivity of Ro2/O2 radicals in aqueous solution. J Phys Chem Ref Data 14:1041–1100

Bogdan C, Rollinghoff M, Diefenbach A (2000) Reactive oxygen and reactive nitrogen intermediates in innate and specific immunity. Curr Opin Immunol 12:64–76

Boveris A (1984) Determination of the production of superoxide radicals and hydrogen peroxide in mitochondria. Methods Enzymol 105:429–435

Bredt D, Hwang P, Glatt C, Lowenstein C, Reed R, Synder S (1991) 450 reductase. Nature 351:714–718

Brune B, Gotz C, Messmer UK, Sandau K, Hirvonen MR, Lapetina EG (1997) Superoxide formation and macrophage resistance to nitric oxide-mediated apoptosis. J Biol Chem 272:7253–7258

Bruskov V, Malakhova L, Masalimov Z, Chernikov A (2002) Heat induced formation of reactive oxygen species and 8-oxoguanine, a biomarker of damage to DNA. Nucleic Acids Res 30:1354–1363

Bunn H, Poyton R (1996) Oxygen sensing and molecular adaptation to hypoxia. Physiol Rev 76:839–885

Butler AR, Glidewell C, Li MS (1988) Nitrosyl complexes of iron sulfur cluster. Adv Inorg Chem 32:335–392

Carney JM, Starke-Reed PE, Oliver CN, Landum RW, Cheng MS et al (1991) Reversal of age-related increase in brain protein oxidation, decrease in enzyme activity, and loss in temporal and spatial memory by chronic administration of the spin-trapping compound N-tert-butyl-α phenylnitrone. Proc Natl Acad Sci U S A 88:3633–3636

Carr A, Mccall MR, Frei B (2000) Oxidation of LDL by myeloperoxidase and reactive nitrogen species reaction pathways and antioxidant protection. Arterioscler Thromb Vasc Biol 20:1716–1723

Castedo M, Hirsch T, Susin SA, Zamzami N, Marchetti P, Macho A, Kroemer G (1996) Sequential acquisition of mitochondrial and plasma membrane alterations during early lymphocyte apoptosis. J Immunol 157:512–521

Castro L, Rodriguez M, Radi R (1994) Aconitase is readily inactivated by peroxynitrite, but not by its precursor, nitric oxide. J Biol Chem 269:29409–29415

Chance B, Sies H, Boveris A (1979) Hydroperoxide metabolism in mammalian organs. Physiol Rev 59:527–605

Coles B, Ketterer B (1990) The role of glutathione and glutathione transferases in chemical carcinogenesis. Crit Rev Biochem Mol Biol 25:47–70

Czapski G, Goldstein S (1995) The role of the reactions of no with superoxide and oxygen in biological systems: a kinetic approach. Free Radic Biol Med 19:785–794

Darley-Usmar V, Wiseman H, Halliwell B (1995) Nitric oxide and oxygen radicals: a question of balance. FEBS Lett 369:131–135

Davidson CA, Kaminski PM, Wolin MS (1997) Am J Physiol 273:L437–L444

Davis K (1987) Protein damage and degradation by oxygen radicals. I. General aspects. J Biol Chem 262:9895–9901

De Grey AD (2002) HO2*: the forgotten radical. DNA Cell Biol 21(4):251–257

De Vos K, Goossens V, Boone E, Vercammen D, Vancompernolle K, Vandenabeele P, Haegeman G, Fiers W, Grooten J (1998) The 55-Kda tumor necrosis factor receptor induces clustering of mitochondria through its membrane-proximal region. J Biol Chem 273:9673–9680

Decuyper-Debergh D, Piette J, Van De Vorst A (1987) Singlet oxygen-induced mutations in M13 lacZ phage DNA. EMBO J 6(10):3155–3161

Demple B, Amabile-Cuevas CF (1991) Redox redux: the control of oxidative stress response. Cell 67:837–840

Di Mascio P, Bechara E, Medeiros M, Briviba K, Sies H (1994) Singlet molecular oxygen production in the reaction of peroxynitrite with hydrogen peroxide. FEBS Lett 355(3):287–289

Di Mascio P, Briviba K, Sasaki S, Catalani L, Medeiros M, Bechara E, Sies H (1997) The reaction of peroxynitrite with tert-butyl hydroperoxide produces singlet molecular oxygen. Biol Chem 378(9):1071–1074

Dröge W (2002) Free radicals in the physiological control of cell function. Physiol Rev 82:47–95

Dumont A, Hehner S, Hofmann T, Ueffing M, Dro¨ Ge W, Schmitz M (1999) Hydrogen peroxide-induced apoptosis is CD95-independent, requires the release of mitochondria-derived reactive oxygen species and the activation of NF-Kb. Oncogene 18:747–757

Enomoto A, Itoh K, Nagayoshie, Haruta J, Kimura T, Harada T, O'connort T, Yamamoto M (2001) High sensitivity of Nrf2 knockout mice to acetaminophen hepatotoxicity associated with decreased expression of Are regulated drug metabolizing enzymes and antioxidant genes. Toxicol Sci 59:169–177

Epe B, Ballmaier D, Roussyn I, Briviba K, Sies H (1996) DNA damage by peroxynitrite characterized with DNA repair enzymes. Nucleic Acids Res 24(21):4105–4110

Esteve J, Mompo J, De La Asuncion J, Sastre J, Asensi M, Boix J, Vina J, Vina J, Pallardo' F (1999) Oxidative damage to mitochondrial DNA and glutathione oxidation in apoptosis: studies in vivo and in vitro. FASEB J 13:1055–1064

Fandrey J, Frede S, Jelkmann W (1994) Role of hydrogen peroxide in hypoxia-induced erythropoietin production. Biochem J 303:507–510

Fang YZ, Yang S, Wu G (2002) Free radicals, antioxidants, and nutrition. Nutrition 18:872–879

Faraci FM, Didion SP (2004) Superoxide dismutase isoforms in the vessel wall. Arterioscler Thromb Vasc Biol 24:1367

Floyd R (1991) Oxidative damage to behavior during aging. Science 254:1597–97

Forman H, Torres M (2001) Redox signaling in macrophages. Mol Aspects Med 22(4–5):189–216

Frenette PS, Wagner DD (1996) Adhesion molecules. Part I. N Engl J Med 334:1526–1529

Garrison W (1987) Reaction mechanisms in the radiolysis of peptides, polypeptides, and proteins. Chem Rev 87:381–398

Gedik CM, Boyle SP, Wood SG, Vaughan NJ, Collins AR (2002) Oxidative stress in humans: validation of biomarkers of DNA damage. Carcinogenesis 23:1441–1446

Ghafourifar P, Cadenas E (2005) Mitochondrial nitric oxide synthase. Trends Pharmacol Sci 26:190–195

Gopaul NK, Halliwell B, Anggård EE (2000) Measurement of plasma F2-isoprostanes as an index of lipid peroxidation does not appear to be confounded by diet. Free Radic Res 33(2):115–127

Gottschling B, Maronpot R, Hailey J, Peddada S, Moomaw C, Klaunig J, Nyska A (2001) The role of oxidative stress in indium phosphide-induced lung carcinogenesis in rats. Toxicol Sci 64:28–40

Griendling KK, Sorescu D, Lasse'Gue B, Ushio-Fukai M (2000) Modulation of protein kinase activity and gene expression by reactive oxygen species and their role in vascular physiology and pathophysiology. Arterioscler Thromb Vasc Biol 20:2175–2183

Griffiths HR, Moller L, Bartosz G, Bast A, Bertonni-Freddari C, Collins A, Coolen S, Haenen G, Hoberg AM, Loft S, Lunec J, Olinski R, Parry J, Pompella A, Poulsen H, Verhagen H, Astley SB (2002) Biomarkers. Mol Aspects Med 23:101–209

Guzik TJ, Mussa S, Gastaldi D, Sadowski J, Ratnatunga C, Pillai R, Channon KM (2002) Mechanisms of increased vascular superoxide production in human diabetes mellitus: role of NAD(P)H oxidase and endothelial nitric oxide synthase. Circulation 105:1656–1662

Haber F, Weiss J (1934) The catalytic decomposition of hydrogen peroxide by iron salts. Proc R Soc Lond Ser A 147:332–351

Halliwell B (2000) Lipid peroxidation, antioxidants and cardiovascular disease: how should we move forward? Cardiovasc Res 47(3):410–418

Halliwell B (2006) Oxidative stress and neurodegeneration: where are we now? J Neurochem 97(6):1634–1658

Halliwell B (2009) The wanderings of a free radical. Free Radic Biol Med 46:531–542

Halliwell B, Gutterdgem (1999) Free radicals in biology and medicine, 3rd edn. Oxford University Press, Midsomer Norton

Halliwell B, Whiteman M (2004) Measuring reactive species and oxidative damage in vivo and in cell culture: how should you do it and what do the results mean? Br J Pharmacol 142:231–255

Halliwell B, Clement M, Long L (2000) Hydrogen peroxide in the human body. FEBS Lett 486:10–13

Hamuro J, Murata Y, Suzuki M, Takatsuki F, Suga T (1999) The triggering and healing of tumor stromal inflammatory reactions regulated by oxidative and reductive macrophages. Gann Monogr Cancer Res 48:153–164

Hauptmann N, Cadenas E (1997) The oxygen paradox: biochemistry of active oxygen. In: Oxidative stress and the molecular biology of antioxidant defenses. Csh monographs vol 34. Cold Spring Harbor Laboratory Press, Plainview

Hawkins C, Brown B, Davies M (2001) Hypochlorite- and hypochlorite-mediate D radical formation and its role in cell lysis. Arch Biochem Biophys 395:137–145

Henderson C, Mclaren A, Moffat G, Bacon E, Wolf C (1998) Pi-class glutathione S-transferase: regulation and function. Chem Biol Interact 111–112:69–82

Hennet T, Richter C, Peterhans E (1993) Tumor necrosis factor-a induces superoxide anion generation in mitochondria of L929 cells. Biochem J 289:587–592

Henry Y, Lepoivre M, Drapier J, Ducrocq C, Boucher J, Guissani A (1993) Epr characterization of molecular targets for NO in mammalian cells and organelles. FASEB J 7:1124–1134

Hockenbery D, Oltvai Z, Yin X, Milliman C, Korsmeyer S (1993) Bcl-2 functions in an antioxidant pathway to prevent apoptosis. Cell 75:241–251

Huang L, Arany Z, Livingston D, And Bunn F (1996) Activation of hypoxia-inducible transcription factor depends primarily upon redox-sensitive stabilization of its alpha subunit. J Biol Chem 271:32253–32259

Huang L, Gu J, Schau M, Bunn H (1998) Regulation of hypoxia-inducible factor 1a is mediated by it oxygen-dependent degradation domain via the

ubiquitin-proteasome pathway. Proc Natl Acad Sci U S A 95:7987–7992
Hug H, Enari M, Nagata S (1994) No requirement of reactive oxygen intermediates in Fas-mediated apoptosis. FEBS Lett 351:311–313
Hur G, Ryu Y, Yun H, Jeon B, Kim Y, Seok J, Lee J (1999) Hepatic ischemia/reperfusion in rats induces iNOS gene transcription by activation of NF-kappaB. Biochem Biophys Res Commun 261:917–922
Ignarro LJ, Kadowitz PJ (1985) The pharmacological and physiological role of cyclic GMP in vascular smooth muscle relaxation. Ann Pharmacol Toxicol 25:171–191
Iida T, Furuta A, Kawashima M, Nishida J, Nakabeppu Y, Iwaki T (2001) Accumulation of 8-Oxo-2-deoxyguanosine and increased expression of Hmth1 protein in brain tumors. Neuro-Oncol 3:73–81
Immenschuh S, Ramadorig (2000) Gene regulation of heme oxygenase-1 as a therapeutic target. Biochem Pharmacol 60:1121–1128
Jacobson M, Burne J, Raff M (1994) Programmed cell death and Bcl-2 protection in the absence of a nucleus. EMBO J 13:1899–1910
Janssen Y, Van Houten B, Borm P, Mossman B (1993) Cell and tissue responses to oxidative damage. Lab Invest 69:261–274
Jeong J, Juedes M, Wogan G (1998) Mutations induced in the supF gene of pSP189 by hydroxyl radical and singlet oxygen: relevance to peroxynitrite mutagenesis. Chem Res Toxicol 11(5):550–556
Johnson T, Yu Z, Ferrans V, Lowenstein R, Finkel T (1996) Reactive oxygen species are downstream mediators of P53-dependent apoptosis. Proc Natl Acad Sci U S A 93:11848–11852
Jones DP (2002) Redox potential of GSH/GSSG couple: assay and biological significance. Methods Enzymol 348:93–112
Jones D (2008) Radical-free biology of oxidative stress. Am J Physiol Cell Physiol 295(4):C849–C868, Epub 2008 Aug 6
Jones SA, O'donnell VB, Wood JD, Broughton JP, Hughes EJ, Jones OT (1996) Expression of phagocyte NADPH oxidase components in human endothelial cells. Am J Physiol Heart Circ Physiol 271:H1626–H1634
Joseph J, Cutler R (1994) The role of oxidative stress in signal transduction changes and cell loss in senescence. Ann N Y Acad Sci 738:37
Jungermann K, Kietzmann T (1997) Role of oxygen in the zonation of carbohydrate metabolism and gene expression in liver. Kidney Int 51:402–412
Kanofsky J (1989) Singlet oxygen production by biological systems. Chem Biol Interact 70(1–2):1–28
Keisari Y, Braun L, Flescher E (1983) The oxidative burst and related phenomena in mouse macrophages elicited by different sterile inflammatory stimuli. Immunobiology 165:78–89
Kilhovd B, Juutilainen A, Lehto S, Rönnemaa T, Torjesen P, Hanssen K, Laakso M (2007) Increased serum levels of advanced glycation end products predict total, cardiovascular and coronary mortality in women with type 2 diabetes: a population-based 18 year follow-up study. Diabetologia 50(7):1409–1417, Epub 2007 May 4
Klatt P, Lamas S (2000) Regulation of protein function by s glutathiolation in response to oxidative and nitrosative stress. Eur J Biochem 267:4928–4944
Klebanoff S, Vadas M, Harlan J, Sparks L, Gamble J, Agosti J, Waltersdorph A (1986) Stimulation of neutrophils by tumor necrosis factor. J Immunol 136:4220–4225
Klotz Lo, Briviba K, Sies H (2000) Signalling by singlet oxygen in biological systems. Section: reactive species as intracellular messengers. Chapter 1. In: Chandan KS, Helmut S, Bauerle PA (eds) Antioxidant and redox regulations of genes. Academic Press, San Diego
Knight T, Kurtz A, Bajt M, Hinson J, Jaeschke H (2001) Vascular and hepatocellular peroxynitrite formation during acetaminophen toxicity: role of mitochondria L oxidant stress. Toxicol Sci 62:212–220
Kohen R, Nyska A (2002) Oxidation of biological systems: oxidative stress phenomena, antioxidants, redox reactions, and methods for their quantification. Toxicol Pathol 30(6):620–650
Korsmeyer S (1995) Regulators of cell death. Trends Genet 11:101–105
Lamas S, Marsden P, Li G, Tempst P, Michel T (1992) Endothelial nitric oxide synthase: molecular cloning and characterization of a distinct constitutive enzyme isoform. Proc Natl Acad Sci U S A 89:6348–6352
Lands WEM et al (1984) In: Pryor W (ed) Free radicals in biology, vol 6. Academic, New York, pp 39–61
Levine R, Stadtman E (2001) Oxidative modification of proteins during aging. Exp Gerontol 36:1495–1502
Liochev SI, Fridovich I (2002) The Haber–Weiss cycle—70 years later: an alternative view. Redox Rep 7:55–57
Liu Y, Zhu B, Luo L, Li P, Paty D, Cynader M (2001) Heme oxygenase-1 plays an important protective role in experimental autoimmune encephalomyelitis. Neuroreport 12:1841–1845
Lo'Pez-Barneo J, Pardal R, Montoro R, Smani T, Garcı'A-Hirschfeld J, Urena J (1999) K1 and Ca21 channel activity and cytosolic [Ca21] in oxygen-sensing tissues. Respir Physiol 115:215–227
Los M, Droge W, Stricker K, Baeuerle PA, Schulze-Osthoff K (1995) Hydrogen peroxide as a potent activator of T lymphocyte functions. Eur J Immunol 25:159–165
Mckersie BD Oxidative stress by, University Of Guelph (Posted on the internet in 1996)
Mehta JL, Rasouli N, Sinha AK, Molavi B (2006) Oxidative stress in diabetes: a mechanistic overview of its effects on atherogenesis and myocardial dysfunction. Int J Biochem Cell Biol 38:794–803
Meier B, Radeke H, Selle S, Younes M, Sies H, Resch K, Habermehl G (1989) Human fibroblasts release reactive oxygen species in response to interleukin-1 or tumor necrosis factor-a. Biochem J 263:539–545

Mendez JI, Nicholson WJ, Taylor WR (2005) Sod isoforms and signaling in blood vessels: evidence for the importance of Ros compartmentalization. Arterioscler Thromb Vasc Biol 25:887–888

Milne G, Musiek E, Morrow J (2005) F2-isoprostanes as markers of oxidative stress in vivo: an overview. Biomarkers 10(Suppl 1):S10–S23

Miyata T, Maeda K, Kurokawa K, Van Ypersele De Strihou C (1997) Oxidation conspires with glycation to generate noxious advanced glycation end products in renal failure. Nephrol Dial Transplant 12:255–258

Mohora M, Greabu M, Muscurel C, Duţă C, Totan A (2007) The sources and the targets of oxidative stress in the etiology of diabetic complications. Romanian J Biophys 17(2):63–84, Bucharest

Moncada S, Higgs A (1993) The L-arginine-nitric oxide pathway. N Engl J Med 329:2002–2012

Montuschi P, Barnes P, Roberts L 2nd (2004) Isoprostanes: markers and mediators of oxidative stress. FASEB J 18(15):1791–1800

Murphy M, Packer M, Scarlet J, Martin S (1998) Peroxynitrite: a biologically significant oxidant. Gen Pharmacol 31:179–186

Neumcke I, Schneider B, Fandrey J, Pagel H (1999) Effects of pro and antioxidative compounds on renal production of erythropoietin. Endocrinology 140:641–645

O'donnell V, Spycher S, Azzi A (1995) Involvement of oxidants and oxidant-generating enzyme(S) in tumor necrosis factor-a-mediated apoptosis: role for lipoxygenase pathway but not mitochondrial respiratory chain. Biochem J 310:133–141

Ohshima H, Sawa T, Akaike T (2006) 8-nitroguanine, a product of nitrative DNA damage caused by reactive nitrogen species: formation, occurrence, and implications in inflammation and carcinogenesis. Antioxid Redox Signal 8(5–6):1033–1045

Pastor N, Weinstein H, Jamison E, Brenowitz M (2000) A detailed interpretation of OH radical footprints in a TBP-DNA complex reveals the role of dynamics in the mechanism of sequence-specific binding. J Mol Biol 304(1):55–68

Patel R, Mollering D, Murphy-Ullrich J, Jo H, Beckman J, Darley-Usmar V (2000) Cell signaling by reactive nitrogen and oxygen species in atherosclerosis. Free Radic Biol Med 28:1780–1794

Piette J (1991) Biological consequences associated with DNA oxidation mediated by singlet oxygen. J Photochem Photobiol B 11(3–4):241–260

Poss W, Huecksteadt T, Panus P, Freeman B, Hoidal J (1996) Regulation of xanthine dehydrogenase and xanthine oxidase activity by hypoxia. Am J Physiol Lung Cell Mol Physiol 270:L941–L946

Preiser J-C (2012) Oxidative stress. J Parenter Enter Nutr 36(2):147–154

Richter C, Gogvadze V, Laffranchi R, Schlapbach R, Schwezer M, Suter M, Walter P (1995) Oxidants in mitochondria: from physiology to diseases. Biochim Biophys Acta 1271:67–74

Ridnour LA, Thomas DD, Mancardi D, Espey MG, Miranda KM, Paolocci N et al (2004) The chemistry of nitrosative stress induced by nitric oxide and reactive nitrogen oxide species. Putting perspective on stressful biological situations. Biol Chem 385:1–10

Saran M, Bors W (1989) Oxygen radicals acting as chemical messengers: a hypothesis. Free Rad Res Commun 7:3–6

Schafer F, Buettnergr (2001) Redox environments of the cell as viewed through the redox state of the glutathione disulfide/glutathione couple. Free Rad Biol Med 30:1191–1212

Schaller MD, Borgman CA, Cobb BS, Vines RR, Reynolds AB, Parsons JT (1992) Pp125fak a structurally distinctive protein-tyrosine kinase associated with focal adhesions. Proc Natl Acad Sci U S A 89:5192–5196

Schulze-Osthoff K, Bakker A, Vanhaesebroeck B, Beyaert R, Jacob W, And Fiers W (1992) Cytotoxic activity of tumor necrosis factor is mediated by early damage of mitochondrial functions. Evidence for the involvement of mitochondrial radical generation. J Biol Chem 267:5317–5323

Semenza G (2000) Hif-1: mediator of physiological and pathophysiological responses to hypoxia. J Appl Physiol 88:1474–1480

Sen CK (2000) Cellular thiols and redox-regulated signal transduction. Curr Top Cell Regul 36:1–30

Shacter E (2000) Quantification and significance of protein oxidation in biological samples. Drug Metab Rev 32(3–4):307–326

Shalaby M, Aggarwal B, Rinderknecht E, Svedersky L, Finkle B, Palladino M Jr (1985) Activation of human polymorphonuclear neutrophil functions by interferon-G and tumor necrosis factor. J Immunol 135:2069–2073

Slater A, Stefan C, Novel I, Van Den Dobbelsteen D, Orrenius S (1995) Signalling mechanisms and oxidative stress in apoptosis. Toxicol Lett 82–83:149–153

Sohal R, Allen R (1990) Oxidative stress as a causal factor in differentiation and aging: a unifying hypothesis. Exp Gerontol 25:499–522

Steinbeck MJ, Khan AU, Karnovsky MJ (1992) Intracellular singlet oxygen generation by phagocytosing neutrophils in response to particles coated with a chemical trap. J Biol Chem 267(19):13425–13433

Steinbeck MJ, Khan AU, Karnovsky MJ (1993) Extracellular production of singlet oxygen by stimulated macrophages quantified using 9,10-diphenylanthracene and perylene in a polystyrene film. J Biol Chem 268(21):15649–15654

Takahashi S, Hirose M, Tamano S, Ozaki M, Orita S, Ito T, Takeuchi M, Ochi H, Fukada S, Kasai H, Shirai T (1998) Immunohistochemical detection of 8-hydroxy-2-deoxyguanosinE in paraffin embedded sections of rat liver after carbon tetrachloride treatment. Toxicol Pathol 26:247–252

Thannickal VJ, Fanburg BL (1995) Activation of an H2O2-generating NADH oxidase in human lung

fibroblasts by transforming growth factor beta 1. J Biol Chem 270:30334–30338
Townsend DM, Tew KD, Tapiero H (2003) The importance of glutathione in human disease. Biomed Pharmacother 57:145–155
Um H, Orenstein J, Wahl S (1996) Fas mediates apoptosis in human monocytes by a reactive oxygen intermediate dependent pathway. J Immunol 156:3469–3477
Upham B, Wagner J (2001) Toxicant-induced oxidative stress in cancer. Toxicol Sci 64:1–3
Valko M, Rhodes CJ, Moncol J, Izakovic M, Mazur M (2006) Free radicals, metals and antioxidants in oxidative stress-induced cancer. Chem Biol Interact 160:1–40
Valko M, Leibfritz D, Moncol J, Mark TD, Cronin C, Milan M, Telser J (2007) Free radicals and antioxidants in normal physiological functions and human disease. Int J Biochem Cell Biol 39:44–84
Vincent J, Zhang H, Szabo C, Preiser J (2000) Effects of nitric oxide in septic shock. Am J Respir Crit Care Med 161:1781–1785
Von Sonntag C (1987) The chemical basis of radiation biology. Taylor & Francis, London
Wang G, Jiang B, Rue E, Semenza G (1995) Hypoxia-inducible factor 1 is a basic helix-loop-helix-Pas heterodimer regulated by cellular O2 tension. Proc Natl Acad Sci U S A 92:5510–5514
Weinstein D, Mihm M, Bauer J (2000) Cardiac peroxynitrite formation and left ventricular dysfunction following doxorubicin treatment in mice. J Pharmacol Exp Ther 294:396–401
Williams M, Henkart P (1996) Role of reactive oxygen intermediates in Tcr-induced death of T cell blasts and hybridomas. J Immunol 157:2395–2402
Wolin MS, Burke-Wolin TM, Mohazzab-H KM (1999) Roles of NADPH oxidases and reactive oxygen species in vascular oxygen sensing mechanisms. Respir Physiol 115:229–238
Wu J (1993) Advanced glycosylation end products: a new disease marker for diabetes and aging. J Clin Lab Anal 7(5):252–255
Wyllie A, Kerr J, Currie A (1980) Cell death: the significance of apoptosis. Int Rev Cytol 68:251–306
Xie Q, Cho H, Calaycay J, Mumford R, Swiderek K, Lee T, Ding A, Troso T, Nathan C (1992) Cloning and characterization of inducible nitric oxide synthase from mouse macrophages. Science 256:225–228
Zamzami Marchetti N, Marchetti P, Castedo M, Decaudin D, Macho A, Hirsch T, Susin S, Petit P, Mignotte B, Kroemer G (1995) Sequential reduction of mitochondrial transmembrane potential and generation of reactive oxygen species in early programmed cell death. J Exp Med 182:367–377
Zhu H, Bunn H (1999) Oxygen sensing and signaling: impact on the regulation of physiologically important genes. Respir Physiol 115:239–247

2 Oxidative Stress in Pathogenesis

Oxidative stress has been well implicated in the pathogenesis of various human diseases. Presently mechanistic considerations of the oxidative stress pathogenesis in most vital organ systems, e.g., nervous system, cardiovascular system, male/female reproductive system, and autoimmune disease-related systems, will be discussed.

Neurodegenerative Diseases: Parkinson's and Alzheimer's Diseases

The brain with major neurons and astrocytes is especially sensitive to the oxidative stress because of the lipid peroxidation in membranes containing high level of polyunsaturated fatty acids (PUFA). Oxidation of lipids, proteins, and DNA in neurons generates many by-products such as peroxides, alcohols, aldehydes, ketones, and cholesterol oxides which are toxic to the blood lymphocytes and macrophages, influencing the in vivo defense system (Ferrari 2000). ROS attacks proteins, oxidizing both the backbone and side chains, which in turn reacts with the amino acid side chain to form carbonyl functions. ROS attacks nucleic acids, causing DNA-protein cross-links and strand breaks, and modifies purine and pyrimidine bases resulting in the DNA mutations (Mattson 2003). ROS are particularly active in the brain and neuronal tissues as the excitatory amino acids and neurotransmitters, whose metabolism produces ROS, which serve as the sources of oxidative stress and result in neural damage. Most significant ill effect on the neurons takes place by dysregulation of the intracellular calcium signaling pathways initiated by the ROS in neuronal cell death (Ermak and Davies 2002). Excitotoxic effects initiated by the ROS induce intracellular calcium influx, leading to the activation of glutamate receptors and apoptosis in the neurodegeneration. All these insults ultimately reflect into the specific disorders.

Oxidative stress has been linked to a range of chronic neurodegenerative disorders, including Alzheimer's disease (AD), Parkinson's disease (PD), Huntington's disease (HD), multiple sclerosis (MS), and amyotrophic lateral sclerosis (ALS). In these conditions, nerve cells in the brain and spinal cord are damaged or lost, leading to either functional loss (ataxia) or sensory dysfunction (dementia). Mitochondrial dysfunctions and excitotoxicity and finally apoptosis result into the pathological conditions in each disease (Gandhi and Abramov 2012). Neurodegeneration mediates a number of factors including the environmental and genetic predisposition. Oxidative stress and additional free radical generation catalyzed by the redox metals play important role in the neurodegeneration. AD and PD being the main neurodegenerative disorders will be the special focus in the present write-up.

M. Bansal and N. Kaushal, *Oxidative Stress Mechanisms and their Modulation*,
DOI 10.1007/978-81-322-2032-9_2, © Springer India 2014

Role of Pathogenesis in AD and PD

Alzheimer's Disease (AD): AD is characterized by the loss of neurons or their synapses in the cerebral cortex and certain subcortical regions and in turn the progressive cognitive decline. Both amyloid plaques and neurofibrillary tangles, as clearly visible microscopically in the AD-affected brain (Tiraboschi et al. 2004), are due to the insoluble deposits of the extracellular amyloid (Aβ) peptide around the neurons. This small amyloid-β protein (39–43 amino acids) originates from a larger protein called amyloid precursor protein (APP), a transmembrane protein that penetrates through the neuron's membrane. APP is critical to the growth, survival, and post-injury repair of the neurons (Priller et al. 2006). Another protein named tau normally stabilizes the microtubules (supporting structures of the neurons guiding nutrients and molecules from the body of the cell to the end of the axon and back) on phosphorylation. In AD, tau proteins get hyperphosphorylated and then pair with other threads, creating neurofibrillary tangles and disintegrating the neuron's transport system (Hemandez and Avila 2007). Amyloid fibrils disrupt the cell's calcium ion homeostasis and induce apoptosis (Yankner et al. 1990).

Further, AD is characterized by the amyloid plaques deposition by chelating Aβ with the transition metal ions (Cu^{2+}, Zn^{2+}, Fe^{3+}). In Aβ the histidine residues at position 6, 13, and 14 coordinate with the transition metals. Binding of Cu^{2+} and Fe^{3+} results in a chemical reaction altering oxidation state of both the metals, producing H_2O_2 catalytically in the presence of transition metals, and finally giving toxic OH· free radicals (Opazo et al. 2002). AD brains show evidence of ROS-mediated injury. There is an increase in the levels of malondialdehyde and 4-hydroxynonenal in the brain and cerebrospinal fluid of AD patients compared to the controls.

Parkinson's Disease (PD): PD is clinically characterized by the progressive rigidity, bradykinesia, and tremor, whereas pathologically by a progressive degeneration of the dopaminergic neurons with age and deposition of inclusion bodies (Lewy bodies) of the protein α-synuclein in the substantia nigra. In PD brain, the concentration of PUFA in the substantia nigra is reduced, while the levels of lipid peroxidation markers (malondialdehyde and 4-hydroxynonenal) are increased (Dalfo et al. 2005). Protein oxidative products as protein carbonyls are seen at high level in the PD brain compared to the controls, and also nitration and nitrosylation of certain proteins due to RNS in the PD brain are also observed (Brown and Borutaite 2004). Oxidative stress in the PD brain results in the increased levels of 8-hydroxydeoxyguanosine and also increase in the common deletions in mitochondrial DNA of the dopaminergic neurons in PD substantia nigra (Bender et al. 2006). Further, dopamine (neurotransmitter) is also a very good metal chelator and electron donor to generate toxic-free radicals. It has high tendency to coordinate with Cu^{2+} and Fe^{3+} and reduce metals to generate H_2O_2 (Gerard et al. 1994). Mutations in the α-synuclein protein modulate negatively the substantia dopamine activity that initiates neuronal cytoplasmic accumulation and interaction of dopamine with iron, producing ROS (Lotharius and Brundin 2002).

Oxidative Stress in the Pathogenesis of PD and AD: Neurons with long axons and multiple synapses require more energy for the axonal transport or long-term plasticity, resulting in mitochondrial dysfunction and further neurodegeneration. These features in different neuronal groups exhibit different degrees of oxidative stress. For example, in the hippocampus, CA1 neurons generate higher levels of superoxide anion than the CA3 neurons and exhibit higher levels of expression of both the antioxidant and ROS-producing genes (Wang and Michaelis 2010). Various sources of ROS production and their influence are shown in Fig. 2.1.

Mitochondria dysfunction and activation of ROS-producing enzymes (as discussed in Chap. 1), e.g., NADPH oxidase, xanthine oxidase, and monoamine oxidase, have been implicated in generating ROS and in turn neurodegeneration. Oxidative damage and the associated mitochondrial dysfunction may result in the energy depletion, accumulation of the cytotoxic mediators, and the cell death. Autophagic activity helps in the mitochondrial turnover, in which membrane autophagosomes

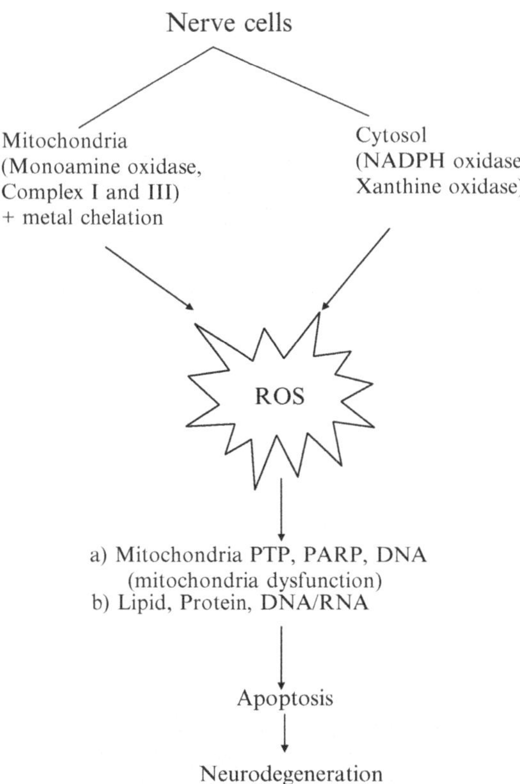

Fig. 2.1 ROS production and effects in nerve cells

sequester damaged/oxidized or dysfunctional intracellular components and organelles and direct them to the lysosomes for degradation. The absence of the autophagy (or mitophagy) may result in an abnormal mitochondrial function and oxidative or nitrative stress. The mitochondrial dysfunction includes the respiratory chain dysfunction and oxidative stress, reduced ATP production, calcium dysregulation, mitochondrial permeability transition pore (PTP) opening, and many more.

Mitochondrial pathology is evident in many neurodegenerative diseases including AD and PD. Mitochondrial dysfunction in the human brain is involved in the pathogenesis of PD and degeneration of dopaminergic neurons. The substantia nigra of PD patients shows reduced activity of the complex I. Complex I inhibitors such as rotenone, 1-methyl-4-phenyl-1,2,3,6-tetrahydropyridine (MPTP), and pesticides cause neurological changes similar to PD (Shapira 2008). Rotenone or MPP^+ (active metabolite of MPTP) produces superoxide anions in the submitochondrial particles, and the neurotoxic effects (Lothiarius and O'Malley 2000) of these is due to the production of oxidative stress as antioxidants prevents these changes. α-Synuclein, inner mitochondrial membrane-associated protein, interacts with the mitochondrial complex I function (Chinta et al. 2010). In the transgenic mice, overexpression of α-synuclein impairs mitochondrial function, increases oxidative stress, and enhances nigral pathology induced by MTPP (Song et al. 2004). Another protein, parkin, associated with the outer mitochondrial membrane, prevents cell death by inhibiting the mitochondrial swelling, cytochrome c release, and caspase activation (Darios et al. 2003). Parkin deficiency causes oxidative stress and mitochondrial impairment (Muftuoglu et al. 2004). Further, a protein PINK1 (phosphatase and tensin homologue, PTEN-induced kinase 1) is a mitochondrial kinase, and its deficiency results in impaired respiration with inhibition of complex I activity, reduced substrate availability, and rotenone-like increased production of ROS in mitochondria (Gandhi et al. 2009). PINK1 deficiency also results in an inability to handle cytosolic calcium challenges due to an impairment of the mitochondrial calcium overload. A combination of ROS production and mitochondrial Ca^{2+} initiates opening of the mitochondrial permeability transition pore (PTP), which allows translocation of the proapoptotic molecules from the mitochondria to the cytosol and that triggers apoptotic cell death.

A reduction in the complex IV activity has been demonstrated in mitochondria from the hippocampus and platelets of AD patients, as well as in the AD animal models. Accumulation of the Aβ leads to oxidative stress, mitochondrial dysfunction, and energy failure prior to the development of the plaque pathology. Deregulation of the calcium homeostasis has been demonstrated in AD, with Aβ causing increased cytoplasmic calcium levels and mitochondrial calcium overload, resulting in an increase in ROS production and opening of the PTP (Abramov et al. 2003). In addition to the alterations in mitochondrial bioenergetics, dysregulation of calcium homeostasis, excitotoxicity, oxidative stress (inflammation), and other mechanisms involve also the protein misfolding leading to the aggregates, proteasome dysfunction, and neuroinflammation in PD (Hirsch et al. 2013).

In AD, the NADPH oxidase has been shown to contribute toward oxidative stress. Activation of NOX2 has been demonstrated in brains of AD patients, and also its deficiency has been shown to improve the AD in a mouse model (Park et al. 2008). At cellular level, amyloid-β induced activation of NADPH oxidase in rat primary culture of microglial cells and human phagocytes, through B-class scavenger receptor, CD36 (Wilkinson et al. 2006). Aβ also activates NOX by including calcium entry into astrocytes (Abramov et al. 2003) and induces opening of the mitochondrial permeability transition pore, mPTP (Abramov et al. 2004). This oxidative stress signal is passed on to the neighboring neurons, which is more damaging than to the astrocytes. In PD, in both the rotenone- and MPTP-induced toxin models, activation of NOX2 in microglia occurs (Gao et al. 2003). Genetic models of PD also exhibit increased oxidative stress. In one such model, loss of PINK1 function is associated with the increased ROS production by NADPH oxidase in the midbrain neurons. The NADPH oxidase is activated by the high cytosolic calcium concentration, leading to the overproduction of superoxide which inhibits the plasmalemmal glucose transporter resulting in the deregulation of the mitochondria metabolism (Gandhi et al. 2009). The oxidative stress response by the microglial cells due to the NADPH oxidase plays a central role in the pathology of PD. This response in microglia occurs through the activation of the ERK signaling pathway by proinflammatory stimuli, leading to the phosphorylation and translocation of the p47 (phox) and p67 (phox) cytosolic subunits, the activation of membrane-bound PHOX, and the production of ROS (Peterson and Flood 2012).

Aβ is able to activate production of H_2O_2 in the cytosol of neocortical neurons (Kaminsky and Kosenko 2008). Inhibitor of XO, allopurinol, significantly suppressed OH^* generation in rat striatum of toxic models of PD induced by paranonylphenol and MPP^+ (Obata et al. 2001), suggesting a potential role for xanthine oxidase in the oxidative stress associated with PD. Monoamine oxidase A (MAOA) and monoamine oxidase B (MAOB), flavoenzymes, are located in the outer membrane of the mitochondria. They have a role in the oxidative catabolism of important amine neurotransmitters, including serotonin, dopamine, and epinephrine (Edmondson et al. 2009).

Electrical and Biological Effects

Direct electrical excitatory effect using low-frequency stimulation of the spinal cord or of the thalamus has been used for the diagnostic or even therapeutic applications. However, high-frequency stimulations (HFS) are considered for damaging and inactivating the neuronal structures, such as nuclei of the basal ganglia and also thalamus/subthalamic nucleus. Intracerebral recordings in the human patients tend to show the arrest of electrical firing in the recorded places. More recent data from the in vitro biological studies show that HFS profoundly affects the cellular functioning and particularly the protein synthesis, suggesting that it could alter the synaptic transmission by reducing the production of neurotransmitters (Benabid et al. 2005). Similarly, repetitive transcranial magnetic stimulation (rTMS) and transcranial direct current stimulation (tDCS), noninvasive cortical stimulation methods, have been successfully employed for the treatment of movement disorders (Wu et al. 2008). Studies show beneficial effects on the clinical symptoms in PD and support the effects on motor and nonmotor symptoms. Rebalancing of the distributed neural network activity and induction of dopamine release occur.

While exploring the oxidative signaling and inflammatory pathways in AD (Anderson et al. 2001), it was shown that activation of microglia with the beta-amyloid peptide activates the production of cyclooxygenase-2, iNOS, and TNF-α. These are considered as key mediators of the pathological cascade of AD. ox-LDL caused a sustained activation of the JNK that resulted in the phosphorylation of the transcription factor c-jun, which was abolished in neurons pretreated with flavonoids. Furthermore, ox-LDL induced the cleavage of procaspase-3 and increased caspase-3-like protease activity in neurons and leading to apoptosis. Dietary flavonoids protect against neuronal apoptosis through selective actions within stress-activated cellular responses, including protein kinase signaling cascades (Schroeter et al. 2001). Guanosine protects human neuroblastoma cells against mitochondrial oxidative stress by inducing heme oxygenase-1 via PI3K/Akt/GSK-3β pathway (Dal-Cim et al. 2012).

Cdk5 (cyclin-dependent kinase 5), a proline-directed serine/threonine kinase, plays multiple roles in neurons development, survival, phosphorylation of cytoskeletal proteins, and synaptic plasticity (Smith and Tsai 2002). Uncontrolled phosphorylation activity of Cdk5 has been closely associated well with AD and PD. Under oxidative stress condition, mitochondrial dysfunctions, excitotoxicity, Aβ exposure, calcium dyshomeostasis, and inflammation lead to rise in the intracellular Ca^{2+}, activating calpain which cleaves p35 (activator of Cdk5) to p25 (Fig. 2.2) forming a more stable yet hyperactive Cdk5/p25 complex which aberrantly hyperphosphorylates various cytoskeletal proteins leading to neurodegeneration (Lee et al. 2000). Activation of Cdk by oxidative stress in AD causes hyperphosphorylation of τ, neurofilament, and other cytoskeletal proteins (Lee et al. 2000). Accumulation of Aβ in cortical neurons induces cleavage of p35 to p25 resulting in activation of kinases and inhibition of phosphatases proceeding NFT (neurofilament tangles) formation, primary markers of AD (Lee et al. 2000). Cdk5-mediated phosphorylation of peroxidases substrates reduces their enzymatic activities resulting in the ROS accumulation within cells (Sun et al. 2008).

HtrA2, a serine protease, was identified to be involved in the neuroprotection, and mutations adjacent to the two phosphorylation sites (S142 and S400) have been found in the PD patients. Cdk5 phosphorylates the HtrA2 at S400 in a p38-dependent manner in humans and mouse cell lines and brain (Fitzgerald et al. 2012). This phosphorylation is involved in maintaining mitochondrial membrane potential under stress conditions.

The activation of JNK pathways is critical for the naturally occurring cell death during development as well as for the pathological death associated with neurodegenerative diseases. Several in vitro and in vivo studies have reported alterations of JNK pathways potentially associated with the neuronal death in PD and AD. Also, Nrf2-ARE signaling pathway is an attractive therapeutic target for neurodegenerative diseases.

Cascades Leading to Dopamine Cell Degeneration

Metabolism of dopamine by the monoamine oxidase generates H_2O_2 and the auto-oxidation of dopamine generates superoxides. Thus, endogenous dopamine as well as exogenous treatment with levodopa (used in PD) may contribute additional oxidative stress insult, like mitochondrial dysfunction (Muller 2011). Also the monoamine oxidase (MAO)-induced metabolism of dopamine and production of H_2O_2 have an important role in the physiological calcium signaling in astrocytes (Vaarmann et al. 2010). In PD, adult substantia nigra pars compacta dopaminergic neurons create intracellular calcium oscillations through L-type calcium channels. This metabolic stress is counterbalanced by the ATP demanding pumps to restore the calcium concentration. It has been demonstrated that the opening of these L-type ion channels results in higher levels of oxidative stress in the mitochondria of such neurons (Surmeier et al. 2011).

Antioxidants Link in Neurodegenerative Disorders

The aim of using antioxidants in any pathology is to neutralize ROS and other kinds of free radicals produced as a consequence of the oxidative stress (Uttara et al. 2009). Brain cells and especially neurons require effective antioxidant protection because of the higher consumption of oxygen

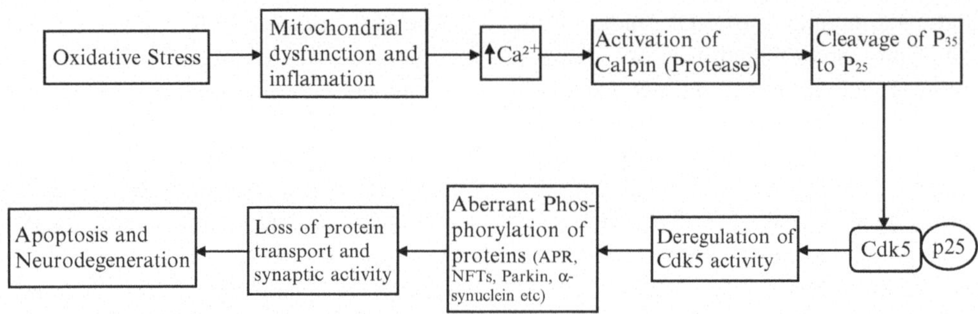

Fig. 2.2 Oxidative stress and Cdk5 in pathophysiology of neurons

(about tenfold), having long life duration and a prominent role of the nitric oxide to form RNS such as peroxynitrite. Glutathione peroxidase is known to localize primarily in glial cells, in which its activity is tenfold higher than in the neurons (Margis et al. 2008). Reduced glutathione (GSH, nonprotein thiol) is the main antioxidant in CNS (Dringen and Hirrlinger 2003) and non-enzymatically acts directly with free radicals. Glutathione peroxidase and glutathione reductase can act enzymatically to remove H_2O_2 and maintain glutathione in a reduced state (Dringen and Hirrlinger 2003).

Widely studied antioxidant therapies have been vitamin E (α-tocopherol, the major scavenger of lipid peroxidation in the brain), vitamin C (intracellular reducing molecule), and coenzyme Q10 (transfers electrons from the complexes I and II to complex III in the respiratory chain). Vitamin E supplementation in an AD mouse model resulted in the improved cognition and reduced Aβ deposition (Conte et al. 2004). The reduction of the amyloid deposition was particularly noted in young AD mice (Sung et al. 2004). Coenzyme Q10 has been shown to have multiple protective effects within the mitochondria. Administration of CoQ10 protects MPTP-treated mice from dopaminergic neuronal loss and also attenuated α-synuclein aggregation. Neuroprotection by CoQ10 in an MPTP-primate model has also been reported (Du and Yan 2010).

However, no antioxidant benefits of vitamin E and/or vitamin C in either AD or PD from large randomized controlled trials have been observed (Dumont et al. 2010). Furthermore, a large meta-analysis of vitamin E clinical trials, CoQ10 trials, and a glutathione trial in PD concluded that there were only minor treatment benefits in the CoQ10 trials that may have been due to the improvement in the respiratory chain deficit rather than a direct antioxidant action (Weber and Ernst 2006). Animal experiments show that antioxidants are effective in the early stages of the disease. Other considerations are to regulate the bioavailability and the effective targeting of the antioxidants. This aspect is discussed in detail in Chap. 6.

Another recent and efficient consideration is of exploiting signaling pathways to mimic the antioxidant activity. Very recently, guanosine have been found protective against mitochondrial oxidative stress in human neuroblastoma cells by a signaling pathway that implicates P13K/Akt/GSK-3β proteins and induction of antioxidant gene enzyme, heme oxygenase-1 (HO-1) (Dal-Cim et al. 2012). The importance of Nrf2-ARE signaling pathway has been well reviewed to be an attractive therapeutic target for neurodegenerative diseases with chemo-preventive agents (vanMuiswinkel and Kuiperij 2005). Nrf2, a key redox regulatory factor, induces endogenous cytoprotective genes of antioxidant- and anti-inflammatory proteins. Dopamine-induced mPTP opening and dopamine-induced cell death could be prevented by inhibition of ROS production by provision of respiratory chain substrates and by alteration in calcium signaling, which suggest potential therapeutic strategies for neuroprotection in PD (Gandhi et al. 2012).

Cardiovascular Diseases

Hypercholesterolemia and Atherosclerosis

Events like intake of high-fat diet (HFD), hypercholesterolemia in the blood, and cholesterol deposition in the arterial wall are accepted as high risk factors for the development of atherosclerosis. This risk has been positively correlated with low-density lipoproteins (LDL), total cholesterol, and total cholesterol/high-density lipoprotein (HDL) ratio (Castelli 1986). In its initial stages, atherosclerosis lesions in the intima of the large, elastic, and muscular arteries consist of the fatty streak that is characterized by the lipid (principally cholesterol and its esters) accumulation in macrophages, T lymphocytes, and smooth muscle cells in addition to the ingested lipoprotein–proteoglycan complexes in more complex foam cells (Ross 1991). This further leads to the fibrous plaques resulting from the synthesis of collagen, elastin, and proteoglycans by smooth muscle cells and macrophages migrated to the intima (Sowers 1992). Qualitative changes in these fibrous plaques, at some later stage, may result in hemorrhage, ulceration, and/or thrombosis, leading subsequently to the arterial occlusions. This results in the ischemic necrosis of vital organs with far-reaching consequences. Peroxidation of polyunsaturated fatty acids (PUFA) gives rise to free radicals and endogenous peroxides, which are highly reactive and have both chemotactic and cytotoxic properties. Hypercholesterolemic atherosclerosis is associated with an increase in the blood and aortic tissue of the MDA content (a LPO product) and OFR producing activity of the polymorphonuclear leukocytes (Prasad and Kalra 1992).

Increased concentration of LDL cholesterol in the plasma constitutes a major risk factor for the atherosclerosis as is demonstrated by various clinical, epidemiological, and genetic studies (Jialal and Devaraj 1996). Both diet-induced hypercholesterolemia and LDL-receptor defective models are characterized by the alteration in the level and composition of the plasma lipoproteins. Lipid-rich LDL and VLDL both have been shown to induce a dose-dependent increase in the monocyte adhesion to the endothelial cells (Endemann et al. 1987). Modified forms of LDL such as acetyl LDL or oxidized LDL are taken up by the scavenger receptor mechanism, resulting in cholesterol accumulation and subsequent foam cell formation (Brown and Goldstein 1983). Clinical and epidemiological studies show that increased levels of LDL cholesterol promote the atherosclerosis. LDL can be oxidatively modified by all major cells of the arterial wall and play a significant role in atherosclerosis in vivo. Macrophages play the role of scavengers as these cells have a large capacity to store altered LDL and diet-induced β-VLDL (Goldstein et al. 1980). Studies indicate that macrophages have only a limited number of receptors for the specific uptake of native LDL, but these can avidly take up certain chemically modified forms of LDL via an alternative specific, saturable receptor – the acetyl-LDL receptor (Parathasarathy et al. 1986).

Minimally oxidized LDL (MM-LDL), initially formed in the subendothelial space, can be taken up by the classical LDL receptor through the apoB and does not associate with the macrophages as normal LDL. However, a significant proportion of the unsaturated acyl chains of the cholesteryl esters and phospholipids in mid-oxidized LDL have been oxidized to hydroperoxides, isoprostanes, and short-chain aldehydes that have potent biological effects. This LDL stimulates production of the monocyte chemotactic protein-1 (MCP-1), resulting in monocyte binding to the endothelium and its subsequent migration into the subendothelial space where monocyte colony-stimulating factor (M-CSF) is also formed (Berliner et al. 1995). M-CSF promotes the differentiation and proliferation of monocytes into macrophages. These macrophages can in turn modify MM-LDL into a more oxidized form and are not recognized by the LDL receptor but become foreign and thus are taken up by the scavenger receptors pathway in macrophages leading to appreciable cholesterol ester accumulation and foam cell formation (Witzum and Steinberg 1991) and resulting in cholesterol accumulation (Fig. 2.3). Extensive oxidation up to 50 % of the cholesterol is

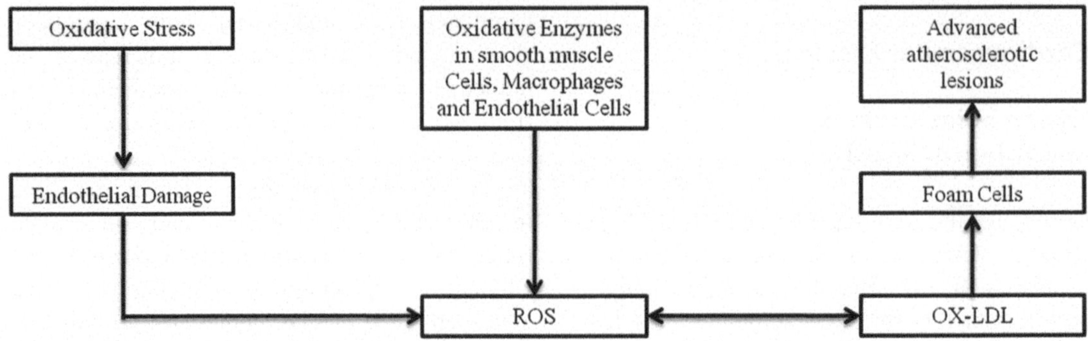

Fig. 2.3 Interaction of oxidative stress and atherosclerosis

converted into 7-ketocholesterol and other oxysterol and binds to the scavenger receptors such as SR-A and/or CD36. Further, most of the unsaturated fatty groups are oxidized to a complex mixture of products, and apoB is extensively fragmented, derivatized, and cross-linked and there is a substantial aldehyde modification to result in the products such as malondialdehyde and hydroxynonenal. Oxidatively modified LDL (ox-LDL) present novel properties, e.g., it is a potent chemoattractant for the monocytes, a potent inhibitor of macrophages mobility in the arterial wall (Jialal and Devaraj 1996), thus can promote its retention in the wall, is cytotoxic, which could promote endothelial dysfunction and atherogenesis by altering the expression of genes in the arterial wall (Jialal and Devaraj 1996).

Another lipoprotein, high-density lipoprotein (HDL), is known for its negative risk factor for the development of atherosclerosis. HDL apolipoprotein, ApoA1, promotes cholesterol efflux from the peripheral deposits within the vasculature and with subsequent transport to the liver for excretion (reverse cholesterol transport). HDL possesses the antioxidant activity that is primarily mediated via inhibition of the oxidation of LDL with a subsequent reduction of the cellular uptake by the monocyte–macrophage system and hence antiatherogenic effect (Nicholls et al. 2005). Antioxidant mechanism of HDL involves its chelation properties due to the presence of proteins such as ceruloplasmin on the surface of the lipoprotein (Kunitake et al. 1992), and also HDL has been demonstrated to accept hydroperoxides from oxidized membranes in in vitro studies, which would potentially provide a pathway for the excretion or detoxification (Klimov et al. 2001). Additionally, Apo A-I has been demonstrated to reduce lipid hydroperoxides into redox-inactive compounds, which thus terminates the chain reactions of lipid peroxidation (Garner et al. 1998). The function of HDL is also due to its associated enzyme proteins exhibiting antioxidant activity. Paraoxonase 1 (PON 1) is an HDL-associated esterase/lactonase that exhibits the anti-inflammatory and antioxidant activity (Kaur and Bansal 2009; Precourt et al. 2011). It degrades oxidized fatty acids within the LDL particle which in turn exhibits an inhibitory effect on a variety of pro-atherosclerotic functions, including a decrease in binding of the circulating monocytes to the endothelium.

It is clear that ROS are responsible for the endothelial dysfunction and development of atherogenesis. Several enzymatic pathways contribute within the vessel wall for the production of different oxidants, such as NADPH oxidase, nitric oxide synthase, myeloperoxidase, xanthine oxidase, lipoxygenase/cyclooxygenase, and mitochondrial respiratory chain/oxidative phosphorylation.

Endothelial NAD(P)H oxidase is a major source of ROS in the vasculature and can be activated by the stimuli such as angiotensin II, thrombin, platelet-derived growth factor, TNF-α, IL-1, and VEGF. C-reactive protein (CRP), a cardiovascular risk marker, has been reported to induce the superoxide production in human aortic endothelial and smooth muscle cells (Venugopal et al. 2003) and may regulate NAD(P)H oxidase through the various activation pathways involving protein kinases and nitric oxide synthases (NOS). In active eNOS, reductase domain (containing the binding sites for NADPH, FAD, and FMN) and an oxygenase domain (containing Zn, tetrahydrobiopterin, BH_4, heme, and L-arginine) are linked by a hinge region to which calmodulin binds (Stocker and Keaney 2001). Under normal conditions, these enzymes transfer electrons from a heme group in the oxygenase domain to the substrate L-arginine to form L-citrulline and NO; BH_4 serves as a cofactor in this process (Bevers et al. 2006). If the availability of either BH_4 or L-arginine decreases, eNOS switches from a coupled state (generate NO) to an uncoupled state (generate O_2^{*-}) because the electrons from the heme reduce oxygen to form O_2^{*-}. NO* reacts rapidly with O_2^{*-} to generate $ONOO^-$ which causes vascular dysfunction. iNOS is found in the vascular smooth muscle cells and also in activated macrophages in the atherosclerotic lesions. It is induced by the microbial endotoxins or cytokine stimulation (Murthy et al. 2004).

Myeloperoxidase (MPO), a heme-containing enzyme, catalyzes the conversion of Cl^- to the hydrochlorous acid (HOCl). Chlorinated biomolecules are considered specific markers of the oxidation reactions catalyzed by the enzyme. The $MPO/H_2O_2/Cl^-$ system can give rise to 3-chlorotyrosine, chlorohydrins such as those of cholesterol and fatty acids, α-chloro fatty acid aldehydes, and free amino acid or protein-bound tyrosyl radicals. Tyrosyl radicals themselves may participate in the secondary oxidation reactions, including the oxidation of LDL. $MPO/H_2O_2/Cl^-$ system and HOCl also oxidize nitrite to the nonradical oxidant, nitryl chloride (NO_2Cl) and the radical $^*NO_2$, both of which promote nitration and can covert tyrosine into 3-nitrotyrosine. MPO plays a major role in the generation of nitrating species in vivo, and that formation of 3-nitrotyrosine is strictly dependent on the availability of $^*NO_2$ (Carr and Frei 2001). MPO has been shown to co-localize with macrophages in the human artery wall, and its characteristic oxidation products have been detected in atherosclerotic lesions (Malle et al. 2000). In hypercholesterolemic rabbits, atherosclerosis resulting from the diet was ascribed to the xanthine oxidase-induced oxidative stress (Ohara et al. 1993). Lipoxygenase (LPO) is another important source of ROS production in the vascular wall and these nonheme-containing dioxygenases oxidize PUFA to hydroperoxy fatty acids derivatives (Kuhn et al. 2005). The mitochondrial ROS have also been shown to be associated with the enhanced susceptibility to the atherosclerosis.

Hypoxia and Stroke

Hypoxia basically refers to a reduced supply of oxygen to the part of a tissue or organ, and when brain is involved, it is called the cerebral hypoxia. It is caused by any event that severely interferes with the brain's ability to receive or process oxygen. Prolonged hypoxia induces the neuronal cell death via apoptosis resulting in a hypoxic brain injury (Malhotra et al. 2001). When the brain is traumatized by the low oxygen levels by choking off the blood supply, this condition is called brain stroke. The widespread self-destruction takes place for days or even a week after the initial stroke.

Oxidative stress plays an important role in the acute ischemic stroke pathogenesis. Free radical formation and subsequent oxidative damage may be a factor in the stroke severity. In one of the studies, serum NO, MDA, and GSH levels were significantly elevated in the acute stroke patients compared to the control within 48 h of stroke (Ozkul et al. 2007). The "neurological deficit score" was negatively correlated with both MDA and NO levels; however, GSH levels were taken as an adaptive mechanism during this period. In ischemic stroke, the cerebral vasculature is a major target of the oxidative stress playing a critical role in the pathogenesis of ischemic brain injury following a cardiovascular attack. Superoxide and its derivatives have been shown to cause the vasodilation via opening of K^+ channels and altered vascular reactivity, breakdown of the blood–brain barrier (BBB), and focal destructive lesions in the animal models of the ischemic stroke (Allen and Bayraktutan 2009).

Among the several stress factors known to induce BBB breakdown, hypoxia is probably the most represented. Evidence of the oxidative stress occurring during hypoxia/ischemic situation raises its possible contribution to the barrier breakdown (Lochhead et al. 2010). Oxygen deprivation injury constitutes one of the most important pathophysiological mechanisms leading to the BBB breakdown. Oxidative stress occurring under O_2 deprivation insult (Chandel et al. 1998) raises the possible contribution of ROS signaling to the BBB breakdown. Involvement of ROS in the RBE4 ECs barrier function disruption during hypoxia was evidenced (AI Ahmad et al. 2009). Further, it was demonstrated that the oxidative stress significantly contributes to the barrier breakdown because artificial generation of the ROS decreased EC integrity (AI Ahmad et al. 2012) and also treatment of RBE4 monolayers with antioxidants during O_2 deprivation stress resulted in overall improvement of both barrier function and cell survival. This provides insight into the effect of oxidative stress on the BBB function during hypoxic insult.

Further, hypoxia-inducible factor 1 (HIF-1), a master regulator of hypoxia-responsive genes, regulates the expression of a broad range of genes that facilitate adaptation to the low O_2 conditions. Its targets include genes that code for the molecules that participate in the vasomotor control, angiogenesis, erythropoiesis, cell proliferation, and energy metabolism. All of these genes may potentially contribute to the recovery of neuronal cells following cerebral ischemia and reperfusion, and hence regulating HIF-1 induction and accumulation is a highly promising therapeutic approach for the cerebral ischemia. A number of mechanisms have been proposed to account for the neuroprotective effect of the HIF-1 (Guo et al. 2009): expression of its downstream gene product erythropoietin has been found to protect cells from hypoxic/ischemic injuries; VEGF expression (another downstream gene of HIF-1) counteracts detrimental ischemic injuries; prevents apoptotic cell death through inhibition of cytochrome c release, caspase activation, and PARP cleavage; suppresses p53 activation; and thereby maintains cell survival.

HIF-1 may contribute to the cellular and tissue damage. It has been reported that the HIF-1 may mediate apoptosis during hypoxia/ischemia. HIf-1-induced apoptosis has been observed in the embryonic stem (ES) cells under hypoxic conditions (Carmeliet et al. 1998). The study indicates that in response to hypoxia, HIF-1α (an inducible subunit) accumulates, associates, and stabilizes the active wild-type p53. It is possible that this increase in the p53 protein is responsible for the apoptosis reported in the hypoxia ES cells. The experimental observations support that HIF-1α may induce cell death in a severe and prolonged ischemia and promote cell survival following mild ischemic insults (Baranova et al. 2007). Thus, HIF-1 plays an important role in the fate of ischemic insults with a double-edged sword effect. Its effects possibly depend on the degree of severity of the insult. Further explanations of the mechanism of the HIF-1 induction in ischemic neurons and its effect on the ischemic brain tissue are well reviewed by Shi (2009).

Further, considering the effective therapeutic targeting of the acute stroke, NOX4 is the most abundant vascular isoform, induced in stroke. Upon ischemia, NOX4 was induced in the human and mouse brain (Kleinschnitz et al. 2010). Mice deficient in the NOX4 ($Nox4^{-/-}$) of either sex were largely protected from the oxidative stress, blood–brain barrier leakage, and neuronal apoptosis, after both transient and permanent cerebral ischemia. Restoration of the oxidative stress reversed the stroke-protective phenotype in $Nox4^{-/-}$ mice. NOX4 therefore represents a major molecular source of oxidative stress in cerebral ischemia including some cases of human stroke and novel class of drug target for stroke therapy.

ROS and Myocardial Infarction

Myocardial infarction (MI), commonly known as heart attack, results from the interruption of the blood supply to a part of the heart, causing heart cells to die. This is most common due to the occlusion of the coronary artery following rupture of atherosclerotic plaque in the wall of an artery. The resulting ischemia and ensuing oxygen shortage, if left untreated for a sufficient period of time, can cause damage or death (infarction) of heart muscle tissue (myocardium). Further, hypoxia and hypoxia–reoxygenation (H/R) are components of the tissue ischemia and reperfusion implicated in the myocardial infarction. Reperfusion (or reoxygenation) injury is the tissue damage caused when blood supply returns to the tissue after a period of ischemia or lack of oxygen.

The inflammatory response is partially responsible for the damage of the reperfusion injury. White blood cells, carried to the area by newly returning blood, release a host of inflammatory factors such as the cytokines (reviewed in Neri et al. 2013) as well as free radicals in response to the tissue damage. Such reactive species may also act indirectly in the redox signaling to turn on apoptosis. White blood cells may also bind to the endothelium of small capillaries, obstructing them and leading to more ischemia.

In prolonged ischemia (60 min or more), the hypoxanthine is formed as breakdown of ATP metabolism. The enzyme xanthine oxidase results in molecular oxygen being converted into the highly reactive superoxide and hydroxyl radicals. Xanthine oxidase also produces uric acid, which may act both as a prooxidant and as a scavenger of the reactive species such as peroxynitrite. Excessive nitric oxide produced during the reperfusion reacts with superoxide to produce the potent reactive species peroxynitrite. Such radicals attack the cell membrane lipids, proteins, and glycosaminoglycans, causing further damage. They may also initiate specific biological processes by the redox signaling. In the first few minutes after the reperfusion, a cascade of biochemical changes results in the opening of the mitochondrial permeability pore (MPT pore) in the mitochondrial membrane of cardiac cells, water enters into the mitochondria to make it dysfunctional and collapse, and calcium released to overwhelm the next mitochondria causes mitochondria energy to reduce or stop completely, resulting in cell death. So protecting the mitochondria is a viable cardioprotective strategy (Hausenloy and Yellon 2008). Cyclophilin D is a protein induced by the excessive calcium flow to interact with other pore components and help in opening the MPT pore. Inhibiting cyclophilin D with the cyclosporine has been shown to prevent the opening of the MPT pore and protect the mitochondria and cellular energy production from the excessive calcium inflows (Javadov and Karmazyn 2007).

A number of studies have described the transplantation of the mesenchymal stem cells (MSCs) from the bone marrow as a strategy for the cardiac repair following myocardial infarction (Huang et al. 2010). However, the therapeutic efficacy of this procedure is greatly limited by the poor survival of the donor MSCs in the infarcted heart, especially because of the oxidative stress environment. It is widely reported that the HDL lowers the risks associated with the ischemic diseases (Duffy et al. 2012), especially because of the reverse cholesterol transport characteristic. In another study (Xu et al. 2012), preconditioning with the HDL resulted in the higher MSC survival rates, improved cardiac remodeling, and better myocardial function than in the MSC control group.

Studies have shown that the heat shock factor-1 (HSF1), a transcription factor for the heat shock proteins (HSPs), confers protection against the cardiovascular diseases, such as ischemia/reperfusion injury and myocardial infarction. HSF1 can prevent cardiomyocytes from apoptosis induced by the various stimulations and cytotoxic oxidative stress leads to apoptosis as a final event (Matsuzawa and Ichijo 2008). JNKs regulate the apoptosis of the H_2O_2-stimulated human pulmonary vascular endothelial cells and play an important role in regulating the left ventricular remodeling by promoting apoptosis (Yamaguchi et al. 2003). HSF1 and HSPs are protective against the oxidative damage (Yan et al. 2005), but also alleviates ischemia/reperfusion injury by prohibiting JNK activity (Zou et al. 2003). Thus, HSF1 may prevent the cardiomyocytes from apoptosis under the various stimulations via inhibition of the intracellular ROS production and then JNK activity. In a recent study, cultured cardiomyocytes of the neonatal rats were transfected with HSF1, ASK1, or both of them before exposure to the H_2O_2 and ROS generation, and JNK activity and apoptosis were examined (Zhang et al. 2011).

H_2O_2 increased intracellular ROS generation and apoptotic cells as expected, and all these cellular events were greatly inhibited by overexpression of HSF1.

Further, protein tyrosine phosphatase (PTP), important regulator in the cell signaling (as detailed in Chap. 5), serves as a molecular target for the ROS. Intermittent oxygenation of the cardiac tissue where reperfusion following ischemia is known to be an etiological factor for the tissue damage associated with the ischemic disease (Brandes et al. 2010). To explore the mechanism, both the respiratory system of mitochondria and NADPH oxidases have been implicated as sources of elevated ROS levels in situations of reperfusion or reoxygenation. To explain the impact of ROS in hypoxia–reoxygenation and ischemia/reperfusion on PTP activity/oxidation and its consequences for tyrosine signaling, a recent study was performed to investigate the potential effects of reoxygenation or reperfusion on PTP-oxidation and tyrosine kinase signaling using cell culture models and an ex vivo model of isolated perfused rat hearts (Sandin et al. 2011). This study demonstrated that the cultured cells exposed to hypoxia followed by reoxygenation and heart tissue subjected to ischemia/reperfusion are characterized by the increased oxidation of PTPs. Further analysis revealed that both cytosolic and receptor-like PTPs are susceptible to H/R-induced PTP-oxidation. Enhanced Erk1/2 phosphorylation was identified as PTP-oxidation sensitive signaling component, which was inactivated in a ROS-sensitive manner after treatment with antioxidant NAC. These findings have the general implication of hypoxia or ischemia affecting signaling processes under pathophysiological conditions.

Antioxidants and CVD

Seeing the association of oxidative stress with various CVDs, antioxidants were used to prevent these diseases in clinical trials with different formulations but produced mixed results. These are well reviewed (Singh and Jialal 2006; Vogiatzi et al. 2009). In contrast to the positive outcomes from various trials with different formulations with vitamins, e.g., vitamins E and C, other antioxidant supplementation studies did not show any positive effect on the primary endpoints related to the cardiovascular events. One apparent reason considered the unexplored threshold doses of the type of the antioxidant with its formulation. Another important parameter considered was the knowledge of the redox reactions in in vivo conditions. For instance, vitamin C supplementation exerts prooxidant and antioxidant effects and at high doses exhibit DNA damage.

Complex informations on the experimental studies regarding antioxidant influence exist. In these studies, the statins increase catalase and BH_4 levels and in turn increase NO production and inhibit LDL oxidation while at the same time restoring vitamin C and E levels and endogenous antioxidants such as ubiquinone and glutathione. Vitamins C and E can inhibit the oxidative process for the prevention of atherosclerotic lesions. Vitamin C stimulates the increase of BH_4 levels and the activity of NO synthase and improves endothelial dysfunction (Lonn et al. 2001). Also vitamin C administration in patients with coronary syndromes, arterial hypertension, and hypercholesterolemia increases NO bioavailability. Vitamin E administration also reduces LDL oxidation and improves NO bioactivity and endothelial dysfunction owing to the malnutrition. Co-administration of vitamins C and E seems to improve endothelial function in hyperlipidemic patients (Engler et al. 2003). Various investigators have related the ability of dietary antioxidant to prevent the formation of highly oxidized LDL. Natural antioxidants such as polyphenols, which are found in fruits and vegetables, seem to be extremely useful, can improve lipid metabolism, and reduce ox-LDL (Wassmann et al. 2001).

Reproductive Systems Disorders (Male and Female)

The reproductive system in an organism, whether in male or female, works for the purpose of reproduction, which is a fundamental characteristic of life. Apart from the external organs of the reproductive system, major internal organs include the gamete producing gonads (testicles or ovaries). Human reproduction takes place as internal fertilization of the female ovum with male sperms. Upon successful fertilization and implantation, gestation of the fetus then occurs within the female's uterus and finally birth of the child. The male reproductive system has one function, production of sperms, whereas the female reproductive system has two: the first is to produce egg cells and the second is to protect and nourish the offspring until birth.

Male Reproduction

The reproductive ability of sexually mature males is dependent upon the capacity of testes to produce large number of structurally and functionally active spermatozoa and maintenance of adequate levels of androgens (male sex hormones). Spermatogenesis is a precisely controlled process, occurring in the seminiferous tubules of testis, which gives rise to mature spermatozoa through a complex sequence of events that result in marked changes in the nuclei of the developing germ cells and finally formation of mature spermatozoa. The cycle of seminiferous epithelium of the testis is a dynamic and time-scaled phenomenon that forms well-defined cellular associations (or stages) within each tubule showing the various cell types in specific ratios to one another. Any alteration in these ratios indicates disturbance in the normal progression of spermatogenesis which can lead to male infertility.

Endocrinology and Gonadotoxicity

The principal androgen, testosterone, a steroid, is manufactured by the interstitial (Leydig) cells of the testes. Secretion of the testosterone increases sharply at puberty, and apart from the development of secondary sexual characteristics of men, testosterone is also essential for the production of sperms. Production of testosterone is controlled by the release of luteinizing hormone (LH) also called interstitial cell-stimulating hormone (ICSH) from the anterior pituitary gland, which in turn is controlled by the release of the gonadotropin-releasing hormone (GnRH) from the hypothalamus. The level of testosterone is under the negative-feedback control from hypothalamus:

$$\text{Hypothalamus} \rightarrow \text{GnRH} \rightarrow \text{Pituitary} \rightarrow \text{LH} \rightarrow \text{Testes} \rightarrow \text{Testosterone}$$

Of the many causes of gonadotoxicity in males, oxidative stress has been identified as one factor that affects fertility status and has been extensively studied. The generation of ROS can be exacerbated by a multitude of environmental, infectious, and lifestyle-related etiologies. A wide range of the industrial by-products and waste chemicals (polychlorinated biphenyls, nonylphenol, or dioxins) causes male infertility, both directly and indirectly. Increasing the presence of the by-products of manufacturing, such as lead, mercury, or cadmium in the environment, has been suggested to pose a serious threat to reproductive health. Lead has been reported to be gonadotoxic with a tendency of suppressing the LH and testosterone levels in animals (Taiwo et al. 2010). Sovol (a commercial mixture of polychlorinated biphenyls) was found to be gonadotoxic in male rat testis (decreased testis weight, sperm cell numbers in ejaculation, testicular weight, testosterone/estradiol in blood and increase in peroxidation) (Agletdinov et al. 2008). These results suggest that the disorders may play an important role in pathogenesis of the male infertility caused by the persistent organic pollutants. Also, with the advent of the modern cancer treatment, survival rates have improved substantially raising new concerns toward quality of life issues such as future fertility and offspring welfare. Chemopreventing agents act by hindering rapidly proliferating cells, hence exerting their gonadotoxic effect also (Ragheb and Sabanegh 2010). The extent of the damage to the germ cells and eventual fecundity depend on the class of chemotherapeutic agents, dosage, spermatogenetic stage targeted, as well as the original pretreatment fertility potential of the patients. In a study (Bahadur et al. 2005), semen quality from patients with leukemia, lymphoma, testicular cancer, and other malignant neoplasms before and after gonadotoxic treatment were monitored. All categories of the patients displayed varying degrees of azoospermia and oligospermia, and recovery of the gonadal function was not significant. This highlighted the importance of ensuring sperm banking before treatment.

Infertility

Infertility has been a major medical and social preoccupation; however, the past few decades have witnessed a remarkable decline in the fertility rates in the industrialized world. Out of the many well-known causes of male infertility, about 40–90 % of the cases are due to deficient and defective sperm production of unidentifiable origin. Oxidative stress is a common pathology seen in approximately half of all the infertile men. Oxidative injury to spermatozoa is considered as a major cause of the sperm dysfunction and the incidence of male infertility. Increased levels of ROS have been correlated with decreased sperm motility, increased sperm DNA damage, sperm cellular membrane lipid peroxidation, and decreased efficacy of oocyte–sperm fusion. All the cellular components, including lipids, proteins, nucleic acids, and sugars, are the potential targets of oxidative stress. The extent of oxidative stress-induced damage depends on the nature, amount, and the duration of the exposure of ROS and also on the extracellular factors such as temperature, oxygen tension, and the composition of the surrounding environment (Aitken and Fisher 1994). The following influences of ROS are observed:

Lipid Peroxidation (LPO): ROS attacks PUFA in sperm plasma membrane, leading to a cascade of chemical reactions called lipid peroxidation (Halliwal 1984). The free radicals react with fatty acid chains and release reactive lipid species, which further react with molecular oxygen to form the lipid peroxyl radical. Peroxyl radicals can react with fatty acids to produce lipid free radicals. Thus, lipid peroxidation in the spermatozoa is a self-propagating reaction.

Sperm Motility: The increased formation of ROS has been correlated with reduction of sperm motility (Armstrong et al. 1999). Decrease in motility is explained that H_2O_2 diffusion across the membranes into cells inhibits the activity of vital enzymes such as glucose-6-phosphatase dehydrogenase (G6PD) that control the rate of glucose flux via hexose monophosphate shunt and in turn control the intracellular availability of NADPH. Another hypothesis involves a series of interrelated events resulting in a decrease in axonemal protein phosphorylation and sperm immobilization, both of which are associated with the reduction in membrane fluidity that is necessary for sperm–oocyte fusion (deLamirande and Gagnon 1995).

DNA Damage: Exposing the sperm to artificially produced ROS causes DNA damage in the form of modification of all the bases, production of base-free sites, deletions, frame shifts, DNA cross-links, and chromosomal rearrangements. Oxidative stress also is associated with the high frequencies of single- and double-strand DNA breaks (Aitken and Krausz 2001). DNA bases and phosphodiester backbones are other sites that are susceptible to the peroxidative damage by ROS. High levels of ROS mediate the DNA fragmentation that is commonly observed in the spermatozoa of infertile individuals. Also, mutations in the mitochondrial DNA, which is also susceptible to oxidative damage, may cause defect of mitochondrial energy metabolism, and therefore lower levels of mutant DNA may compromise sperm motility in vivo (Spiropoulos et al. 2002).

Oxidative Damage to Protein: Oxidative attack on proteins results in the site-specific amino acid modifications, fragmentation of the peptide chain, aggregation of cross-linked reaction products, altered electric charge, and increased susceptibility or extreme tolerance to proteolysis. Primary, secondary, and tertiary protein structures alter the relative susceptibility of certain amino acids. Sulfur-containing amino acids and, specifically, thiol groups are very susceptible (Farr and Kogama 1991).

Apoptosis: ROS may also initiate a chain of reactions that ultimately lead to apoptosis. Apoptosis may help to remove abnormal germ cells and prevent their overproduction during spermatogenesis (Sakkas et al. 1999), thus maintaining the nursing capacity of the Sertoli cells. High levels of ROS cause DNA damage and disrupt the inner and outer mitochondrial membranes, releasing cytochrome c and activating the caspases and at least apoptosis.

Free Radicals and Sperm Functions

ROS are generated mainly by the sperm and seminal leukocytes within semen (Garrido et al. 2004) and produce infertility by two key mechanisms. First, they damage the sperm membrane, decreasing sperm motility and its ability to fuse with the oocyte. Second, ROS can alter the sperm DNA, resulting in the passage of defective paternal DNA on the conceptus. Several studies show positive/negative correlation between seminal leukocytes numbers and ROS production. Activation state of the leukocytes was considered to play an important role in determining final ROS output. This is supported by the observations of a positive correlation between seminal ROS production and proinflammatory seminal plasma cytokines such as interleukins (IL-6, IL-8), and TNFα.

Small amounts of ROS produced by the spermatozoa are essential to many of the physiological processes such as fertilization, capacitation, hyperactivation, motility, and sperm–oocyte fusion (Agarwal et al. 2004). ROS such as nitric oxide or superoxide anion have also shown to promote capacitation and the acrosome reaction (Griveau et al. 1995). They also act as second messenger molecules and transmit signals by increasing the influx of calcium ions, which leads to increased production of ATP through a series of chain reactions. Capacitation has been shown to occur in the female genital tract, a process carried out to prepare the spermatozoa for interaction with oocyte. During this process, the levels of intracellular calcium, ROS, and tyrosine kinase increase, leading to an increase in cAMP (Aitken 1995). This facilitates hyperactivation of the spermatozoa, a condition in which they are highly motile. However, only capacitated spermatozoa exhibit hyperactivated motility and undergo a physiological acrosome reaction, thereby acquiring the ability to fertilize (deLamirande et al. 1997).

Most semen specimens contain variable number of the leukocytes, with neutrophils as the predominant type, and are considered potential sources of ROS (Aitken 1995). Activated neutrophils generate and release ROS in high concentrations to form cytotoxic reactions against nearby cells and pathogens. Leukocytospermia has long been associated with decreased sperm concentration, motility, and morphology as well as decreased hyperactivation and defective fertilization (Moskovstev et al. 2007). Spermatozoa's own production of ROS is independent of the leukocytes and depends on the maturation level of the sperm.

During spermatogenesis, a defect of the cytoplasmic extrusion mechanism results in release of spermatozoa from germinal epithelium carrying surplus residual cytoplasm, and these cytoplasmic droplets are a major source of ROS (Gomez et al. 1996). The resulting spermatozoa are immature and functionally defective, and residual cytoplasm by spermatozoa is positively correlated with ROS generation via mechanisms that may be mediated by the cytosolic enzyme glucose-6-phosphate dehydrogenase (G6PD) (Aitken 1999). G6PD (NADPH oxidase, NOXs) at the sperm plasma membrane controls the glucose flux and intracellular production of β-nicotinamide adenine dinucleotide phosphate (NADPH) through the hexose monophosphate shunt. NADPH is used to fuel the generation of ROS via NADPH oxidase located within the sperm membrane. NADPH-dependent oxidoreductase (diaphorase) at the mitochondrial level also contributes ROS (Gavella and Lipovac 1992). As a result, teratozoospermic sperm produces increased amounts of ROS compared with morphologically normal sperm.

Further, one group investigated that NOX 5 enzyme of sperm is a calcium-dependent NADPH oxidase and is quite distinct from leukocyte NADPH oxidase, with NOX 5 activity not being controlled by protein kinase C as occurs in the leukocyte. While intrinsic (by sperm) and extrinsic (by leukocyte, 1000×) ROS production is negatively correlated with sperm DNA integrity, the relationship is significantly stronger for the intrinsic ROS production. The close proximity between intrinsic ROS production and sperm DNA makes it a more important variable in terms of fertility potential. Also spermatozoa are rich in mitochondria for the constant supply of energy for their motility. Unfortunately, when spermatozoa contain dysfunctional mitochondria, increased production of ROS occurs, affecting further mitochondrial function. Such a relationship could be due to two mutually interconnected phenomena: ROS causing damage to the mitochondrial membrane and the damaged mitochondrial membrane further causing an increase in ROS production. Increased ROS levels have been correlated with decreased sperm motility. One hypothesis suggests that H_2O_2 diffuses across the membrane into the cells and inhibits the activity of some vital enzymes. Another theory involves a series of interrelated events resulting in a decrease in axonemal protein phosphorylation and sperm immobilization, both of which are associated with a reduction in membrane fluidity that is necessary for sperm–oocyte fusion. Loss of motility observed when spermatozoa are incubated overnight is highly correlated with the lipid peroxidation status of the spermatozoa.

Varicocele patients (dilatation of testis veins) have increased ROS in serum, testis, and semen samples. Increased nitric oxide also has been demonstrated in the spermatic veins of patients with varicocele, which could be responsible for the spermatozoa dysfunction (Ozbek et al. 2000). ROS in patients with varicocele is due to the excessive presence of xanthine oxidase, a source of superoxide anion from the substrate xanthine and nitric oxide in dilated spermatic veins. On the other hand, it has been recorded that varicocelectomy increases the concentration of antioxidants such as SOD, catalase, GPx, and vitamin C, in seminal plasma as well as improves sperm quality (Mostafa et al. 2001). Patients with varicocele had increased 8-hydroxy-2-deoxyguanosine (8-OHdG), indicating oxidative DNA damage (Smith et al. 2006). Analysis conclude that oxidative stress significantly increased in infertile patients with varicocele as compared with normal sperm donors and antioxidant concentrations were significantly lower in infertile patients compared with controls.

Antioxidants' Role

Increased ROS generation in males with suboptimal sperm quality has been elucidated, offering multiple targets for a potential therapy. The high rate of mitosis and metabolic activity during spermatogenesis in the seminiferous tubules makes the germ cells highly sensitive to the free radicals, thus creating a need for an effective antioxidant system. The germinal cells in the testis as well as the epididymal spermatozoa are equipped with enzymatic and nonenzymatic scavenger systems to prevent lipoperoxidative damage. Seminal plasma and sperm themselves also have an array of the protective antioxidants.

Three basic endogenous antioxidant enzymes (superoxide dismutase, SOD; catalase and glutathione peroxidase, GPx) play a significant role (Tremellen 2008). The cytosolic Cu/Zn-SOD is a remarkably dominant SOD isoenzyme in the seminal plasma as well as in spermatozoa (Peeker et al. 1997). Addition of SOD to sperm in culture has been confirmed to protect them from oxidative attack. The majority of evidence does support a link between deficient catalase activity and male infertility. Catalase with Cu/Zn-SOD removes O_2^- and may play an important role in decreasing lipid peroxidation and protecting spermatozoa during genitourinary inflammation (Sikka et al. 1995). Glutathione peroxidases (GPx 1–5) are a family of enzymes. This enzyme is located and is active in almost all the reproductive organs. Male factor infertility has been linked with a reduction in seminal plasma and spermatozoa GPx activity. The classic intracellular GPx1 is expressed in sperm/genital tract and a direct relationship has been demonstrated with sperm motility (Dandekar et al. 2002). More significantly, a direct relationship has been reported between male fertility and phospholipid hydroperoxide glutathione peroxidase (PHGPx or GPx4), a selenoprotein that is highly expressed in testicular tissue. In addition coordinated activity of GPx, glutathione reductase (GR, regenerate glutathione), and glutathione clearly plays a pivotal role in protecting sperm from oxidative attack. Other enzymes, such as glutathione-S-transferases, ceruloplasmin, or heme oxygenase-1, may also participate in the enzymatic control of oxygen radicals and their products (Tremellen 2008).

The nonenzymatic antioxidants related to the male reproductive system include ascorbic acid (vitamin C), α-tocopherol (vitamin E), glutathione, amino acids (taurine, hypotaurine), albumin, carnitine, carotenoids, flavonoids, urate, coenzyme

Q-10, resveratrol, and prostasomes. These agents principally act by directly neutralizing free radical activity. Coenzyme Q-10 is an antioxidant that is related to low-density lipoproteins and protects against peroxidative damage. Since it is an energy-promoting agent, it also enhances sperm motility (Lewin and Lavon 1997). It is present in the sperm midpiece and recycles vitamin E and prevents its prooxidant activity (Aitken et al. 1993). Albumin also helps neutralize lipid peroxide-mediated damage to the sperm plasma membrane and DNA (Twigg et al. 1998). Extracellular organelles (prostasomes) secreted by the prostate have been shown to fuse with leukocytes within semen and reduce their production of free radicals (Saez et al. 1998). A significant reduction in nonenzymatic antioxidant activity in seminal plasma of infertile compared with fertile men has been reported.

Vitamin E (tocopherol) is a major antioxidant in the sperm membranes and appears to have a dose-dependent effect and plays a vital role in protecting cell membranes from oxidative damage by scavenging all the three major types of free radicals (Suleiman et al. 1996). Vitamin C is an important water-soluble antioxidant, neutralizes hydroxyl superoxide and hydrogen peroxide radicals, and prevents sperm agglutination (Agarwal et al. 2004). It prevents lipid peroxidation, recycles oxidized vitamin E, and protects against DNA damage induced by H_2O_2 radicals (Kodama et al. 1997). Resveratrol is a potential lipid-soluble antioxidant that is commonly found in many plants. It inhibited lipid peroxidation of ram semen most effectively even when applied in low concentrations (Sarlos et al. 2002).

Female Reproduction

In mammals, oogenesis starts in the germinal epithelium in the development of the ovarian follicles, the functional unit of the ovary. Oogenesis consists of several subprocesses with final maturation to form an ovum. Folliculogenesis is a separate subprocess that accompanies and supports all oogenetic subprocesses.

Endocrinology and Gonadotoxicity

Ovaries of the sexually mature females secrete a mixture of estrogens (17β-estradiol is the most abundant and potent) and progesterone. Apart from the development of secondary sexual characteristics of the female, estrogens (steroids) are responsible for the monthly preparation of body for a possible pregnancy and its maintenance if it occurs. Progesterone is also a steroid and has a role in the menstrual cycle and pregnancy. Estrogens and progesterones are small hydrophobic molecules that are transported in the blood bound to a serum globulin. The hormone-receptor complex enters the nucleus (if it is formed in the cytoplasm) and binds to the specific sequences of DNA, called the estrogen (or progesterone) response elements. Response elements are located in the promoters of genes. The hormone-receptor complex acts as a transcription factor (often recruit other transcription factors for help) which turns on (or sometimes off) the transcription of the target genes.

The synthesis and secretion of estrogens are stimulated by FSH, which in turn is controlled by the hypothalamic gonadotropin-releasing hormone (GnRH). Progesterone production is stimulated by the LH, which is also stimulated by GnRH:

$$\text{Hypothalamus} - \text{GnRH} - \text{Pituitary} - \text{FSH} \\ -\text{Follicle} - \text{Estrogen}\left(\text{negative feedback}\right)$$

$$\text{Hypothalamus} - \text{GnRH} - \text{Pituitary} - \text{LH} - \text{Corpus} \\ \text{luteum} - \text{progesterone}\left(\text{negative feedback}\right)$$

About every 28 days, some blood and other products of the disintegration of the inner lining of the uterus, endometrium, are discharged from the uterus, a process called menstruation. During this time, a new follicle begins to develop in one of the ovaries. After menstruation ceases, the follicle continues to develop, secreting an increasing amount of estrogen which causes the endometrium to become thicker and more richly supplied with blood vessels and glands. A rising level of LH causes the developing egg within the follicle to complete the first meiotic division (meiosis 1), forming a secondary oocyte. After about 2 weeks, there is a sudden surge in the production of LH which triggers ovulation: the release of the secondary oocyte into a corpus luteum. Stimulated by LH, the corpus luteum secretes progesterone which continues the preparation of the endometrium for a possible pregnancy and inhibits the contraction of the uterus and development of a new follicle. If fertilization does not occur, the rising level of progesterone inhibits the release of GnRH which, in turn, inhibits further production of progesterone. As the progesterone level drops, the corpus luteum begins to degenerate and the endometrium begins to break down via apoptosis. The inhibition of the uterine contraction is lifted and the bleeding and cramps of menstruation begin.

Aggressive chemotherapy and radiotherapy used for the treatment of some cancers and autoimmune disorders are the most common causes of gonadotoxicity and subsequent infertility. Patients receiving chemotherapy are at risk of developing "premature ovarian failure" (POF, a well-known consequence of the exposure of the female gonad to chemotherapeutic drugs). The well-known gonadotoxic cyclophosphamide-based multiagent cytotoxic chemotherapy is one of the combinations of choice in treating female breast cancer (Kaufmann et al. 2003). Cancer of the cervix is another malignancy that affects the reproductive age of women, and some of those patients receive radiosensitizing chemotherapy which again might affect their gonads. Co-treatment with GnRH agonist may reduce ovarian damage significantly in the female patients treated for Hodgkin lymphoma and is considered in addition to assisted reproduction for women in the reproductive age receiving gonadotoxic chemotherapy (Blumenfeld et al. 2008). Also, in a study (Brougham et al. 2012) anti-Mullerian hormone (AMH), detectable in girls of all ages, falls rapidly during cancer treatment in both the prepubertal and pubertal age. Both fall during the treatment and recovery thereafter varied with the risk of gonadotoxicity. AMH is used as a marker of damage to the ovarian reserve in girls receiving treatment of cancer.

Oxidative Stress and Infertility

Oxidative Stress, Oogenesis, and Folliculogenesis: ROS may have a regulatory role in the oocyte maturation, folliculogenesis, ovarian steroidogenesis, and luteolysis. Mammalian ovulation or follicular rupture results from the vascular changes and the proteolytic cascade. This is mediated by the cytokines, VEGF and ROS (both nitrogen and oxygen radicals). Interleukin-1β causes nitrite to accumulate in the rat ovaries, demonstrating close interaction between the cytokines and NOS (Ben-Shlomo et al. 1994).

There is a delicate balance between the ROS and antioxidant enzymes in ovarian tissues. Expression of various markers of the oxidative stress have been demonstrated in normal cycling ovaries (Suzuki et al. 1999), and their concentrations have been demonstrated to be lower in the follicular fluid than in the blood, suggesting that follicular fluid contains highly active antioxidant system (Jozwik et al. 1999). Enhanced expression of the luteal Cu/Zn-SOD may be due to the hCG which may have an important role in the maintenance of the corpus luteal function in pregnancy. Also nitric oxide radical is one of the local factors involved in the ovarian folliculogenesis and steroidogenesis. NO binds to the heme-containing enzyme guanylate cyclase, which activates the cyclic-GMP (LaPolt et al. 2003). Plasma concentration of nitrate monitored during follicular cycle has revealed peak levels at ovulation (Ekerhovd et al. 2001). NO inhibits the ovarian and corpus luteum steroidogenesis (Seino et al. 2002) and has luteolytic action mediated through the increased prostaglandins and apoptosis (Vega et al. 2000). The preovulatory follicle has a potent antioxidant defense, which can be exhausted by the intense peroxidation (Aten et al. 1992). Transferrin, a blood plasma glycoprotein that binds the iron, is known to suppress ROS generation and has been proven an important factor for the successful development of the follicles.

Oxidative Stress, Endometrium, and Endometriosis: Oxidative stress is involved in the modulation of cyclic changes in the endometrium. There is a cyclical variation in the expression of SOD in the endometrium. Elevated lipid peroxidation and decreased SOD activity in the late secretory phase with increased ROS levels (Sugino et al. 2004) have been linked to be

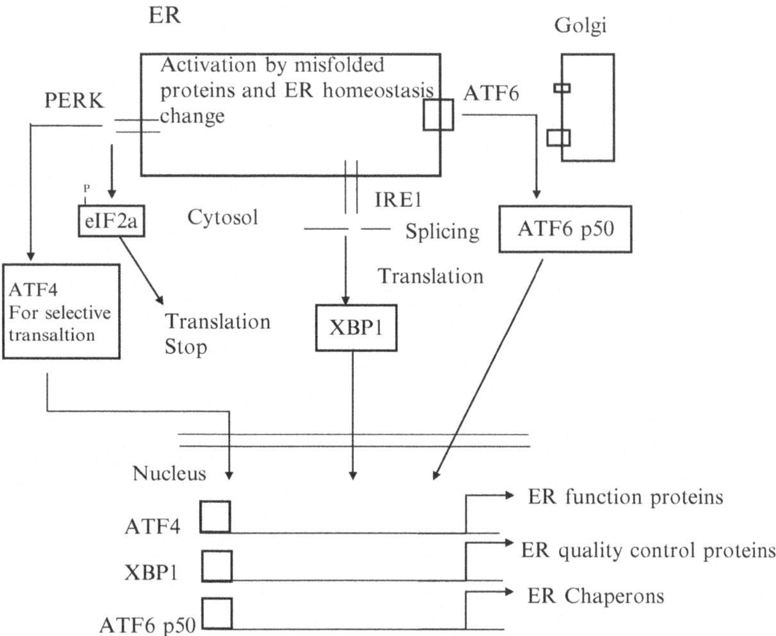

Fig. 2.4 Activation and response of unfolded protein response (UPR) pathways

important in the genesis of menstruation and endometrial shedding. The expression of eNOS and iNOS has been demonstrated in the human endometrium and endometrial vessels (Ota et al. 1998). NO is thought to regulate the microvasculature of endometrium. eNOS is also thought to bring about the changes that prepare endometrium for implantation. Stimulation of the cyclooxygenase enzyme is brought about by the ROS via activation of the NF-kB, suggesting a mechanism for menstruation (Sugino et al. 2004). VEGF and Ang-2, key regulators of endometrial angiogenesis, are induced by hypoxia and ROS (Park et al. 2006), and their expression changes are thought to play an integral role in producing the abnormally distended and fragile vessels. Oxidative stress is thus implicated in the genesis of endometrial pathophysiology (Hickey et al. 2006). Endometriosis, blockage of sperm–egg union, is a complex phenomenon. Women with endometriosis have increased peritoneal fluid, macrophages, cytokines, and prostaglandins. ROS from macrophages may increase growth and adhesion of the endometrial cells in the peritoneal cavity, promoting endometriosis adhesions and infertility (Alpay et al. 2006). However, this etiology is controversial as others.

Further, the concentration of ROS plays a major role both in the implantation and fertilization of eggs (Sharma and Agarwal 2004). More severe attack by the ROS may lead to more extensive and irreparable cell damage, resulting ultimately in death through necrosis or apoptosis. These pathological effects are mediated by the opening of ion channels, lipid peroxidation, protein modifications, and DNA oxidation. ROS activate the calcium release channels in the ER membrane, which include the inositol-1,4,5-trisphosphate receptor, IP_3R, and the ryanodine receptor (Hool and Corry 2007). Ca^{2+} release activates diverse Ca^{2+}-sensitive processes within the cell (Hool and Corry 2007), and loss of chaperone activity results in the accumulation of misfolded proteins within the lumen, leading to further generation of ROS as attempts are made to refold them (Tu and Weissman 2004). Accumulation also stimulates the unfolded protein response (UPR), a highly conserved set of signaling pathways (Fig. 2.4) that aim to restore homeostasis, but if this fails, it will stimulate apoptosis (Ron and Walter 2007). Rise in the cytosolic Ca^{2+} ion concentration will also adversely affect mitochondrial function, including an increase in their own production of the ROS and opening of the permeability transition pore (PTP). As a result, the mitochondrial membrane potential, ATP synthesis, and ionic homeostasis fail and the cell undergoes necrosis or apoptosis (Leist et al. 1997).

A complex cytokine influence at the maternal–fetal interface creates conditions that are necessary to support the embryo implantation in the endometrium (Krussel et al. 2003). Critical changes occur in the vascular system which accompany follicular growth. As endometrium grows in the menstrual cycle, vessel regeneration occurs, i.e., spiral arteries and capillaries (Bausero et al. 1998). Estrogen promotes angiogenesis in the endometrium by controlling the expression of factors such as VEGF. ROS generated from the NADPH oxidase is critical for VEGF signaling in vitro and angiogenesis in vivo (Ushio-Fukai and Alexander 2004). Small amounts of ROS are produced from the endothelial NADPH oxidase activated by growth factors and cytokines.

Oxidative Stress, Pregnancy, and Placental Changes: Oxidative stress plays a role in both the normal development of placenta and in the pathophysiology of the complications such as miscarriage, preeclampsia, intrauterine growth restriction (IUGR), and premature rupture of the membranes. Development of placental hypoxia, reperfusion, and in turn oxidative stress triggers the release of cytokines and prostaglandins, which results in the endothelial cell dysfunction and plays an important role in the development of preeclampsia (Bilodeau and Hubel 2003). Activation of the mononuclear phagocytes can be triggered in the endometriosis by a number of factors including damaged RBCs and the apoptotic endometrial cells. A positive correlation between the concentrations of TNF-α in the peritoneal fluid and endometriosis has been reported (Bedaiwy and Falcone 2003).

The placenta at the start is supported by the secretions from the endometrial glands with low oxygen concentration which is more protective for the developing embryo rather than the maternal circulation (Burton et al. 2003). Maternal arterial blood is prevented from entering the intervillous space of the placenta by the plugs of the extravillous cytotrophoblast cells that invade down the mouths of the uterine spiral arteries. The maternal intraplacental circulation recovers fully toward the end of the first trimester, when these plugs dislocate the circulation in the periphery of the placenta, where trophoblast invasion is least and progressively extends into the central region (Jauniaux et al. 2003). Onset of the circulation is associated with a threefold rise in the oxygen concentration within the placenta, stimulation of ROS generation, particularly in the critical syncytiotrophoblastic layer, which contains low concentrations of the principal antioxidant enzymatic defenses.

NO also regulates the microvasculature of the endometrium and is important in menstruation. Expression of iNOS was highest in patients with preterm pregnancy and not in patients in term labor. The expression of these enzymes decreased by 75 % at the term and was barely detectable in preterm in labor patients or term labor patients (Bansal et al. 1997), reiterating that NO has a role in the maintenance of uterine quiescence. Low levels of NO are important in ovarian function and implantation and cause relaxation of oviduct musculature. High levels of NO are reported as having deleterious effects on sperm motility, are toxic to embryos, and inhibit implantation (Lee et al. 2004). High levels of NO, such as those produced by macrophages, can negatively influence fertility. High levels of NO adversely affect sperm, embryos, implantation, and oviductal function, indicating that reduction in the peritoneal fluid NO production or blocking NO effects may improve fertility in women with endometriosis (Osborn et al. 2002).

Oxidative Stress and Spontaneous Miscarriage: Any imbalance between the cytokines and angiogenesis factors could result in the implantation failure and pregnancy loss (Choi et al. 2003). In cases of miscarriage, onset of the maternal intraplacental circulation is disorganized (Jauniaux et al. 2000), and it starts at an earlier stage and occurs randomly throughout the placenta. In 70 % of these cases, extravillous trophoblast invasion is superficial and consequently plugging of the spiral arteries is less complete. The apoptotic index is increased compared with control placentas of a similar gestation age, and there is morphological evidence of degenerate syncytiotrophoblast sloughing off in some areas. In these cases, it seems that increased oxidative stress causes widespread destruction of the trophoblast. In confirmation of these findings, increased lipid peroxides in villous, decidual tissues and the serum of women undergoing pregnancy loss have been observed (Toy et al. 2010). High increase in the oxidative stress in the placenta takes place at 10–12 weeks of gestation on adapting to the maternal environment which causes increase in the expression and activity of the antioxidant enzymes (Jauniaux et al. 2000). Polymorphisms in the enzymes detoxifying ROS have been linked to an increased risk of miscarriage (Sata et al. 2003). Also the selenium deficiency with reduced activity of glutathione peroxidase is associated with miscarriage (Zachara et al. 2001).

Placental Oxidative Stress in Preeclampsia: Normal pregnancy is said to be a condition of the oxidative stress, as circulating levels of the ox-LDL increase and the total antioxidant capacity in pregnant women decreases compared with nonpregnant women (Belo et al. 2004). Pregnancy is also associated with a systemic inflammatory response, as evidenced by the activation of peripheral granulocytes, monocytes, and lymphocytes during the third trimester, all of which produce ROS. These states are observed to a much greater degree in preeclampsia. There is clear evidence of the placental oxidative stress in cases of the early onset preeclampsia, including increased concentrations of the protein carbonyls, lipid peroxides, nitrotyrosine residues, and DNA oxidation (Burton et al. 2009). The cause of the oxidative stress is thought to be vascular, because early onset of preeclampsia is associated with deficient conversion of the spiral arteries. In particular, the myometrial segments of the arteries are adversely affected. As the myometrial segment contains a highly contractile portion of the artery, it is proposed that failure to convert this section results in intermittent perfusion of the placenta and a low-grade ischemia–reperfusion-type injury (Hung et al. 2001). In support of this hypothesis, it is shown that hypoxia–reoxygenation in vitro is a potent inducer of the oxidative stress in term placental explants, much more than hypoxia alone. Exposure of explants to changes in oxygenation causes generation of the ROS within the nitrotyrosine residues in a pattern matching closely to that seen in preeclamptic placentas. Furthermore, labor, in which the placenta is exposed to repeated episodes of ischemia–reperfusion, induces high levels of oxidative stress (Cindrova-Davies et al. 2007a).

Early onset of preeclampsia is associated with intrauterine growth restriction (IUGR) and high levels of ER stress in these placentas (Yung et al. 2008; Burton and Yung 2011). Induction of similar stress in trophoblast-like cell lines causes a reduction in their proliferation rate. In addition, the high levels of ER stress may contribute to the inflammatory response by stimulating the p38 and NF-kB pathways. Hence, both ER stress and oxidative stress may contribute to the placental pathophysiology in preeclampsia (Burton et al. 2009). Increased phosphorylation of IkB, an inhibitory subunit of NF-kB, is observed in term placental explants subjected to hypoxia–reoxygenation in vitro, which provides a model for malperfusion of the placenta in vivo (Hung et al. 2001). Activation of the pathway is associated with increased tissue levels of the proinflammatory enzyme COX-2 and interleukin-1β, increased secretion of TNF-α, and activation of the apoptotic cascade by the cleavage of caspase 3 (Cindrova-Davies et al. 2007b). Further, increased phosphorylation of p38 is observed in the term placenta after labor compared with control participants delivered by caesarean section (Cindrova-Davies 2009). ASK1 (upstream kinase of p38 and SAPK-JNK) is also activated in explants exposed to either hypoxia–reoxygenation or H_2O_2 (Cindrova-Davies 2009). Activation is associated with increased levels of the soluble receptor for VEGF, which has been implicated in the pathogenesis of preeclampsia.

Role of Antioxidants

Earlier well-known basic enzymatic and nonenzymatic antioxidants were suggested to protect the oocyte and the embryo from oxidative stress by detoxifying and neutralizing the ROS production (Attaran et al. 2000). Considering antioxidants as a potential therapy for preeclampsia, vitamins C and E trials have not been successful (Roberts et al. 2010; Xu et al. 2010). However, in vitro experiments show positive results (Cindrova-Davies 2009). The difference may result from the ability of the vitamins to access the relevant trophoblast cell compartment in the necessary concentration in vivo. It is notable that the multivitamin usage during the preconceptional period is associated with a reduced risk of preeclampsia among lean or normal weight women (Catov et al. 2009). Conversely, women with a low dietary intake of vitamin C have been reported to have a trend toward increased risk (Klemmensen et al. 2009).

Autoimmune Diseases

Autoimmune disorder is a condition that occurs when the immune system mistakenly attacks and destroys the healthy body tissue. In patients with an autoimmune response, result in an hypersensitivity reaction, similar to the response in allergic conditions. In allergies, the immune system reacts to an outside substance, whereas with autoimmune disorders, the immune system reacts to normal body tissues. Organs and tissues commonly affected by the autoimmune disorders include blood vessels, connective tissues, endocrine glands such as the thyroid or pancreas, joints, muscles, red blood cells, and skin. Autoimmune diseases are multifactor diseases to which hereditary dispositions and environmental factors are related. Oxidative stress affects immune systems directly or indirectly. In the present write-up, three such disorders having link with oxidative stress have been discussed: HIV, colitis, and rheumatoid arthritis.

HIV

Long back it was proposed that the oxidative mechanisms are of critical significance in the genesis of AIDS (acquired immune deficiency syndrome) and then further predicted that the mechanisms responsible for AIDS could be reversed by the administration of the reducing agents, especially those containing the sulfhydryl groups. The discovery of the HIV (human immunodeficiency virus) supported these as it considered the oxidative stress as a principal mechanism in both the development of AIDS and expression of HIV (Papadopulos-Eleopulos et al. 1989). In further experimentation, researchers found that the asymptomatic HIV-infected individuals and AIDS patients have decreased sulfhydryl and total glutathione and also the reducing agents suppress the expression of HIV. Since the viral production require thiols, which they obtain from the host, it may be assumed that the decreased SH level in the HIV-positive individuals may be the result of the HIV infection. However, for the HIV expression, oxidative stress is a prerequisite (Papadopulos-Eleopulos et al. 1991). The systemic decrease of the glutathione concentration in the HIV seropositive individuals may result from both decrease in synthesis and increased degradation. The oxidative stress to which the AIDS patients are subjected would lead to the cellular anomalies in many cells, including lymphocytes, resulting in the opportunistic infection, immunological abnormalities, and neoplasia. All these show in favor of the oxidation as being a critical factor in the pathogenesis of the AIDS and HIV expression.

HIV/AIDS patients suffer from several infections because of the poor immune system, especially as CD4-T cell immunodeficiency. Different factors released may trigger apoptosis in $CD4^{+}$ T cell, including viral protein (i.e., gp 120, Tat), inflammatory cytokines from the activated macrophages (i.e., TNF-α), and toxins from microorganisms. In the HIV-infected patients, increased oxidative stress has been implicated in the increased HIV transcription through the activation of the NF-kB (Greenspan and Aruoma 1994). Glutathione (GSH) is a major intracellular thiol, which acts as a free radical scavenger and is thought to inhibit the activation of NF-kB (Sharon et al. 1997). NF-kB is involved in the transcription of HIV-1. Thus, ROS may potentially be involved in the pathogenesis of the HIV infection through direct effects of the cells and through the interactions with the NF-kB and activation of the HIV replication. The viral Tat protein liberated by the HIV-1-infected cells interferes with the calcium homeostasis, activates caspases, and induces mitochondrial generation and accumulation of the ROS, all being important events in the apoptotic cascades of several cell types. $CD4^+$ T cell subset depletion in the HIV/AIDS patients is the most dramatic effect of the apoptosis mediated by redox abnormalities and induction of Fas/APO-1/CD95 receptor expression (Jaworowski and Crowe 1999). The proportion of the lymphocytes expressing Fas was shown to be elevated in the HIV-infected individuals. Some micronutrients play an essential role in maintaining the normal immune function and may protect immune effector cells from the oxidative stress (Meydani and Beharka 1998). Thus, infection by the HIV causes the persistent chronic inflammation through the intracellular increase of ROS, thus increasing the apoptotic index, mostly the one mediated by FAS/CD95, and depleting CD4+ T lymphocytes.

In the HIV/AIDS patients (Gil et al. 2003), the redox-related parameters and various types of the T lymphocytes load were studied and compared to the healthy subjects. Reduction of GSH levels and an increase in the MDA and total hydroperoxides levels were detected in the plasma of HIV^+ patients. These patients also showed an increase of the DNA fragmentation in the lymphocytes as well as a significant reduction of the glutathione peroxidase and an increase in the SOD activity in erythrocytes. These results also show that the substantial oxidative stress occurs during HIV infection.

Since the discovery of HIV infection, numerous antiretroviral drugs that control the disease when administered in a potent combination are referred to as the highly active antiretroviral therapy (HAART). This therapy reduces the viral load and improves immune system reconstitution, leading to a significant reduction of the HIV-related morbidity and mortality (McArthur and Brew 2010). However, the HAART does not completely eliminate HIV and the treatment must continue for long, which has been related to the long-term adverse events that can compromise the patient health. The prevalence of the HIV-associated neurocognitive disorder (HAND) is increasing as the HIV-infected individuals are living longer. HAND is manifested by the enhanced neuroinflammation, reactive astrocytes, formation of multinucleated giant cells, blood–brain barrier (BBB) damage, formation of microglial nodules, and neural apoptosis associated with increased viral replication and deterioration of the immune responses (Kanmogne et al. 2007). HAND has been characterized by the development of cognitive, behavioral, and motor abnormalities and occurs in about 50 % of the HIV-infected individuals (McArthur and Brew 2010).

As presented above, the oxidative stress is associated with HIV infection and therefore is true for the neurological disorders too. Studies have reported that HIV-1 viral proteins including gp120 and tat, released from the infected cells, induce oxidative stress in the CNS either directly or indirectly (Mollace et al. 2001). In vitro studies also demonstrated an increased ROS generation by gp120 exposure to the astrocytes (Reddy et al. 2012). Nrf2, ARE, and antioxidant genes, a well-known oxidative stress protective system, may have link with the inflammation and their role in the HAND. Further, the NOX2 subunit of NADPH oxidase was found involved in the HIV-1-mediated ROS generation (Williams et al. 2010). Inhibition of the NADPH oxidase with NOX2 knockdown and with pharmacological inhibitors, e.g., diphenyleneiodonium (DPI) and apocynin, significantly attenuated the HIV-1. Tat protein induces production of the inflammatory mediators such as TNF-α, IL-6, and MCP-1 in the microglia and macrophages (Mollace et al. 2001). Activation of NADPH oxidase in neurons can contribute to the cell death under stress stimuli.

Activation of the NF-kB signaling has been shown to play an important role in inducing the oxidative stress and increased inflammation response mediated by the HAND (Shah and Kumar 2010), and also upregulation of the NF-kB results in stimulating several inflammatory genes that play a vital role in the HAND (Williams et al. 2009). Upregulation of MMP-9 in the astrocytes treated with HIV-1 and its proteins such as gp120 and Tat has been observed (Ju et al. 2009). MMP-9 has been reported to be increased in the CSF of HIV-infected neurologically impaired patients as well (Sporer et al. 1998). In addition, increased expression of the NF-kB has been demonstrated to be mediated by the Nrf2 pathway, thereby resulting in increased MMP-9 expression (Mao et al. 2011). These findings suggest that the Nrf2 may be beneficial in preventing the oxidative stress and inflammatory cascades induced in the HAND and may provide insight in the development of novel therapeutic strategies against HAND.

Colitis

Colitis (pl. colitides) refers to an inflammation of the colon and is often used to describe an acute or chronic inflammation of the large intestine (colon, cecum, and rectum). The signs and symptoms of the colitides are quite variable and dependent on the etiology of the given colitis and factors that modify its course and severity. There are many types of colitis and classified by etiology: autoimmune inflammatory bowel disease (IBD, a group of chronic colitides), ulcerative colitis (UC, a chronic colitis that affects the large intestine), Crohn's disease (a type of IBD often leads to a colitis), idiopathic (microscopic colitis (a colitis diagnosed by microscopically), lymphocytic colitis, collagenous colitis), iatrogenic (diversion colitis, chemical colitis), vascular disease (ischemic colitis), and infectious colitis. Ulcerative colitis (UC) is idiopathic, chronic, and relapsing inflammatory bowel disease, which elicits the risk of colorectal cancer, the third most common malignancy in humans. Studies in the animal models of UC have helped to shed light on the mechanisms of the inflammation-driven colorectal carcinogenesis. The available evidence suggests that the DNA damage caused by the oxidative stress in the characteristic damage–regeneration cycle is a major contributor to colorectal cancer development in UC patients. Based on this concept, the dietary antioxidants are considered as the protective factors for the UC and associated carcinogenesis.

The colons of individuals with IBD are infiltrated with the neutrophils and activated macrophages that are capable of producing the high levels of ROS and RNS. In addition, inflammatory cytokines such as TNF-α and IFN-γ, which are overproduced in the IBD, are potent inducers of ROS and NO. Excessive production ROS and RNS leads to tissue damage of the host via oxidation of lipids, proteins, and DNA. Moreover, chronic inflammatory bowel diseases, both UC and Crohn's disease, are significant risk factors for the development of colon cancer. The cytokine, IL10 with potent anti-inflammatory and immune regulatory activity, inhibits the production of the inflammatory cytokines, such as IL1 and TNF-α, which stimulate the production of ROS. IL10 also inhibits the production of ROS in neutrophils and human monocytes (Kuga et al. 1996). It was reported that IL10-deficient mice ($IL10^{-/-}$) develop a spontaneous inflammatory bowel disease 3–6 months after birth (Berg et al. 1996). Further, study (Narushima et al. 2003) showed the presence of oxidative stress in the inflammatory bowel disease in nonsteroidal anti-inflammatory drug-(NSAID)-treated $IL10^{-/-}$ mice and suggested a role for the oxidative stress in the pathophysiology of this model of the inflammatory bowel disease. The potential pathogenicity of the free radicals may have a pivotal role in the ulcerative colitis (UC). Fish oil omega-3 fatty acids exert anti-inflammatory effects on the patients with UC (Barbosa et al. 2003), by acting as free radical scavenger.

In further understanding the pathophysiology of the inflammatory diseases, endoplasmic reticulum (ER) stress has been linked. The synthesis, folding, and processing of the secreted and membrane proteins by the ER involve ER chaperones, maintenance of ER calcium pools, and an oxidative environment. A variety of stimuli, including the virus infections, endogenous imbalances in the cell, accumulation of the unfolded or misfolded proteins, loss of the calcium homeostasis, and glucose deprivation, can increase stress to the ER through a battery of UPR molecular pathways (shown in Fig. 2.4 in section "Female Reproduction"). In a recent study (Bogaert et al. 2011), involvement of the ER stress in IBD was studied at molecular level, and different implications of these were observed in colonic and ileal disease which were related to the differences in the development of ileal or colonic disease.

The development of new diagnostic modalities at an early or precancerous stage is crucial to improve the prognosis of the UC-associated neoplasia (Fujii et al. 2008). Advanced oxidation protein products (AOPPs) are new protein markers of oxidative stress with proinflammatory properties, which accumulate in many pathological conditions (Wykretowicz et al. 2007). Being the products of oxidative imbalance themselves, AOPPs further participate in the potentiation and perpetuation of both oxidative stress and inflammation (Peng et al. 2006). Metallothioneins (MTs) have highly conserved number and position of cysteine residues, enabling them to incorporate monovalent and divalent metal atoms and to reduce ROS and RNS. MTs are known to participate in fundamental cellular processes such as cell proliferation and apoptosis (Cioffi et al. 2004). P53, tumor suppressor gene, mutations are the most frequently reported somatic gene alterations in human cancer, leading to accumulation of p53 gene products in tumor cells that can initiate an immune response with generation of circulating anti-p53 antibodies (p53Abs) (El-Sayed et al. 2003). Based on these informations, a recent study (Hamouda et al. 2011) was to exploit the use of p53Abs, MTs, and some oxidative stress markers in the early detection of dysplasia in chronic UC patients. Elisa of p53 antibodies (Abs) and MTs and spectroscopic analysis of AOPPs and GSH were carried. There was a positive correlation between AOPPs and both MTs and p53 Abs, and also between p53Abs and MTs. There was a negative correlation between AOPPs and GSH, and also between GSH and both MTs and p53Abs. In conclusion, oxidative stress and oxidative cellular damage play an important role in the pathogenesis of chronic UC and the associated carcinogenic process. P53Abs levels could help in early detection of dysplasia in these conditions.

Rheumatoid Arthritis

Rheumatoid arthritis (RA) is a chronic, systemic inflammatory disorder that may affect many tissues and organs, but principally attacks flexible (synovial) joints. The process involves an inflammatory response of the capsule around the joints (synovium), secondary to the swelling (hyperplasia) of synovial cells, excess synovial fluid, and the development of fibrous tissue (pannus) in the synovium. The pathology of the disease process often leads to the destruction of the articular cartilage and ankylosis (fusion) of the joints. RA can also produce diffuse inflammation in the lungs, membrane around the heart (pericardium), the membrane of the lung (pleura), and white of the eye (sclera), and also nodular lesions, most common in subcutaneous tissue. Although the cause of RA is unknown, autoimmunity plays a pivotal role in both its chronicity and progression, and it is considered a systemic autoimmune disease.

RA is a chronic multisystem disease with an unknown etiology. Increased oxidative stress and decreased antioxidant status are the hallmarks in patients of RA as compared to healthy individuals. A study (Karatas et al. 2003) indicates that increased oxidative stress and/or defective antioxidant status contributes to the pathology of RA. MDA levels in patients with RA were found to be significantly higher than controls, whereas levels of vitamins A, E, and C and activities of glutathione peroxidase and SOD were lower in the patients compared to controls. Plasma catalase had also been reported to be significantly lower in patients with RA (Kamanli et al. 2004). An epidemiological study (Knekt et al. 2002) suggested that low selenium status may be a risk factor for rheumatoid factor-negative RA. This shows that there is increased state of oxidative stress in RA, which proposes the use of antioxidants supplementation in such patients. In view of the animal studies strongly suggesting anti-inflammatory role of antioxidants like superoxide dismutase (Salvemini et al. 2001) and vitamin E (Behaska et al. 2002) in experimentally induced arthritis, antioxidant therapy strategies have been proposed for the prevention and treatment of RA (Cerhan et al. 2003). Antioxidant implications and complications are reviewed by Mahajan and Tandon (2004).

RA being dependent on environmental factors and highly influenced by genetic composition,

Vasanthi et al. (2009) designed a study to generate data of the disease condition and the biochemical aspects in peripheral blood of population. Statistically significant changes were observed in the levels of MDA, vitamin E, total NO, and ESR in the patient group. Significant differences were also observed in ESR and vitamin E levels in patients with active disease. Increased oxidative stress status existed, which may lead to the connective tissue degradation leading to the joint and periarticular deformities in RA. In another study (Desai et al. 2010), oxidative stress was evaluated by measuring MDA and enzymatic antioxidant status by estimating the SOD and GR in the patients of RA. This study revealed that there was an increased oxidative stress and a decreased antioxidant defense in patients with RA as compared with healthy individuals. In extension to these, another study (Biniecka et al. 2011) was to assess the levels and spectrum of mitochondrial DNA mutations in synovial tissue from patients with inflammatory arthritis and to link these with oxidative stress status as assessed by analyzing in vivo tissue hypoxic status, lipid peroxidation, and cytochrome c oxidase expression levels. The effect of antioxidant treatment on the above processes was also examined. The findings demonstrate that hypoxia-induced mitochondrial dysfunction drives mitochondrial genome mutagenesis and antioxidants significantly rescue these events in synovial tissue from patients with inflammatory arthritis.

Further, seeing the increasing evidence that oxidative stress may play a key role in joint destruction in RA, the role of Nrf2, a transcription factor that maintains the cellular defense against oxidative stress, was studied (Wruck et al. 2011) in the synovial tissue from patients with RA using immunohistochemistry (IHC). Antibody-induced arthritis (AIA) was induced in Nrf2-KO (knock out) and Nrf2-WT (wild type) control mice. Nrf2 was activated in the joints of arthritic mice and of patients with RA. Nrf2-KO mice had more severe cartilage injuries and more oxidative damage, and the expression of Nrf2 target genes was enhanced in Nrf2-WT but not in KO mice during AIA. Both VEGF-A mRNA and protein expression was upregulated in Nrf2-KO mice during AIA. An unexpected finding was the number of spontaneously fractured bones in Nrf2-KO mice with AIA. These results provide strong evidence that oxidative stress is significantly involved in cartilage degradation in experimental arthritis and indicate that the presence of a functional Nrf2 gene is a major requirement for limiting cartilage destruction.

To establish the correlation of the redox status in peripheral blood and the oxidative status at the site of inflammation in RA patients, spectrophotometry and/or flow cytometry analysis was carried (Kundu et al. 2012). The basal levels of total ROS, superoxide, and hydroxyl radicals were significantly raised in neutrophils sourced from peripheral blood and synovial infiltrate. However, there was no major increase in the RNS generated in monocytes from both sources. Furthermore, raised levels of superoxide in neutrophils of synovial infiltrate showed a positive correlation with NADPH oxidase activity in synovial fluid. Therefore, peripheral blood analysis directly correlates the inflammation status and in turn joint damage in RA patients and hence easy diagnosis. Staron et al. (2012) also generated biochemical analysis data in erythrocytes from RA patients. The level of the lipid peroxidation, antioxidant enzyme activities (CAT, SOD, GPx), level of the –SH groups, and GSH and $Na^{+}K^{+}$ ATPase activity in erythrocytes from patients with RA were estimated. There were no significant differences in CAT and GSH-Px activities. SOD activity is lower in RA patients than in the control group. Increase in the lipid peroxidation is observed in RA patients. Levels of the GSH and –SH groups are significantly lower in RA patients than in the control groups. Total ATPase and $Na^{+}K^{+}$ ATPase activities decreases in RA patients.

Inflamed synovium is infiltrated by neutrophils, macrophages, T cells, and B cells, which release a variety of proinflammatory mediators. Persistent inflammation results in destruction of the cartilage and bone. This occurs through a number of mechanisms, including oxidative and proteolytic breakdown of the collagen and proteoglycans (Wright et al. 2010). Once sequestered within the joint space, neutrophils degranulate and release a variety of potentially harmful enzymes and peptides (Edwards and Hallett 1997). They may also undergo a respiratory burst and generate several ROS, including superoxide, H_2O_2, hypohalous acids, and possible hydroxyl radical (Wright et al. 2010).

RA is a heterogeneous disease, in which MPO may play a role in the pathogenesis, severity, and/or outcomes. Indeed, MPO is present at high concentrations in SF of patients with RA. Also MPO is a marker of cardiovascular risk as discussed in CVD section in this chapter, and CVD is recognized as an important cause of death in patients with RA. Given the potential of MPO to contribute to both the pathology, a study was designed to determine whether MPO is active and promotes oxidative stress in SF through the production of hypochlorous acid and its relation with inflammatory activity in RA (Stamp et al. 2012). Plasma or SF was collected from RA patients and control individuals for analysis. Detection of 3-chlorotyrosine confirms that hypochlorous acid is produced in SF and reacts with proteins. There is no other known biological reaction that produces 3-chlorotyrosine and MPO is the only human enzyme capable of generating hypochlorous acid (Winterboum and Kettle 2000). Hence, the strong correlation between 3-chlorotyrosine and the levels of MPO indicates that this enzyme catalyzes the production of hypochlorous acid in SF. Furthermore, the association of protein carbonyls with both MPO and 3-chlorotyrosine suggests that hypochlorous acid (strongest two-electron oxidant) is a major driver of the oxidative damage that occurs to proteins in the inflamed joint. It readily oxidizes cysteine and methionine residues as well as cross-linking and fragmenting proteins, inactivating α_1-antiprotease inhibitor and adversely affecting the functions of LDL and HDL (Nicholls and Hazen 2009). Given its extreme and diverse reactivity, it is likely that hypochlorous acid contributes to the tissue damage that occurs in patients with RA. Thus, targeting specific inhibitors, such as 2-thioxanthines, against MPO would be expected to lower oxidative stress within the inflamed joint (Tiden et al. 2011).

References

Neurodegeneration Diseases

Abramov AY, Canevari L, Duchen MR (2003) Changes in intracellular calcium and glutathione in astrocytes as the primary mechanism of amyloid neurotoxicity. J Neurosci 23:5088–5095

Abramov AY, Canevari L, Duchen MR (2004) Beta-amyloid peptides induce mitochondrial dysfunction and oxidative stress in astrocytes and death of neurons through activation of NADPH oxidase. J Neurosci 24:565–575

Anderson I, Adinolfi C, Doctrow S, Huffman K, Joy KA, Malfrov B, Soden P, Rupniak HT, Bames JC (2001) Oxidative signalling and inflammatory pathways in Alzheimer's disease. Biochem Soc Symp 67:141–149

Benabid AL, Waliace B, Mitrofanis J, Xia C, Piallat B, Fraix V, Batir A, Krack P, Poliak P, Berger F (2005) Therapeutic electrical stimulation of the central nervous system. C R Biol 328:177–186

Bender A, Krishnan KJ, Morris CM, Taylor GA, Reeve AK, Perry RH, Jaros E, Hershoson JS, Betts J, Klopstock T, Taylor RW, Turnbull DMJ (2006) High levels of mitochondrial DNA deletions in substantia nigra neurons in aging and Parkinson disease. Nat Genet 38:515–517

Brown GC, Borutaite V (2004) Inhibition of mitochondrial respiratory complex I by nitric oxide, peroxynitrite and S-nitrosothiols. Biochim Biophys Acta 1658:44–49

Chinta SJ, Mallajosyula JK, Rane A, Andersen JK (2010) Mitochondrial alpha-synuclein accumulation impairs complex I function in the dopaminergic neurons and results in increased mitophagy in vitro. Neurosci Lett 486:235–239

Conte V, Uryu K, Fujimoto S, Yao Y, Rokach J, Longhi L, Trojanowski JQ, Lee VM, McIntosh TK, Pratico D (2004) Vitamin E reduces amyloidosis and improves cognitive function in Tg2576 mice following repetitive concussive brain injury. J Neurochem 90:758–764

Dal-Cim T, Motz S, Egea J, Parada E, Romero A, Budni J, Martin de Saavedra MD, Barrio LD, Tasca CL, Lopez MG (2012) Guanosine protects human neuroblastoma SH-SY5Y cells against mitochondrial oxidative stress by inducing Heme oxigenase-1 via PI3K/Akt/GSK-3β pathway. Neurochem Int 61:397–404

Dalfo EP, Portero-Otin MMP, Ayala VP, Martinez A, Pamplona M, Ferrer IM (2005) Evidence of oxidative stress in the neocortex in incidental lewy body disease. J Neuropath Exptl Neurol 64:816–830

Darios F, Corti O, Lucking CB, Hampe C, Muriel MP, Abbas N, Gu WJ, Hirsch EC, Rooney T, Ruberg M, Brice A (2003) Parkin prevents mitochondrial swelling and cytochrome release in mitochondria-dependent cell death. Hum Mol Genet 12:517–526

Dringen R, Hirrlinger J (2003) Glutathione pathways in the brain. Biol Chem 384:505–516
Du H, Yan SS (2010) Mitochondrial permeability transition pore in Alzheimer's disease cyclophilin D and amyloid. Biochim Biophys Acta 1802:198–204
Dumont M, Lin MT, Beal MF (2010) Mitochondria and antioxidant targeted therapeutic strategies for Alzheimer's disease. J Alzheim Dis 20:5633–5643
Edmondson DE, Binda C, Wang J, Upadhyay AK, Mattevi A (2009) Molecular and mechanistic properties of the membrane-bound mitochondrial monoamine oxidases. Biochemistry 48:4220–4230
Ermak G, Davies KJ (2002) Calcium and oxidative stress from cell signaling to cell death. Mol Immunol 38:713–721
Ferrari CKB (2000) Free radicals, lipid peroxidation and antioxidants in apoptosis: implication in cancer, cardiovascular and oxidants in apoptosis: implications in cancer, cardiovascular and neurological diseases. Biologia 55:581–590
Fitzgerald JC, Camprubi MD, Dunn L, Wu HC, Ip NY, Kruger R, Martins LM, Wood NW, Plun-Favreau H (2012) Phosphorylation of HtrA2 by cyclin-dependent kinase-5 is important for mitochondrial function. Cell Death Differ 19:257–266
Gandhi S, Abramov AY (2012) Mechanism of oxidative stress in neurodegeneration. Oxidative Med Cell Longev, PMID 22685618
Gandhi S, Wood-Kaczmar A, Yao Z, Plun-Favreau H, Deas E, Klupsch K, Downwara J, Latchman DS, Tabrizi SJ, Wood NW, Duchen MR, Abramov AY (2009) PINK1-associated Parkinson's disease is caused by neuronal vulnerability to calcium-induced cell death. Mol Cell 33:627–639
Gandhi S, Vaarmann A, Yao Z, Duchen MR, Wood NW, Abramov AY (2012) Dopamine induced neurodegeneration in a PINK1 model of Parkinson's disease. Plos ONE 7:e37565
Gao HM, Liu B, Hong JS (2003) Critical role for microglial NADPH oxidase in rotenone-induced degeneration of dopaminergic neurons. J Neurosci 23:6181–6187
Gerard C, Chehal H, Hugel RP (1994) Complexes of iron (III) with ligands of biological interest dopamine and 8-hydroxyquinine-5-sulfonic acid. Polyhedron 13:591–597
Hemandez F, Avila J (2007) Tauopathies. Cell Mol Life Sci 64:2219–2233
Hirsch EC, Jenner P, Przedborski S (2013) Pathogenesis of Parkinson's disease. Mov Disord 28:24–30
Kaminsky YG, Kosenko EA (2008) Effects of amyloid-beta peptides on hydrogen peroxide-metabolizing enzymes in rat brain in vivo. Free Rad Res 42:564–573
Lee MS, Kwon YT, Li M, Peng J, Friedlander RM, Tsai LH (2000) Neurotoxicity induces cleavage of p35 to p25 by calpain. Nature 405:360–364
Lotharius J, Brundin P (2002) Impaired dopamine storage resulting from αsynuclein mutations may contribute to the pathogenesis of Parkinson's disease. Hum Mol Genet 11:2395–2407
Lothiarius J, O'Malley KL (2000) The Parkinsonism-inducing drug 1-methyl-4-phenylpyridinium triggers intracellular dopamine oxidation: a novel mechanism of toxicity. J Bio Chem 275:38581–38588
Margis R, Dunand C, Teixeira FK, Margis-Pinheiro M (2008) Glutathione peroxidase family-an evolutionary overview. FEBS J 275:3859–3970
Mattson MP (2003) Will caloric restriction and folate protect against AD and PD? Neurology 60:690–695
Muftuoglu M, Elibol B, Dalmizrak O, Ercan A, Kulaksiz G, Ogus H, Dalkara T, Ozer N (2004) Mitochondrial complex I and IV activities in leukocytes from patients with parkin mutations. Mov Disord 19:544–548
Muller T (2011) Motor complications, levodopa metabolism and progression of Parkinson's disease. Expert Opin Drug Metab Toxicol 7:847–855
Obata T, Kubota S, Yamanaka Y (2001) Allopurinol suppresses para-nonylphenol and 1-methyl-4-phenylpyridinium ion (MPP+)-induced hydroxyl radical generation in rat striatum. Neurosci Lett 306:9–12
Opazo C, Huang X, Chemy R, Chemy R (2002) Metalloenzyme-like activity of Alzheimer's disease β-amyloid. Cu-dependent catalytic conversion of dopamine cholesterol, and biological reducing agents to neurotoxic H_2O_2. J Biol Chem 277:40302–40308
Park L, Zhou P, Pitstick R, Carpone C, Anrather J, Norris EH, Younkin L, Youakin S, Carlson G, McEwen BS, Ladecola C (2008) Nox2-derived radicals contribute to neurovascular and behavioral dysfunction in mice overexpressing the amyloid precursor protein. Proc Natl Acad Sci U S A 105:1347–1352
Peterson LJ, Flood PM (2012) Oxidative stress and microglial cells in Parkinson's disease. Mediators Inflamm 2012, 401264
Priller C, Bauer T, Mitteregger G, Krebs B, Kretzschmar HA, Herms J (2006) Synapse formation and function is modulated by the amyloid precursor protein. J Neurosci 26:7212–7221
Schroeter H, Spencer JP, Rice-Evans C, Williams RJ (2001) Flavonoids protect neurons from oxidized low-density-lipoprotein-induced apoptosis involving c-jun N-terminal kinase (JNK), c-jun and caspase-3. Biochem J 358:547–557
Shapira AH (2008) Mitochondria in the aetiology and pathogenesis of Parkinson's disease. Lancet Neurol 7:97–109, 68
Smith DS, Tsai LH (2002) Cdk5 behind the wheel: a role in trafficking and transport? Trends Cell Biol 12:28–36
Song DD, Shults CW, Sisk A, Rockenstein E, Masliah E (2004) Enhanced substantia nigra mitochondrial pathology in human α-synuclein transgenic mice after treatment with MPTP. Exp Neurol 186:158–172
Sun KH, DePablo Y, Vincent F, Shah K (2008) Deregulated Cdk5 promotes oxidative stress and mitochondrial dysfunction. J Neurochem 107:265–278
Sung S, Yao Y, Uryu K, Yang H, Lee VM, Trajanowaki JQ, Pratico D, Faseb J (2004) Early vitamin E supplementation in young but not aged mice reduces Abeta levels and amyloid deposition in a transgenic model of Alzheimer's disease. FASEB J 18:323–325

Surmeier DJ, Guzman JN, Sanchez-Padilla J, Goldberg JA (2011) The origin of oxidant stress in Parkinson's disease and therapeutic strategies. Antioxid Redox Signal 14:1289–1301

Tiraboschi P, Hansen LA, Thal IJ, Corey-Bloom J (2004) The importance of neuritic plaques and tangles to the development and evolution of AD. Neurology 62:1984–1989

Uttara B, Singh AV, Zamboni P, Mahajan RT (2009) Oxidative stress and neurodegenerative diseases: a review of upstream and downstream antioxidant therapeutic options. Curr Neuropharmacol 7:65–74

Vaarmann A, Gandhi S, Abramov AY (2010) Dopamine induces Ca^{2+} signaling in astrocytes through reactive oxygen species generated by monoamine oxidase. J Biol Chem 285:25018–25023

van Muiswinkel FL, Kuiperij HB (2005) The Nrf2-ARE signaling pathway: promising drug target to combat oxidative stress in neurodegenerative disorders. Curr Drug Targets CNS Neuro Disord 4:267–281

Wang X, Michaelis EK (2010) Selective neuronal vulnerability to oxidative stress in the brain. Front Aging Neurosci 2:12

Weber CA, Ernst ME (2006) Antioxidants supplements and Parkinson's disease. Ann Pharmacother 40:935–938

Wilkinson B, Koenigsknecht-Taboo C, Grommes C, Lee CYD, Landreth (2006) Fibrillar β-amyloid-stimulated intracellular signaling cascades require Vav for induction of respiratory burst and phagocytosis in monocytes and microglia. J Biol Chem 281:20842–20850

Wu AD, Fregni F, Simon DK, Deblieck C, Pascual-Leone A (2008) Noninvasive brain stimulation for Parkinson's disease and dystonia. Neurotherapeutics 5:345–361

Yankner BA, Duffy LK, Kirschner DA (1990) Neurotrophic and neurotoxic effects of amyloid beta protein reversal by tachykinin neuropeptide. Science 250:279–282

Cardiovascular Diseases

Al Ahmad A et al (2009) Maintaining blood-brain barrier integrity pericytes perform better astrocytes during prolonged oxygen deprivation. J Cell Physiol 218:612–622

Al Ahmad A, Gassmann M, Ogunshola OO (2012) Involvement of oxidative stress in hypoxia-induced blood-brain barrier breakdown. Microvasc Res 84:222–225

Allen CL, Bayraktutan U (2009) Oxidative stress and its role in the pathogenesis of ischaemic stroke. Int J Stroke 4:461–470

Baranova O, Miranda LF, Pichiule P, Dragatsis I, Johnson RS, Chavez JC (2007) Neuron-specific inactivation of the hypoxia inducible factor 1α increases brain injury in a mouse model of transient focal cerebral ischemia. J Neurosci 23:6320–6332

Berliner JA, Navab M, Fogelman AM, Frank JS, Dermer LL, Edwards PA, Watson AD, Lusis AJ (1995) Atherosclerosis: basic mechanisms, oxidation, inflammation and genetics. Circulation 91:2488–2496

Bevers LM, Braam B, Post JA, Zonneveld AJ, rbelink TJ, Koomans HA, Verhaar MC, Joles JA (2006) Tetrahydrobiopterin but not L-arginine, decreases NO synthase uncoupling in cells expressing high levels of endothelial NO synthase. Hypertension 47:87–94

Brandes RP, Weissmann N, Schroder K (2010) NADPH oxidases in cardiovascular diseases. Free Radic Biol Med 49:687–706

Brown MS, Goldstein JL (1983) Lipoprotein metabolism in the macrophages. Ann Rev Biochem 52:223–261

Carmeliet P, Dor Y, Herbert JM, Fukumura D, Brusselmans K, Dewerchin M, Neeman M, Bono F, Abramovitch R, Maxwell P, Koch CJ, Ratcliffe P, Moons L, Jain RK, Collen D, Keshorte E (1998) Role of HIF-1 alpha in hypoxia-mediated apoptosis, cell proliferation and tumour angiogenesis. Nature 394:485–490

Carr AC, Frei B (2001) The nitric oxide congener nitrite inhibits myeloperoxidase/ H_2O_2/Cl^--mediated modification of low density lipoprotein. J Biol Chem 276:1822–1828

Castelli WP (1986) The triglyceride issue: a view from Framingham. Am Heart J 112:432–437

Chandel NS, Maitape E, Goldwasser E, Mathieu CE, Simon MC, Schumacbu PT (1998) Mitochondrial reactive oxygen species trigger hypoxia –induced transcription. Proc Natl Acad Sci U S A 95:11715–11720

Duffy D, Holmes DN, Roe MT, Peterson ED (2012) The impact of high-density lipoprotein cholesterol levels on long-term outcomes after non-ST-elevation myocardial infarction. Am Heart J 163:705–713

Endemann G, Pronzcuk A, Freidman G, Lindsey S, Alderson L, Hayes KC (1987) Monocyte adherence to endothelial cells in vitro is increased by β-VLDL. Am J Pathol 126:1–6

Engler MM, Engler MB, Malloy MJ, Chiu EY, Schlother MC, Paul SM, Shiehinger M, Lin KY (2003) Antioxidant vitamin C and E improve endothelial function in children with hyperlipidemia. Endothelial Assessment of Risk from lipid in Youth (EARLY) trial. Circulation 108:1059–1063

Garner B, Cooke JP, Morrow JD, Ridker PM, Rifai N, Miller L, Witzhum JL, Mietus-Snycler (1998) Oxidation of high density lipoproteins. II, evidence for direct reduction of lipid hydroperoxides by methionine residues of apolipoproteins AI and AII. J Biol Chem 273:6088–6095

Goldstein JL, Ho YK, Brown MS, Innerarity TL, Mahley RW (1980) Cholesteryl ester accumulation in macrophages resulting from receptor mediated uptake and degradation hypercholesterolemic canine β-VLDL. J Biol Chem 225:1839–1848

Guo S, Miyake M, Liu KJ, Shi H (2009) Specific inhibition of hypoxia inducible factor exaggerates cell injury induced by in vitro ischemia through deteriorating cellular redox environment. J Neurochem 5:1309–1321

Hausenloy DJ, Yellon DM (2008) Time to take myocardial reperfusion injury seriously. N Eng J Med 359:518–520

Huang J, Zhang Z, Guo J, Ni A, Deb A, Zhang L, Mirotsou M, Pratt RE, Dzau VJ (2010) Genetic modification of mesenchymal stem cells overexpressing CCR1 increases cell viability, migration, engraftment and capillary density in the injured myocardium. Circ Res 106:1753–1762

Javadov S, Karmazyn M (2007) Mitochondrial permeability transition pore opening as an endpoint to initiate cell death and as a putative target for cardioprotection. Cell Physiol Biochem 20:1–22

Jialal I, Devaraj S (1996) Low density lipoprotein oxidation, antioxidants and atherosclerosis: a clinical biochemistry perspectives. Clin Chem 42:498–506

Kaur HD, Bansal MP (2009) Studies on associated enzymes under experimental hypercholesterolemia: possible modulation on selenium supplementation. Lipids Health Dis 8:1–16

Kleinschnitz C, Grund H, Wingler K, Armitage ME, Jones E, Mittal M, Barit D, Schwarz T, Geis C, Kraft P, Barthel K, Schuhmann MK, Herrmann AM, Meuth SG, Stoll G, Meurer S, Schrewe A, Becker L, Gailus-Durner V, Fuchs H, Klopstock T, de Angelis MH, Jandeleit-Dahm K, Shah AM, Weissmann N, Schmidt HH (2010) Post-stroke inhibition of induced NADPH oxidase type 4 prevents oxidative stress and neurodegeneration. Plos Biol 8:e1000479

Klimov AN, Kozheynikoya KA, Kuzmin AA, Kuzetrov AS, Belora EV (2001) On the ability of high density lipoproteins to remove phospholipid peroxidation products from erythrocyte membranes. Biochemistry (Mosc) 66:300–304

Kuhn H, Romisch J, Belkner J (2005) The role of lipoxygenase-isoforms in atherogenesis. Mol Nutr Food Res 49:1014–1029

Kunitake ST, Jarvis MR, Hamilton RL, Kane JP (1992) Binding of transition metals by apolipoprotein A-1-containing plasma lipoproteins: inhibition of oxidation of low density lipoproteins. Proc Natl Acad Sci U S A 89:6993–6997

Lochhead JJ, Mccaffrey G, Quigley CE, Finch J, DeMarco KM, Nametz N, Davis TP (2010) Oxidative stress increases blood-brain barrier permeability and induces alterations in occluding during hypoxia-reoxygenation. J Cereb Blood Flow Metab 30:1625–1636

Lonn EM, Yusuf S, Dzavik V, Doris C, Yi Q, Smith S, Moore Cox A, Bosch J, Riley W, Teo K (2001) Effects of ramipril and vitamin E on atherosclerosis: the Study to Evaluate Carotid Ultrasound changes in patients treated with Ramipril and vitamin E (SECURE). Circulation 103:919–925

Malhotra R, Lin Z, Vinconz C, Brosius FC 3rd (2001) Hypoxia induces apoptosis via two independent pathways in Jurkat cells: differential regulation by glucose. Am J Physiol Cell Physiol 281:C1596–C1603

Malle E, Waeg G, Schreiber R, Grone EF, Sattler W, Grone HJ (2000) Immunohistochemical evidence for the myeloperoxidase/H_2O_2/halide system in human atherosclerotic lesions: colocalization of myeloperoxidase and hypochlorite-modified proteins. Eur J Biochem 267:4495–4503

Matsuzawa A, Ichijo H (2008) Redox control of cell fate by MAP kinase: physiological roles of ASK1-MAP kinase pathway in stress signaling. Biochim Biophys Acta 1780:1325–1336

Murthy KG, Szabo C, Salzman AI (2004) Cytokines stimulate expression of inducible nitric oxide synthase in DLD-1 human adenocarcinoma cells by activating poly(A) polymerase. Inflamm Res 53:604–608

Neri M, Fineschi V, Di Paolo M, Pomara C, Riezzo I, Tunilazzi E, Cerretani D (2013) Cardiac oxidative stress and inflammatory cytokines response after myocardial infarction. Curr Vasc Pharmacol, PMID 23716180

Nicholls SJ, Dustina GJ, Cutri B, Bao S, Deummond GR, Rye KA, Barter PJ (2005) Reconstituted high-density lipoprotein inhibit the acute pro-oxidant and proinflammatory vascular changes induced by a periarterial collar in normocholesterolemic rabbits. Circulation 111:1543–1550

Ohara Y, Peterson TE, Harrison DG (1993) Hypercholesterolemia increases endothelial superoxide anion production. J Clin Invest 91:2546–2551

Ozkul A, Akyol AL, Yenisey C, Arpaci E, Kyloglu N, Tataroglu C (2007) Oxidative stress in acute ischemic stroke. J Clin Neurosci 14:1062–1066

Parathasarathy S, Printz DJ, Boyd D, Joy L, Steinberg D (1986) Macrophage oxidation of low density lipoprotein generates a modified form recognized by scavenger receptor. Atherosclerosis 6:505–510

Prasad K, Kalra J (1992) Oxygen free radicals and hypercholesterolemic atherosclerosis: effect of vitamin E. Am Heart J 125:958–961

Precourt LP, Amre D, Denis MC, Lavoie JC, Delvin E, Seidman E, Levy E (2011) The three-gene paraoxonase family: physiologic roles, actions and regulation. Atherosclerosis 214:20–36

Ross R (1991) The pathogenesis of atherosclerosis. In: Braunward E (ed) Heart disease. W.B. Saunders Co., Philadelphia, pp 1135–1152

Sandin A, Dagnell M, Gonon A, Pernow J, Stangl V, Aspenstrom P, Kappert K, Ostman A (2011) Hypoxia followed by re-oxygenation induces oxidation of tyrosine phosphatases. Cell Signal 23:820–826

Shi H (2009) Hypoxia inducible factor 1 as a therapeutic target in ischemic stroke. Curr Med Chem 16: 4593–4600

Singh U, Jialal I (2006) Oxidative stress and atherosclerosis. Pathophysiology 13:129–142. Review

Sowers JR (1992) Insulin resistance, hyperinsulinemia, dyslipidemia, hypertension and accelerated atherosclerosis. J Clin Pharm 32:529–535

Stocker R, Keaney JF (2001) Role of oxidative modifications in atherosclerosis. Physiol Rev 84:1381–1478

Venugopal SK, Devaraj S, Jialal I (2003) C-reactive protein decreases prostacyclin release from human aortic endothelial cells. Circulation 108:1676–1678

Vogiatzi G, Tousoulis D, Stefanadis C (2009) The role of oxidative stress in atherosclerosis. Hellenic J Cardiol 50:402–409. Review

Wassmann S, Laufs U, Baumer AT, Muller K, Ahibary K, Linz W, Itter G, Rosen R, Bohm M, Nickenig G (2001) HMG-CoA reductase inhibitors improve endothelial dysfunction in normocholesterolaemic hypertension via reduced production of reactive oxygen species. Hypertension 37:1450–1457

Witzum JL, Steinberg D (1991) Role of oxidized low-density lipoprotein in atherogenesis. J Clin Invest 88(1):785–1792

Xu J, Qian J, Xie X, Lin L, Zou Y, Fu M, Huang z, Zhang G, Su Y, Ge J (2012) High density lipoprotein protects mesenchymal stem cells from oxidative stress-induced apoptosis via activation of the PI3K/Akt pathway and suppression of reactive oxygen species. Int J Mol Sci 13:17104–17120

Yamaguchi O, Higuchi Y, Hirotani S, Kashiwase K, Nakayama H, Hikoso S, Takeda T, Watanabe T, Asahi M, Taniike M, Matsumura Y, Tsujimoto I, Hongo K, Kusakari Y, Kurihara S, Nishida K, Ichijo H, Hori M, Otsu K (2003) Targeted deletion of apoptosis signal-regulating kinase 1 attenuates left ventricular remodeling. Proc Natl Acad Sci U S A 100:15883–15888

Yan LJ, Rajasekaran NS, Sathyanarayanan S, Benjamin J (2005) Mouse HSF1 disruption perturbs redox state and increases mitochondrial oxidative stress in kidney. Antioxid Redox Signal 7:465–471

Zhang L, Jiang H, Gao X, Zou Y, Liu M, Liang Y, Zhu W, Chen H, Ge J (2011) Heat shock transcription factor-1 inhibits H_2O_2-induced apoptosis via down-regulation of reactive oxygen species in cardiac myocytes. Mol Cell Biochem 347:21–28

Zou Y, Zhu W, Sakamoto M, Quin Y, Akazawa H, Toko H, Mizukami M, Takeda N, Minamino T, Takano H, Nagai T, Nakai A, Komuro I (2003) Heat shock transcription factor I protects cardiomyocytes from ischemia/reperfusion injury. Circulation 108:3024–3030

Male Reproduction

Agarwal A, Nallella KP, Allamaneni SS, Said TM (2004) Role of antioxidants in treatment of male infertility: an overview of the literature. Reprod Biomed Online 8:616–627

Agletdinov EF, Kamilov FK, Alekhin EK, Romantsov MG, Bulygin KV, Makasheva LO (2008) Gonadotoxic effects of polychlorinated biphenyls in experiments on male rats. Antibiot Khimioter 53:15–18

Aitken RJ (1995) Free radicals lipid peroxidation and sperm function. Reprod Fertil Dev 7:659–668

Aitken RJ (1999) The Amoroso lecture. The human spermatozoon – a cell in crisis? J Reprod Fertil 115:1–7

Aitken RJ, Fisher H (1994) Reactive oxygen species generation and human spermatozoa: the balance of benefit and risk. Bioessays 16:259–267

Aitken RJ, Krausz C (2001) Oxidative stress, DNA damage and the Y chromosome. Reproduction 122:497–506

Aitken RJ, Harkiss D, Buckingham DW (1993) Analysis of lipid peroxidation mechanisms in human spermatozoa. Mol Reprod Dev 35:302–315

Armstrong JS, Rajasekaran M, Chamulitrat W, Gatti P, Hellstrom WJ, Sikka SC (1999) Characterization of reactive oxygen species induced effects on human spermatozoa movement and energy metabolism. Free Radic Biol Med 26:869–880

Bahadur G, Ozturk O, Muneer A, Wafa R, Ashraf A, Jaman N, Patel S, Oyede AW, Ralph DJ (2005) Semen quality before and after gonadotoxic treatment. Hum Reprod 20:774–781

Dandekar SP, Nandkarni GD, Kulkarni VS, Punekar S (2002) Lipid peroxidation and antioxidant enzymes in male infertility. J Postgrad Med 48:186–189

de Lamirande E, Gagnon C (1995) Impact of reactive oxygen species on spermatozoa: a balancing act between beneficial and detrimental effects. Hum Reprod 10(suppl I):15–21

de Lamirande E, Leclerc P, Gagnon C (1997) Capacitation as a regulatory event that primes spermatozoa for the acrosome reaction and fertilization. Mol Hum Reprod 3:175–194

Farr SB, Kogama T (1991) Oxidative stress responses in *Escherichia coli* and *Salmonella typhimurium*. Microbiol Rev 55:561–585

Gavella M, Lipovac V (1992) NADPH-dependent oxidoreductase (diaphorase) activity and isozyme pattern of sperm in infertile men. Arch Androl 28:135–141

Garrido N, Meseguer M, Simon C, Pellieer A, Remohi J (2004) Pro-oxidative and anti-oxidative imbalance in human semen and its relation with male fertility. Asian J Androl 6:59–65

Gomez E, Buckingham DW, Brindle J, Lanzafame F, Irvine DS, Aitken RJ (1996) Development of an image analysis system to monitor the retention of residual cytoplasm by human spermatozoa: correlation with biochemical markers of the cytoplasmic space, oxidative stress and sperm function. J Androl 17:276–287

Griveau JF, Renard P, Le Lannou D (1995) Superoxide anion production by human spermatozoa as a part of the ionophore-induced acrosome reaction in vitro. Int J Androl 18:67–74

Halliwal B (1984) Tell me about free radicals, doctor: a review. J Roy Soc Med 82:747–752

Kodama H, Yamaguchi R, Fukuda J, Kasai H, Tanaka T (1997) Increased oxidative deoxyribonucleic acid damage in the spermatozoa of infertile male patients. Fertil Steril 68:519–524

Lewin A, Lavon H (1997) The effect of coenzyme Q10 on sperm motility and function. Mol Asp Med 18(Suppl):S213–S219

Moskovstev SI, Willis J, White J, Mullen BM (2007) Leukocytospermia: relationship to sperm deoxyribonucleic acid integrity in patients evaluated for male factor infertility. Fertil Steril 88:737–740

Mostafa T, Anis TH, El-Nashar A, Iman H, Othman IA (2001) Varicocelectomy reduces reactive oxygen species levels and increases antioxidant activity of seminal plasma from infertile men with varicocele. Int J Androl 24:261–265

Ozbek E, Turkoz Y, Gokdeniz R, Davarci M, Ozugurlu F (2000) Increased nitric oxide production in the spermatic vein of patients with varicocele. Eur Urol 37:172–175

Peeker R, Abramson L, Marklund SL (1997) Superoxide dismutase isoenzymes in human seminal plasma and spermatozoa. Mol Hum Reprod 13:1061–1066

Ragheb AM, Sabanegh ES Jr (2010) Male fertility-implications of anticancer treatment and strategies to mitigate gonadotoxicity. Anticancer Agents Med Cem 10:92–102

Saez F, Motta C, Boucher D, Grizard G (1998) Antioxidant capacity of prostasomes in human semen. Mol Hum Reprod 4:667–672

Sakkas D, Mariethoz E, Mnicsirdi G, Bizzaro D, Bianchi PG, Bianchi U (1999) Origin of DNA damage in ejaculated human spermatozoa. Rev Reprod 4:31–37

Sarlos P, Molner A, Kokai M, Gabor GY, Ratky J (2002) Comparative evaluation of the effect of antioxidants in the conservation of ram semen. Acta Vet Hung 50:235–245

Sikka SC, Rajasekaran M, Hellstrom WJ (1995) Role of oxidative stress and antioxidants in male infertility. J Androl 16:464–468

Smith R, Kaune H, Parodi D, Madariaga M, Rios R, Morales I, Castro A (2006) Increased sperm DNA damage in patients with varicocele: relationship with seminal oxidative stress. Hum Reprod 21:986–993

Spiropoulos J, Turnbull DM, Chinnerry PF (2002) Can mitochondrial DNA mutations cause sperm dysfunctions? Mol Hum Reprod 8:719–721

Suleiman SA, Ali ME, Zaki ZM, el-Malik EM, Nasr MA (1996) Lipid peroxidation and human sperm motility: protective role of vitamin E. J Androl 17:530–537

Taiwo AM, Ige SO, Babalola OO (2010) Assessments of possible gonadotoxic effect of lead on experimental male rabbits. Glob Vet 5:282–286

Tremellen K (2008) Oxidative stress and male infertility: a clinical perspective. Hum Reprod Update 14:243–258. Review

Twigg J, Irvine DS, Houston P, Fulton P, Michael L, Aitken RJ (1998) Latrogenic DNA damage induced in human spermatozoa during sperm preparation: protective significance of seminal plasma. Mol Hum Reprod 4:439–445

Female Reproductive System

Alpay Z, Saed GM, Diamond MP (2006) Female infertility and free radicals: potential role in adhesions and endometriosis. J Soc Gynecol Investig 13:390–398

Aten RF, Duarte KM, Behrman HR (1992) Regulation of ovarian antioxidant vitamins, reduced glutathione, and lipid peroxidation by luteinizing hormone and prostaglandin F2 alpha. Biol Reprod 46:401–407

Attaran M, Pasqualotto E, Falcone T, Goldberg JM, Miller KF, Agarwal A, Sharma RK (2000) The effect of follicular fluid reactive oxygen species on the outcome of in vitro fertilization. Int J Fertil Womens Med 45:314–320

Bansal RK, Goldsmith PC, He Y, Zaloudek CJ, Ecker JL, Riemer RK (1997) A decline in myometrial nitric oxide synthase expression is associated with labor and delivery. J Clin Invest 99:2502–2508

Bausero P, Cavaille F, Meduri G, Freitas S, Perrot-Applanat M (1998) Paracrine action of vascular endothelial growth factor in the human endometrium: production and target sites, and hormonal regulation. Angiogenesis 2:167–182

Bedaiwy MA, Falcone T (2003) Peritoneal fluid environment in endometriosis. Clinicopathological implications. Minerva Ginecol 55:333–345

Belo L, Caslake M, Santos-Silva A (2004) LDL size, total antioxidant status and oxidised LDL in normal human pregnancy: a longitudinal study. Atherosclerosis 177:391–399

Ben-Shlomo I, Kokia E, Jackson MJ, Adashi EY, Payne DW (1994) Interleukin-1 beta stimulates nitrite production in the rat ovary: evidence for heterologous cell-cell interaction and for insulin-mediated regulation of the inducible isoform of nitric oxide synthase. Biol Reprod 51:310–318

Bilodeau JF, Hubel CA (2003) Current concepts in the use of antioxidants for the treatment of preeclampsia. J Obstet Gynaecol Can 25:742–750

Blumenfeld Z, Avivi I, Eckman A, Epelbaum R, Rowe JM, Dann EJ (2008) Gonadotropin-releasing hormone agonist decreases chemotherapy-induced gonadotoxicity and premature ovarian failure in young female patients with Hodgkin lymphoma. Fertil Steril 89:166–173

Brougham MF, Crofton PM, Johnson EJ, Evans N, Anderson RA, Wallace WH (2012) Anti-Mullerian hormone is a marker of gonadotoxicity in pre- and postpubertal girls treated for cancer: a prospective study. J Clin Endocrinol Metab 97:2059–2067

Burton GJ, Yung HW (2011) Endoplasmic reticulum stress in the pathogenesis of early-onset pre-eclampsia. Pregnancy Hypertens 1:72–78

Burton GJ, Hempstock J, Jauniaux E (2003) Oxygen, early embryonic metabolism and free radical-mediated embryopathies. Reprod BioMed Online 6:84–96

Burton GJ, Yung HW, Cindrova-Davies T (2009) Placental endoplasmic reticulum stress and oxidative stress in the pathophysiology of unexplained intrauterine growth restriction and early onset preeclampsia. Placenta 30(Suppl A):S43–S48

Catov JM, Nohr EA, Bodnar LM (2009) Association of periconceptional multivitamin use with reduced risk of preeclampsia among normal-weight women in the Danish National Birth Cohort. Am J Epidemiol 169:1304–1311

Choi HK, Choi BC, Lee SH, Kim JW, Cha KY, Baek KH (2003) Expression of angiogenesis- and apoptosis-related genes in chorionic villi derived from recurrent pregnancy loss patients. Mol Reprod Dev 66:24–31

Cindrova-Davies T (2009) From placental oxidative stress to maternal endothelial dysfunction. Placenta 30(Suppl A):S55–S65

Cindrova-Davies T, Yung HW, Johns J (2007a) Oxidative stress, gene expression and protein changes induced in the human placenta during labor. Am J Pathol 171:1168–1179

Cindrova-Davies T, Spasic-Boskovic O, Jauniaux E (2007b) Nuclear factor-kappa B, p38, and stress-activated protein kinase mitogen-activated protein kinase signaling pathways regulate proinflammatory cytokines and apoptosis in human placental explants in response to oxidative stress: effects of antioxidant vitamins. Am J Pathol 170:1511–1520

Ekerhovd E, Enskog A, Caidahl K, Klintland N, Nilsson L, Brannstrom M, Norstrom A (2001) Plasma concentrations of nitrate during the menstrual cycle, ovarian stimulation and ovarian hyperstimulation syndrome. Hum Reprod 16:1334–1339

Hickey M, Krikun G, Kodaman P, Schatz F, Carati C, Lockwood CJ (2006) Long-term progestin-only contraceptive result in reduced endometrial blood flow and oxidative stress. J Clin Endocrinol Metab 9:3633–3638

Hool LC, Corry B (2007) Redox control of calcium channels: from mechanisms to therapeutic opportunities. Antioxid Redox Signal 9:409–435

Hung TH, Skepper JN, Burton GJ (2001) In vitro ischemia-reperfusion injury in term human placenta as a model for oxidative stress in pathological pregnancies. Am J Pathol 159:1031–1043

Jauniaux E, Watson AL, Hempstock J (2000) Onset of maternal arterial blood flow and placental oxidative stress; a possible factor in human early pregnancy failure. Am J Pathol 157:2111–2122

Jauniaux E, Hempstock J, Greenwold N (2003) Trophoblastic oxidative stress in relation to temporal and regional differences in maternal placental blood flow in normal and abnormal early pregnancies. Am J Pathol 162:115–125

Jozwik M, Wolczynski S, Szamatowicz M (1999) Oxidative stress markers in preovulatory follicular fluid in humans. Mol Hum Reprod 5:409–413

Kaufmann M, von Minckwitz G, Smith R, Vekro Y, Gianni L, Eiermann W, Howell A, Costa SD, Beuzeboc P, Untech M, Blohmer JU, Sinn HP, Sittek R, Souchon R, Tulusan AH, Volm T, Semu HJ (2003) International expert panel on the use of primary (preoperative) systemic treatment of operable breast cancer, review and recommendations. J Clin Oncol 21:2600–2608

Krussel JS, Bielfeld P, Polan ML, Simon C (2003) Regulation of embryonic implantation. Eur J Obstet Gynecol Reprod Biol 110:S2–S9

Klemmensen A, Tabor A, Osterdal ML (2009) Intake of vitamins C and E in pregnancy and risk of preeclampsia: prospective study among 57 346 women. BJOG 116:964–974

LaPolt PS, Leung K, Ishimaru R, Tafoya MA, You-hsin Chen J (2003) Roles of cyclic GMP in modulating ovarian functions. Reprod Biomed Online 6:15–23

Lee TH, Wu MY, Chen MJ, Chao KH, Ho HN, Yang YS (2004) Nitric oxide is associated with poor embryo quality and pregnancy outcome in in vitro fertilization cycles. Fertil Steril 82:126–131

Leist M, Single B, Castoldi AF (1997) Intracellular adenosine triphosphate (ATP) concentration: a switch in the decision between apoptosis and necrosis. J Exp Med 185:1481–1486

Osborn BH, Haney AF, Misukonis MA, Weinberg JB (2002) Inducible nitric oxide synthase expression by peritoneal macrophages in endometriosis-associated infertility. Fertil Steril 77:46–51

Ota H, Igarashi S, Hatazawa J, Tanaka T (1998) Endothelial nitric oxide synthase in the endometrium during the menstrual cycle in patients with endometriosis and adenomyosis. Fertil Steril 69:303–308

Park JK, Song M, Dominguez CE, Walter MF, Santanam N, Parthasarathy S, Murthy AA (2006) Glycodelin mediates the increase in vascular endothelial growth factor in response to oxidative stress in the endometrium. Am J Obstet Gynecol 195:1772–1777

Roberts JM, Myatt L, Spong CY (2010) Vitamins C and E to prevent complications of pregnancy-associated hypertension. N Engl J Med 362:1282–1291

Ron D, Walter P (2007) Signal integration in the endoplasmic reticulum unfolded protein response. Nat Rev Mol Cell Biol 8:519–529

Sata F, Yamada H, Yamada A (2003) A polymorphism in the CYP17 gene relates to the risk of recurrent pregnancy loss. Mol Hum Reprod 9:725–728

Sharma RK, Agarwal A (2004) Role of reactive oxygen species in gynecologic diseases. Reprod Med Bio 3:177–199

Seino T, Salto H, Kaneko T, Takahashi T, Kawachi H (2002) Eight-hydroxy-2′-deoxyguanosine in granulose cells is correlated with the quality of oocytes and embryos in an in vitro fertilization-embryo transfer program. Fertil Steril 77:1184–1190

Sugino N, Karube-Harada A, Taketani T, Sakata A, Nakamura Y (2004) Withdrawal of ovarian steroids stimulates prostaglandin F2alpha production through nuclear factor-kappaB activation via oxygen radicals in human endometrial stromal cells: potential relevance to menstruation. J Reprod Dev 50:215–225

Suzuki T, Sugino N, Fukaya T, Sugyamas S, Uda T, Takayh R, Yajima A, Sasano H (1999) Superoxide dismutase in normal cycling human ovaries: immunohistochemical localization and characterization. Fertil Steril 72:720–726

Toy H, Camuzcuoglu H, Camuzcuoglu A (2010) Decreased serum prolidase activity and increased oxidative stress in early pregnancy loss. Gynecol Obstet Invest 69:122–127

Tu BP, Weissman JS (2004) Oxidative protein folding in eukaryotes: mechanisms and consequences. J Cell Biol 164:341–346

Ushio-Fukai M, Alexander RW (2004) Reactive oxygen species as mediators of angiogenesis signaling: role of NAD(P)H oxidase. Mol Cell Biochem 264:85–97

Vega M, Urrutia L, Iniguez G, Gabler F, Devoto L, Johnson MC (2000) Nitric oxide induces apoptosis in the human corpus luteum in vitro. Mol Hum Reprod 6:681–687

Xu H, Perez-Cuevas R, Xiong X, Reyes H, Roy C, Julien P, Smith G, von Dadelszen P, Leduc L, Audibert F, Moutquin JM, Piedboeuf B, Shatenstein B, Parra-Cabrera S, Choquette P, Winsor S, Wood S, Benjamin A, Walker M, Helewa M, Dubé J, Tawagi G, Seaward G, Ohlsson A, Magee LA, Olatunbosun F, Gratton R, Shear R, Demianczuk N, Collet JP, Wei S, Fraser WD, INTAPP Study Group (2010) An international trial of antioxidants in the prevention of preeclampsia (INTAPP). Am J Obstet Gynecol 202:239.e1–239.e10

Yung HW, Calabrese S, Hynx D (2008) Evidence of placental translation inhibition and endoplasmic reticulum stress in the etiology of human intrauterine growth restriction. Am J Pathol 173:451–462

Zachara BA, Dobrzynski W, Trafikowska U (2001) Blood selenium and glutathione peroxidases in miscarriage. BJOG 108:244–247

Autoimmune Diseases

Barbosa DS, Cecchini R, ElKadri MZ, Rodriquez MA, Burini RC, Dichi I (2003) Decreased oxidative stress in patients with ulcerative colitis supplemented with fish oil omega-3 fatty acids. Nutrition 19:837–842

Behaska AA, Wu D, Serafini, Meydani SN (2002) Mechanism of vitamin E inhibition of cyclooxygenase activity in macrophage from old mice: role of peroxynitrite. Free Radic Biol Med 32:503–511

Berg DJ, Davidson N, Kuhn R, Muller W, Menon S, Helland G, Thompson-snipes L, Leach MW, Rennick D (1996) Enterocolitis and colon cancer in interleukin -10-deficient mice are associated with aberrant cytokine production and CD4(+)TH1-like response. J Clin Invest 98:1010–1020

Biniecka M, Fox E, Gao W, Ng CT, Veale DJ, Fearon U, O'Sullivan J (2011) Hypoxia induces mitochondrial mutagenesis and dysfunction in inflammatory arthritis. Arthritis Rheum 63:2172–2182

Bogaert S, De Vos M, Oliever K, Peeters, Elewaut D, Lambrecht B, Poullot P, Laukens D (2011) Involvement of endoplasmic reticulum stress in inflammatory bowel disease: a different implication for colonic and ileal disease. Plos ONE 6, e25589

Cerhan JR, Sagg KG, Merlino LA, Mikuls TR, Criswell LA (2003) Antioxidant micronutrient and risk of rheumatoid arthritis in a cohort of order women. Am J Epidemiol 157:345–354

Cioffi M, riegler G, Vietri MT, Pilla P, Caserta L, Carratu R, Sica V, Molinari AM (2004) Serum p53 antibodies in patients affected with ulcerated colitis. Inflamm Bovel Dis 10:606–611

Desai PB, Manjunath DS, Kadi S, Chetana K, Vanishree J (2010) Oxidative stress and enzymatic antioxidant status in rheumatoid arthritis: a case control study. Eur Rev Med Pharmacol Sci 14:959–967

El-Sayed ZA, Farag DH, Eissa S (2003) Tumor suppressor protein p53 and anti-p53 autoantibodies in pediatric rheumatological diseases. Pediatr Allergy Immunol 14:229–233

Edwards SW, Hallett MB (1997) Seeing the wood for the trees: the forgotten role of neutrophils in rheumatoid arthritis. Immunol Today 18:320–324

Fujii S, Katsumata D, Fujimori T (2008) Limits of diagnosis and molecular markers for early detection of ulcerative colitis-associated colorectal neoplasia. Digestion 77(suppl 1):2–12

Gil L, Martinez G, Gonzalez I, Tarinas A, Alvarez A, Giuliani A, Molina R, Tapanes R, Perez J, Leon OS (2003) Contribution to characterization of oxidative stress in HIV/AIDS patients. Pharmacol Res 47:217–224

Greenspan HC, Aruoma O (1994) Could oxidative stress initiate programmed cell death in HIV infection? A role from plant derived metabolites having synergistic antioxidant activity. Chem Biol Interact 143:145–148

Hamouda HE, Zakaria SS, Ismail SA, Khedr MA, Mayah WW (2011) p53 antibodies, metallothioneins and oxidative stress markers in chronic ulcerative colitis with dysplasia. World J Gastontol 17:2417–2423

Jaworowski A, Crowe SM (1999) Does HIV cause depletion of $CD4^+$ T cells in vivo by the induction of apoptosis? Immunol Cell Biol 77:90–98

Ju SM, Song HY, Lee JA, Lee SJ, Choi SY, Park J (2009) Extracellular HIV-1 Tat up-regulates expression of matrix metalloproteinase-9 via a MAPK-NF-kappaB dependent pathway in human astrocytes. Exp Mol Med 41:86–93

Kamanli A, Naziroglu M, Aydilek N, Hacievliyagil C (2004) Plasma lipid peroxidation and antioxidant levels in patients with rheumatoid arthritis. Cell Biochem Funct 22:53–57

Kanmogne GD, Schall K, Leibhart J, Knipe B, Gendehnan HE, Persidsky Y (2007) HIV-1 gp120 compromise es blood-brain barrier integrity and enhances monocyte migration across blood-brain barrier: implication for viral neuropathogenesis. J Cereb Blood Flow Metab 27:123–134

Karatas F, Ozates I, Canatan H, Halifeoglu I, Karatepe M, Colakt R (2003) Antioxidant status and lipid peroxidation in patients with rheumatoid arthritis. Ind J Med Res 118:178–181

Knekt P, Heliovaara M, Aho K, Aifthan G, Marniemi T, Aromaa A (2002) Serum selenium, serum alphatocopherol and the risk of rheumatoid arthritis. Epidemiology 11:402–405

Kuga S, Otsuka T, Nitro H, Nunoi H, Nemoto Y, Nakano T, Ogo T, Umei T, Nihe Y (1996) Suppression of superoxide anion production by interleukin-10 is accompanied by a downregulation of the genes for subunit proteins of NADPH oxidase. Exp Hematol 24:151–157

Kundu S, Ghosh P, Datta S, Ghosh A, Chattopadhyay S, Chatterjee M (2012) Oxidative stress as a potential biomarker for determining disease activity in patients with rheumatoid arthritis. Free Radic Res 46:1482–1489

Mahajan A, Tandon V (2004) Antioxidants and rheumatoid arthritis. J Ind Rheumatol Assoc 12:139–142

Mao L, Wang H, Qiao L, Wang X (2011) Disruption of Nrf2 enhances the upregulation of nuclear factor-kappa B activity, tumor necrosis factor-alpha and matrix metalloproteinase-9 after spinal cord injury in mice. Mediators Inflamm 2010:238321

McArthur JC, Brew BJ (2010) HIV-associated neurocognitive disorders: is there a hidden epidemic? AIDS 24:1367–1370

Meydani SN, Beharka AA (1998) Recent developments in vitamin E and the immune response. Nutr Rev 56:s49–s58

Mollace V, Nottet HS, Clayette P, Turco MC, Muscoli C, Salvemini D, Perno CF (2001) Oxidative stress and

neuroAIDS: triggers, modulators and novel antioxidants. Trends Neurosci 24:411–416
Narushima S, Spitz DR, Oberley LW, Toyokuni S, Miyata T, Gunnett CA, Buettner GR, Zhang J, Ismail H, Lynch RG, Berg DJ (2003) Evidence for oxidative stress in NSAID-induced colitis in $IL10^{-/-}$ mice. Free Radic Biol Med 34:1153–1166
Nicholls SJ, Hazen SL (2009) Myeloperoxidases, modified lipo-proteins and atherogenesis. J Lipid Res 50(Suppl):S346–S351
Papadopulos-Eleopulos E, Healand-Thomel B, Causer DA, Dufty AP (1989) An alternative explanation for the radiosensitization of AIDS patients. Int J Radiat Oncol Biol Phys 17:695–696
Papadopulos-Eleopulos E, Healand-Thomel B, Causer DA, Turner VF, Papadintrion JM (1991) Changes in thiols and glutathione as consequences of simian immune deficiency virus infection. Lancet 338:1013
Peng KF, Wu XF, Zhao HW, Sun Y (2006) Advanced oxidation protein products induce monocyte chemoattractant protein-1 expression via p38 mitogen-activated protein kinase activation in rat vascular smooth muscle cells. Chin Med J (Engl) 119:1088–1093
Reddy PV, Gandhi N, Samikkannu T, Saiyed Z, Agudelo M, Yndart A, Khatavkar P, Nair MP (2012) HIV-1 gp120 induces antioxidant response element-mediated expression in primary astrocytes: role in HIV associated neurocognitive disorder. Neurochem Int 61:807–814
Salvemini D, Mazzon E, Dugo L, serraino I, De Sarro A, Caputi AP, Cuzzocrea S (2001) Amelioration of joint disease in a rat model of collagen induced arthritis by M40403, a superoxide dismutase mimetic. Arthritis Rheum 44:2909–2921
Shah A, Kumar A (2010) HIV-1 gp120-mediated increases in IL-8 production in astrocytes are mediated through the NF-kappaB pathway and can be silenced by gp120-specific siRNA. J Neuroinflammation 7:96
Sharon LW, Louise MW, Maureen LH, Jack PV, Peter GW (1997) Oxidative stress and thiol depletion in plasma and peripheral blood lymphocytes from HIV-infected patients: toxicological and pathological implications. AIDS 11:1689–1697
Sporer B, Paul R, Koedel U, Grimm R, Wick M, Goebel FD, Pfister HW (1998) Presence of matrix metalloproteinase-9 activity in the cerebrospinal fluid of human immunodeficiency virus-infected patients. J Infect Dis 178:854–857
Stamp LK, Khalilova I, Tarr JM, Senthilmohan R, Turner R, Haigh RC, Winyard PG, Kettle AJ (2012) Myeloperoxidase and oxidative stress in rheumatoid arthritis. Rheumatology 51:1796–1803
Staron A, Makosa G, Koter-Michalak M (2012) Oxidative stress in erythrocyte from patients with rheumatoid arthritis. Rheumatol Int 32:331–334
Tiden AK, Sjogren T, Svesson M, Bemlind A, Senthilmohan R, Auchere F, Norman H, Markgren PD, Gustavrson S, Schmidt S, Landquist S, Forber LV, Maqon NJ, Paton LN, Jamerson GN, Eriksson H, Kettle AJ (2011) 2-Thioxanthines are suicide inhibitors of myeloperoxidase that block oxidative stress during inflammation. J Biol Chem 286:37578–37589
Vasanthi P, Nalini G, Rajasekhar G (2009) Status of oxidative stress in rheumatoid arthritis. Int J Rheum Dis 12:29–33
Williams R, Dhillon NK, Hegde ST, Yao H, Peng F, Callen S, Chebloune Y, Davis RL, Buch SJ (2009) Proinflammatory cytokines and HIV-1 synergistically enhance CXCL10 expression in human astrocytes. Glia 57:734–743
Williams R, Yao H, Peng F, Yang Y, Bethel-Brown C, Buch S (2010) Cooperative induction of CXCL10 involves NADPH oxidase: implications for HIV dementia. Glia 58:611–621
Winterboum CC, Kettle AJ (2000) Biomarkers of myeloperoxidase-derived hypochlorous acid. Free Radic Biol Med 29:403–409
Wright HL, Moots RJ, Bucknall RC, Edwards SW (2010) Neutrophil function in inflammation and inflammatory diseases. Rheumatology 49:1618–1631
Wruck CJ, Fragoulis A, Gurzynski A, Brandenburg LO, Kan YW, Chan K, Hassenpflug J, Freitag-Wolf S, Varoga D, Lippross S, Pufe T (2011) Role of oxidative stress in rheumatoid arthritis: insights from the Nrf2-knockout mice. Ann Rheum Dis 70:844–850
Wykretowicz A, Adamska K, Krauze T, Guzik P, Szczepanik A, Rutkowska A, Wysoki H (2007) The plasma concentration of advanced oxidation protein products and arterial stiffness in apparently healthy adults. Free Radic Res 41:645–649

Oxidative Stress in Metabolic Disorders/Diseases

3

Oxidative stress occupies a central role in a variety of metabolic pathologies such as diabetes, obesity, and other secondary complications associated with them. Reactive oxygen species and the molecular mechanism activated by them are unifying factors in these diseases. In this chapter we will discuss three such dynamic metabolic pathophysiological phenomena which constitute the metabolic syndrome and are regulated in part by the oxidative stress.

Diabetes

Diabetes mellitus constitutes a group of metabolic diseases characterized by hyperglycemia (increased glucose levels) which may be a result of defects in insulin secretion or action or both (Dewanjee et al. 2008). There are two main types of diabetes, Type I and Type II, described below (Michael et al. 2000):

(A) ***Type I diabetes*** (insulin-dependent diabetes)

The major factors that can cause type 1 diabetes are:

1. Autoimmunity (type 1A)
2. Genetic susceptibility, environmental or idiopathic (type 1B)

(B) ***Type 2 diabetes mellitus*** (formerly called non-insulin-dependent diabetes mellitus)

Major causative factors are:

1. A derangement in β-cell secretion of insulin
2. A decrease response of peripheral tissue to respond to insulin (insulin resistance)

Diabetes as a progressive disease can lead to variety of other pathophysiological complications (Dewanjee et al. 2008) ranging from the *acute metabolic complications* which include diabetic ketoacidosis, hyperosmolar nonketonic coma, and hypoglycemia to *late systemic complications* which include: atherosclerosis, diabetic microangiopathy, cerebrovascular disease, coronary artery disease, retinopathy, nephropathy, and peripheral and autonomic neuropathies.

Hyperglycemia acts as a connector between these diabetic complications and diabetes (Brownlee 2001; Rolo and Palmeira 2006). Apart from hyperglycemia, other key players that are critical in diabetic pathogenesis are hyperlipidemia and enhanced oxidative stress. For instance, it has been studied that diabetes can direct toward induction of the oxidative stress due to increased speed of the production of free radicals and reduced antioxidants resistance. Studies indicate that high glucose is one of a major source of reactive oxygen species via its auto-oxidation, metabolism, and the development of advanced glycosylation end products (AGEs) (Ha and Lee 2000). Both hyperglycemia and oxidative stress are interconnected and act as the two most important mediators of diabetes.

M. Bansal and N. Kaushal, *Oxidative Stress Mechanisms and their Modulation*,
DOI 10.1007/978-81-322-2032-9_3,

Major Mechanisms of Hyperglycemia-Induced Damage

In diabetic conditions increased level of glucose are observed, and this glucose acts as a main source of the free radicals via its oxidation. Glucose as enediol in a transition metal-dependent reaction is oxidized to an enediol radical anion that is converted into reactive ketoaldehydes and to superoxide anion radicals. These superoxide anion radicals can then form subsequent free radicals such as peroxynitrite radicals (Halliwell and Gutteridge 1990; Hogg et al. 1993) on reaction with nitric oxide or hydrogen peroxide and extremely reactive hydroxyl radicals if not degraded by antioxidant enzymes (Jiang et al. 1990; Wolff and Dean 1987). These hyperglycemia-induced superoxide radicals are also found to promote the lipid peroxidation of low-density lipoprotein (LDL) further adding to the pool of free radicals in the body (Tsai et al. 1994; Kawamura et al. 1994). Furthermore, in diabetes, glucose interacts with proteins that culminate in the production of advanced glycation end products (AGEs) (Hori et al. 1996; Mullarkey et al. 1990), which via their receptors (RAGEs) not only inactivate enzyme functions (McCarthy et al. 2001) but also promote the free radical formation (Baynes and Thorpe 1999; Baynes 1991). This increase in intracellular oxidative stress by AGEs activates the transcription factor NF-kB, which is a common mediator in major inflammatory processes and oxidative conditions (Mohamed et al. 1999a). Activation of NF-kB via enhanced production of nitric oxide is believed to be a mediator of islet beta cell damage.

Hyperglycemia-induced tissue damage is mediated by four important molecular mechanisms (Rolo and Palmeira 2006) (Fig. 3.1):

1. Increased formation of advanced glycation end products (AGEs)
2. Activation of protein kinase C isoforms
3. Increased glucose flux through polyol pathway
4. Increased overactivity of hexosamine pathway flux

These mechanisms of hyperglycemia-induced damage in turn are a result of the mitochondrial

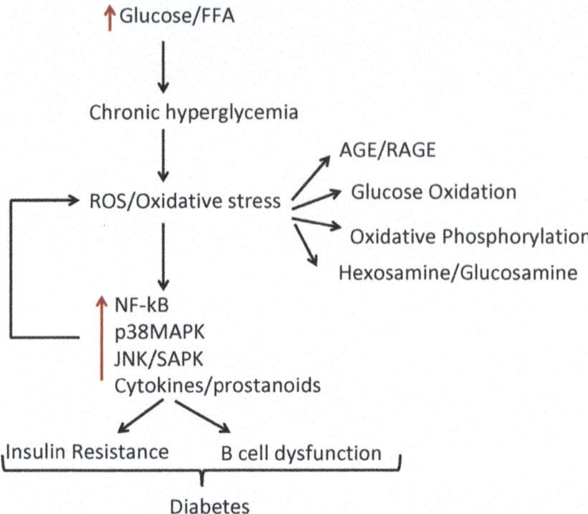

Fig. 3.1 Molecular pathways of ROS-mediated induction of diabetes

overproduction of reactive oxygen species (ROS). While, in the diabetic microvasculature diseases, this is a consequence of intracellular hyperglycemia; in diabetic macrovascular and in the heart, it results from the increased oxidation of fatty acids and in part from pathway specific insulin resistance.

Increased Polyol Pathway Flux

The polyol pathway constitutes a family of aldo-keto reductase enzymes that can reduce variety of the carbonyl compounds using NADPH to their respective sugar alcohols (polyols). In tissues having high levels of aldose reductase such as the nerve, retina, lens, glomerulus, and vascular cells (Ramasamy and Goldberg 2010), glucose uptake is mediated by the insulin-independent GLUTs (glucose transporters) thus increasing intracellular glucose concentrations simultaneously with hyperglycemia. It has been suggested that hyperglycemia-induced increase in the polyol pathway causes tissue damage by an increase in redox stress caused by the consumption of NADPH. Since NADPH is a cofactor in regeneration of the reduced glutathione (GSH), which is an important scavenger of ROS, its increased consumption can induce or exacerbate intracellular oxidative stress. It has also been demonstrated that decreased glutathiolation of the cellular proteins is related to decreased NO availability in diabetic rats. On the contrary, experiments with restoration of NO levels in diabetic animals increase glutathiolation of the cellular proteins and inhibit aldose reductase activity.

These findings are backed by studies indicating that the overexpression of human aldose reductase increased atherosclerosis in diabetic mice and reduced the expression of genes that regulate regeneration of glutathione (Vikramadithyan et al. 2005). Similarly, levels of GSH were decreased in the lens of diabetic mice and transgenic mice overexpressing aldose reductase (Lee and Chung 1999; Chung et al. 2003; Zhang et al. 2000; Ii et al. 2004).

Increased Intracellular AGE Formation

In hyperglycemic conditions, nonenzymatic reaction between proteins and glucose or other glycating compounds formed from the glucose and increased fatty acid oxidation leads to the formation of advanced glycosylation end products (AGEs) (Wautier and Schmidt 2004; Candido et al. 2003). Increased amount of AGEs are observed in extracellular matrix in diabetes (Stitt et al. 1998, 1997; Nishino et al. 1995; Horie et al. 1997; Niwa et al. 1997). Increased AGEs can damage cells in three major ways:

1. Firstly, AGEs modify the intracellular proteins leading to their altered function.
2. Secondly, AGEs modify extracellular matrix components leading to the abnormal interaction between these matrix components among themselves and with their matrix receptors (integrins) expressed on the cell surfaces.
3. Finally, AGEs modify plasma proteins which bind to the receptors of AGE (RAGE) on various cells such as macrophages, vascular endothelial cells, and vascular smooth muscle cells leading to induced production of ROS. These events in turn activate the pleiotropic transcription factor nuclear factor (NF)-κB, causing multiple pathological changes in gene expression (Goldin et al. 2006).

In addition, to these distinct mechanisms, other mechanisms involving AGE-dependent modification of corepressor mSin3A leading to the increased recruitment of *O*-linked *N*-acetylglucosamine (*O*-GlcNAc) transferase, with consequent increased modification of Sp3 by *O*-GlcNAc, have been reported. This decreases the binding to a glucose-responsive GC box in the angiopoietin (Ang)-2 promoter, resulting in the increased Ang-2 expression which in renal endothelial cells can increase the expression of intracellular adhesion molecule-1 and vascular cell adhesion molecule-1 (VCAM)-1. This mechanism sensitizes the microvascular endothelial cells of kidney to the proinflammatory effects of tumor necrosis factor α (Yao et al. 2007).

AGE-dependent modification of various proteins and receptors has been implicated in various other diabetes-related pathologies such as ischemia (Rivard et al. 1999; Schatteman et al. 2000) and impaired wound healing (Gallagher et al. 2007). On the contrary in diabetic mouse models of impaired angiogenesis and wound healing, decreasing mitochondrial ROS formation normalizes both the ischemia-induced new vessel formation and wound healing (Gallagher et al. 2007; Thangarajah et al. 2009).

Increased Protein Kinase C Activation

PKC enzyme phosphorylates various target proteins, and its activity is dependent on both Ca^{2+} ions and phosphatidylserine and is greatly enhanced by the diacylglycerol (DAG) (Geraldes and King 2010). Consistently increased activation of several PKC isoforms can cause tissue injury by diabetes-induced ROS. These effects are a result of the enhanced de novo synthesis of DAG from glucose via triose phosphate induction. Due to diabetes there is increased ROS which inhibits the activity of the glycolytic enzyme GAPDH leading to raised intracellular levels of the DAG precursor triose phosphate (Inoguchi et al. 1992; Craven et al. 1990; Shiba et al. 1993; Scivittaro et al. 2000).

Additionally, enhanced activity of PKC activity as seen in the hyperglycemic conditions can also be enhanced by interaction of AGEs with their cell-surface receptors (Derubertis and Craven 1994). This hyperglycemia-induced activation of protein kinase C and p38α mitogen-activated protein kinase (MAPK) can increase the expression downstream targets of PKC signaling such as SHP-1 (Src homology-2 domain-containing phosphatase-1), a protein tyrosine phosphatase. This signaling cascade culminates in the pericyte apoptosis resulting from dephosphorylation of the platelet-derived growth factor (PDGF) receptor-β and its reduced downstream signaling (Geraldes et al. 2009). Similarly PKC activation by the increased fatty acid oxidation in insulin-resistant arterial endothelial cells and the heart is reported to be a key player in diabetic atherosclerosis and cardiomyopathy.

PKC activation by high glucose induces expression of the permeability-enhancing factor VEGF in vascular smooth muscle cells (Williams et al. 1997) and eNOS in cultured endothelial cells (Kuboki et al. 2000) with a concomitant decrease in the pool of NO (Ganz and Seftel 2000). In cultured mesangial cells (Pugliese et al. 1994; Craven et al. 1997) and glomeruli of the diabetic rats (Kikkawa et al. 1994), this inhibition of NO is a contributor to PKC-mediated accumulation of microvascular matrix protein by inducing expression of transforming growth factor (TGF)-β1, fibronectin, and type IV collagen (Pugliese et al. 1994). Furthermore, increased activity of PKC in the hyperglycemic conditions has also been implicated in the overexpression of the fibrinolytic inhibitor, plasminogen activator inhibitor (PAI)-1 (Feener et al. 1996) and in the activation of NF-κB in cultured endothelial cells and vascular smooth muscle cells (Pieper and Riaz-ul-Haq 1997; Yerneni et al. 1999).

Increased Hexosamine Pathway Flux

Increased fatty acid oxidation as seen in the diabetic conditions is a major contributor to the pathogenesis of diabetic complications. There is an increase in the flux of fructose 6-phosphate into the hexosamine pathway (Sayeski and Kudlow 1996; Kolm-Litty et al. 1998; Chen et al. 1998; Du et al. 2000), converting it to glucosamine 6-phosphate by the enzyme fructose 6-phosphate amidotransferase (GFAT). This glucosamine 6-phosphate is then converted to UDP-N-acetylglucosamine glutamine. This product is used by specific *O*-GlcNAc transferases during the posttranslational modification of various cytoplasmic and nuclear proteins by *O*-GlcNAc at serine and threonine residues and is critical in diabetes-related pathologies. For example, hyperglycemia-induced increase in the *O*-GlcNAcylation of the transcription factor Sp1 activates PAI-1 promoter in vascular smooth muscle cells (Chen et al. 1998) and TGF-β1 and PAI-1 in arterial endothelial cells (Du et al. 2000). Similarly, *O*-GlcNAcylation at the Akt activation site of eNOS protein (Yamagishi et al. 2001; Hart 1997; Musicki et al. 2005) leads to inhibition of eNOS activity in arterial endothelial cells which is important in vascular complication of diabetes. Finally, reduced sarcoplasmic reticulum Ca (2^+) ATPase 2a (SERCA2a) expression and its promoter activity impair the cardiomyocyte calcium cycling through increased nuclear O-GlcNAcylation (Clark et al. 2003) during diabetes (Pang et al. 2004), whereas inhibition of GFAT blocks the hyperglycemia-induced increases in the transcription of both TGF-α (Yerneni et al. 1999) and TGF-β1 (Sayeski and Kudlow 1996).

The above mentioned pathways share some commonality with other metabolic diseases such as obesity in terms of oxidative stress being a common denominator. In diabetic conditions, it is important to understand that oxidative insult not only precedes the activation of these pathways but also is a key player in developing insulin resistance and ß-cell dysfunction. The next section addresses the molecular mechanism which directly aid in developing diabetic complications through induction of oxidative stress.

Oxidative Stress and Diabetes

ROS is an important mediator in diabetes and pathogenesis of diabetic complications (Rösen et al. 2001; Nishikawa et al. 2000b). Chronic elevation of plasma glucose levels called hyperglycemia leads to the ROS formation (Brownlee 2001) and causes major diabetes-associated complications including nephropathy, retinopathy, neuropathy, and macro- and microvascular damages (DeFronzo 1997; The Diabetes Control and Complications Trial Research Group 1993). These microvascular complications are caused by the elevated free fatty acids (FFA) and correlated with insulin resistance (McGarry 2002; Boden 1997) and FFA-induced oxidative stress-mediated deterioration of ß-cell function (Poitout and Robertson 2002; Harmon et al. n.d.). In diabetic conditions ROS not only directly causes macromolecular damage but also activates a variety of cellular stress-sensitive pathways of cellular damage as signaling molecules. The activation of these pathways of cellular damage by ROS is linked to insulin resistance and decreased insulin secretion.

In previous sections we have discussed various hyperglycemia-induced biochemical pathways such as glycation end products (AGEs) and receptors for AGE (RAGE) (Brownlee 1995), protein kinase C (PKC) (Koya and King 1998), and the polyol pathway (Stevens et al. 2000) that play a significant role in the etiology of diabetic complications. However, it is important to understand that oxidative stress caused by hyperglycemia is an earlier event that precedes these pathways of diabetic complications (Rösen et al. 2001; Brownlee 1995; Koya and King 1998). For example, ROS (superoxide and hydroxyl) mediates the nonenzymatic protein glycation, through metal-catalyzed glucose autoxidation as an early account for glucose cytotoxicity (Brownlee 2001; Wolff and Dean 1987).

Hyperglycemia activates biochemical pathways of stress-activated signaling involving nuclear factor-kB (NF-κB), NH_2-terminal Jun kinases/stress-activated protein kinases (JNK/SAPK), p38 mitogen-activated protein (MAP) kinase, and hexosamine (Barnes and Karin 1997; Kyriakis and Avruch 1996; Marshall et al. 1991) (Fig. 3.1). In the next sections we will discuss these pathways separately in detail.

Activation of NF-κB

Out of various signaling pathways mentioned earlier, the activation of NF-κB by hyperglycemia, ROS, and oxidative stress is the most extensively studied (Barnes and Karin 1997; Mohamed et al. 1999b; Bierhaus et al. 2001). NF-κB as a transcription factor regulates expression of a large number of genes, important in inflammatory responses, apoptosis, and those linked to the complications of diabetes (e.g., vascular endothelial growth factor (VEGF) and RAGE) (Mohamed et al. 1999b). Many of the gene products can in turn also activate NF-κB (e.g., VEGF, RAGE) by a positive feedback loop. Increased oxidative stress activates NF-κB via degradation of its inhibitory subunit, inhibitory protein B (IB), due to its phosphorylation by the upstream serine kinase, IB kinase ß (IKK-ß). Hyperglycemia-induced intracellular ROS and hence activation of NF-κB have been reported in bovine endothelial cells (Nishikawa et al. 2000a) with subsequent increase in PKC activity, AGE, and sorbitol levels. These studies also indicated that strategies that disrupt the mitochondrial ROS production also suppressed the NF-κB, PKC, AGE, and sorbitol induction and thus blocked the hyperglycemia-induced increase in ROS production. This implies that the ROS formation and NF-kB activation are the initial signaling events in the diabetic complications.

Activation of Stress Kinases (JNK/SAPK and p38 MAPK)

The JNKs/SAPKs are members of MAP serine/threonine protein kinases superfamily which also includes the p38 MAP kinases (p38 MAPKs) and the extracellular signal-related kinases (ERKs) (Kyriakis and Avruch 1996). Unlike ERKs (also referred to as MAPKs), which are mitogen-activated kinases, JNK/SAPK and p38 MAPK are stress-activated kinases. These stress-activated kinases are responsive to a variety of exogenous and endogenous stress-inducing stimuli, including hyperglycemia, ROS, oxidative stress, osmotic stress, proinflammatory cytokines, heat shock, and ultraviolet irradiation. Hyperglycemia-induced oxidative stress activates JNK/SAPK leading to apoptosis in human endothelial cell (Ho et al. 2000a; Natarajan et al. 1999).

Similarly, p38 MAPK pathway is also activated in response to hyperglycemia and in diabetes (Begum and Ragolia 2000; Igarashi et al. 1999) by increased ROS production. Increases in total levels of JNK/SAPK and p38 MAPK have been reported in the nerve tissue of patients with type 1 and type 2 diabetes (Purves et al. 2001), although a causative role in the pathophysiology has not been established. On the contrary, vitamin C supplementation as an antioxidant not only suppressed the H_2O_2 generation, JNK/SAPK activity but also the subsequent apoptosis induced by hyperglycemia (Ho et al. 2000a).

Activation of Hexosamine Pathway

Hyperglycemia-induced oxidative stress increases the flux of glucose or FFAs into a variety of cell types results in the activation of the hexosamine biosynthetic pathway (Marshall et al. 1991; Boden et al. 1994). This activation of hexosamine pathway in turn leads to insulin resistance and the development of late complications of diabetes (Marshall et al. 1991; Boden et al. 1994; Schleicher and Weigert 2000). For example, hyperglycemia-induced increase in hexosamine pathways is seen in bovine endothelial cells. The fact that this effect was blocked by an inhibitor of electron transport, a mitochondrial uncoupling agent (CCCP), and the expression of either uncoupling protein 1 or MnSOD (Du et al. 2000) indicates the importance of oxidative stress in this mechanism.

Cumulatively, it is evident that hyperglycemia and ROS-mediated activation NF-kB, JNK/SAPK, p38 MAPK, and hexosamine stress-sensitive pathways are critical for diabetes and its associated pathologies. What has become equally intriguing is the growing number of reports linking the activation of these same pathways to insulin resistance and ß-cell dysfunction.

Insulin Resistance and Oxidative Stress

The pathophysiology of type 2 diabetes is characterized by insulin resistance and decreased insulin (DeFronzo 1997; Reaven 2000; Kahn 1994; Grodsky 2000). Due to increased insulin resistance or the compensatory insulin secretory response decrease, an impaired glucose tolerance occurs. This increase in insulin, FFA, and/or glucose levels can increase ROS-mediated oxidative stress as well as activate stress-sensitive pathways which in turn can start a vicious circle of both insulin action and secretion, thereby accelerating the progression to overt type 2 diabetes.

As discussed above, oxidative stress is linked to the etiology of numerous diabetes-associated complications (Rösen et al. 2001; Nishikawa et al. 2000b; West 2000). Studies indicate that these effects of ROS production and oxidative stress are mediated through induction of insulin resistance (Paolisso and Giugliano 1996; Rudich et al. 1997; Ceriello 2000; Yaworsky et al. 2000; Maddux et al. 2001). For example, antioxidant such as lipoic acid (LA), vitamin E, vitamin C, or glutathione supplementation improves insulin sensitivity (Maddux et al. 2001; Rudich et al. 1999; Packer et al. 2000). As mentioned previously, ROS and oxidative stress activate multiple serine kinase cascades (Kyriakis and Avruch 1996) which act on a number of potential targets (substrates), including the insulin receptor (IR) and the family of IR substrate (IRS) proteins. For example, serine phosphorylation of IRS-1 and IRS-2 decreases the extent of tyrosine phosphorylation and thus attenuates insulin action (Paz et al. 1997; Birnbaum 2001). Similarly, activation of IKK-ß, a serine kinase that regulates the NF-kB pathway, also inhibits insulin action (Yuan et al. 2001). These studies indicate the critical role of IKK-ß in the pathogenesis of insulin resistance and need to focus on developing IKK-ß inhibitors as potential therapeutics to increase the insulin sensitivity.

Since obesity and insulin resistance are associated, it is most likely that the adipocyte-derived factor may be the mediator of oxidative stress-induced insulin resistance in the prediabetic stage. Some of these factors include tumor necrosis factor (Hotamisligil and Spiegelman 1994), leptin (Cohen et al. 1996), FFAs (McGarry 2002; Boden 1997; Randle et al. 1988), and, most recently, resistin (Steppan et al. 2001). Out of these, free fatty acids (FFAs) are the most strongest and likely link between obesity and insulin resistance (McGarry 2002; Boden 1997; Shulman 2000). FFAs can decrease the insulin sensitivity by several mechanisms such as inhibition of insulin-stimulated glucose transport (Shulman 2000) and direct interaction with transcription factors to regulate gene expression, especially those involved in lipid and carbohydrate metabolism (Duplus et al. 2000). Activation of NF-kB is also one of the mechanisms by which FFAs cause cellular damage (Hennig et al. 2000, 1999; Dichtl et al. 1999; Lee et al. 2001). This effect might be linked to FFA-induced ROS and activation of PKC-*theta* (Griffin et al. 1999). Most importantly, elevated FFA levels cause mitochondrial dysfunctions leading to the uncoupling of oxidative phosphorylation (Brownlee 1995; Wojtczak and Schonfeld 1993) and the generation of ROS, including superoxide (Koya and King 1998; Bakker et al. 2000a). FFAs, further exacerbate these effects by impairment of endogenous antioxidant defenses by reducing intracellular glutathione (Paolisso and Giugliano 1996; Toborek and Hennig 1994; Hennig et al. 2000). These studies suggest that the alteration in cellular redox status is a contributory component of the proinflammatory effects of FFAs.

ß-Cell Dysfunction

ß cells are central to the pathology of diabetes as sensors of glucose and secreting the appropriate amount of insulin in response to a glucose stimulus (Meglasson and Matschinsky 1986). Oxidative stress is known to target these ß cells and blunt insulin secretion (Maechler et al. 1999). Other factors that add to the ß-cell dysfunction are chronic hyperglycemia, elevated FFA levels, or both (Grodsky 2000; Robertson et al. 2000). These molecules cause damage by increasing the production of ROS and RNS and activation of stress-sensitive pathways. The fact that ß Cells have lower antioxidant enzymes such as catalase, glutathione peroxidase, and superoxide dismutase (Tiedge et al. 1997) makes them more susceptible to oxidative insult caused by ROS and RNS. These results are verified in studies with transgenic mice overexpressing antioxidant enzymes in islets showing protection from deleterious effects noted above (Tiedge et al. 1998; Benhamou et al. 1998).

In ß-cell preparations, oxidative stress increases production of p21 (an inhibitor of cyclin-dependent kinase); decreases insulin mRNA, cytosolic ATP, and calcium flux in cytosol and mitochondria; and causes apoptosis (rev. in Maechler et al. 1999). Moreover, oxidative stress also inhibits the glucose-mediated insulin secretion (Maechler et al. 1999). The findings that lipid peroxidation inhibits insulin secretion and glucose oxidation (Miwa et al. 2000) are consistent with the fact that antioxidants can protect against ß-cell toxicity and the generation of glycation end products and attenuation of NF-kB (Tanaka et al. 1999; Ho and Bray 1999; Tajiri et al. 1997; Ho et al. 1999, 2000b). Additionally, the activation of the hexosamine pathway has also been found to be important in glucose-stimulated insulin secretion and simultaneous increase in H_2O_2 indicating the loss of ß-cell function (Kaneto et al. 2001). There are numerous studies which indicate an inverse relationship between hyperglycemia and insulin release (Grodsky 2000; Boden et al. 1996).

ß-cell impairment in insulin secretion has also been associated with an FFA-induced increase in ROS (Carlsson et al. 1999). FFAs-mediated oxidative damage causes decreased mitochondrial membrane potential and increased uncoupling proteins, leading to the opening of K^+-sensitive ATP channels and selective impairment of glucose-stimulated, but not K^+-stimulated, insulin secretion (Lameloise et al. 2001; Segall et al. 1999). Genetic defects such as leptin production or leptin receptors impairment further amplify these toxic effects of FFAs leading to the poor insulin secretion (Unger and Zhou 2001). As mentioned earlier the combination of increased glucose and FFAs can maximize ß-cell toxicity. In isolated islets or HIT cells exposed to the chronic elevated glucose and FFA levels, a decrease in both insulin mRNA and the activation of an insulin gene reporter construct was reported (Jacqueminet et al. 2000). These studies along with others indicate that ß-cell lipotoxicity is amplified by the concurrent hyperglycemia (Poitout and Robertson 2002; Harmon et al. n.d.).

Obesity

In the earlier chapters it has been discussed that lipids per se are susceptible to damage by ROS-induced oxidative stress. The products of lipid peroxidation such as malondialdehyde (MDA), thiobarbituric acid-reactive substances (TBARS), lipid hydroperoxides, conjugated dienes, 4-hydroxynonenal

(4-HNE), and F2-isoprostanes (8-epiPGF2α) have been widely implicated in obesity. For example, concentration of F2-IsoPs and other lipid peroxidation markers are higher in individuals with obesity and correlate directly with BMI, plasma cholesterol concentration, the percentage of body fat, LDL oxidation, and TG levels (Pihl et al. 2006; Block et al. 2002); in contrast, antioxidant defense markers are lower according to the amount of body fat and central obesity (Chrysohoou et al. 2007; Hartwich et al. 2007). A research showed that high-fat diet induces a significant increase in OS stress and inflammation (Patel et al. 2007).

Obesity is a prevalent metabolic disorder which is associated with increased rates of the morbidity and mortality worldwide (Bray 2000). A direct relation between obesity and other metabolic diseases such as diabetes, hypertension, dyslipidemia, insulin resistance, coronary artery disease, and osteoarthritis has been found in obese individuals (Bray 2000; Davi et al. 2003). At molecular level oxidative stress has been indicated as a unifying mechanism between obesity and associated metabolic disorders (Higdon and Frei 2003). Numerous factors contribute to the obesity-associated oxidative stress. These factors may not be mutually exclusive, but differentially lead to systemic oxidative stress depending on the metabolic and physical status of the obese individual. Here we have compiled such sources and mechanisms of oxidative stress critical in obesity (Fig. 3.2).

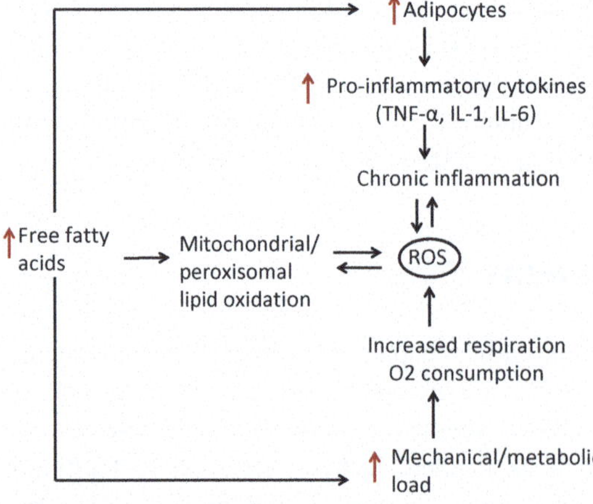

Fig. 3.2 Central role of ROS in obesity

Adipose Tissue/Adipocyte

Excessive adipocytes as observed in obesity as sources of proinflammatory cytokines, including TNF-α, IL-1, and IL-6, are directly linked to the obesity-associated OS. These inflammatory cytokines stimulate the production of reactive oxygen and nitrogen species by macrophages and monocytes leading to increased OS. TNF-α particularly, can lead to the generation of superoxide anions by inhibiting the activity of PCR and thus increasing the interaction of electrons with oxygen to generate superoxide anion (Fonseca-Alaniz et al. 2007). In addition to this pathway, adipose tissue having the secretory capacity of angiotensin II stimulates nicotinamide adenine dinucleotide phosphate (NADPH) oxidase activity. NADPH oxidase comprises the major route for ROS production in adipocytes (Morrow 2003).

Lipid Oxidation

Obesity is characterized by the increased dietary fat intake, increased fat storage, and excessive intracellular triglycerides and dyslipidemia (Davi et al. 2002; Vincent et al. 2004). Obesity-associated increased accumulation of lipid molecules may be more susceptible targets for the oxidative modification by ROS (Olusi 2002; Vincent et al. 2001). Numerous studies have validated these findings suggesting a positive correlation exists between levels of lipids and oxidative stress (Vincent et al. 2001; Furukawa et al. 2004). White adipose tissues are the preferable site of accumulation of oxidation products of lipids (Furukawa et al. 2004). Increased lipid oxidation in turn increases the risk for thrombosis, endothelial dysfunction, and atherosclerosis (Rodriguez-Porcel et al. 2002; Lyon et al. 2003). The hallmark of obesity includes elevated triglycerides, lowered high-density lipoproteins (HDL), and elevated LDLs (Dobrian et al. 2000). Obese individuals have a shorter log phase of LDL and rapid lipid peroxidation in the polyunsaturated fatty acids of LDL particles (Van Gaal et al. 1998; Ozata et al. 2002). Decreased antioxidant defenses and increased concentrations of 4-HNE per unit intramuscular triglycerides in obese people indicate increased susceptibility of lipids to oxidative modification with obesity (Russell et al. 2003). Also consumption of certain lipids by obese men such as conjugated linolenic acid dramatically increases urinary concentrations of products of lipid oxidation such as 8-epiPGF2α (Basu et al. 2000).

Obesity-associated oxidative stress may be due to the metabolic impact of intracellular triglycerides (Bakker et al. 2000b). For example, mitochondrial and peroxisomal oxidation of fatty acids (excessive triglycerides) are capable of producing free radicals (•O_2^-) within the mitochondrial electron transport chain via decrease in intramitochondrial adenosine diphosphate and, therefore, OS. On the contrary some studies suggest possibility of preexisting mitochondrial abnormalities that allow for overproduction of ROS (Duvnjak et al. 2007). Similarly, enhanced free fatty acid (FFA) levels as seen in abdominal or visceral adiposity can produce nitroxide radicals particularly in the smooth vascular and endothelial cells via a protein kinase C mechanism (Inoguchi et al. 2000). Furthermore, these FFA can also induce the oxidative respiratory burst in white cells and acutely increase ROS formation (•O_2^-, hypochlorous acid, ONOO•) in culture (Inoguchi et al. 2000).

We can conclude that the ROS and lipid peroxidation biomarkers enter systemic circulation and initiate a vicious cycle of systemic oxidative stress in obesity.

Oxygen Overconsumption and Increased Metabolic ROS

Obesity is linked to the increased mechanical load and myocardial metabolism, thus increased oxygen consumption (Fig. 3.2). Obese individuals are known to have high cell respiration rates and oxygen consumption due to the additive mechanical load of carrying excessive body weight which may be further exacerbated during physical activity (Vincent et al. 2004; Salvadori et al. 1999). The negative consequence of this increased oxygen consumption and respiration is activation of metabolic pathways that form free radicals by increasing electron transport chain activity, increased chances of electron leakage that can partially reduce oxygen leading to the formation of •O_2^- and subsequently H_2O_2 (Ji 1995, 1996). Studies indicate mechanically less efficiency during exercise is an additive contributor to increased energy expenditure for a given exercise load in obese people (Vincent et al. 2005a). This accelerated and inefficient mitochondrial respiration is associated with increased lipid hydroperoxide production in the obese (Vincent et al. 2005b).

Additionally, increased respiration and oxygen consumption in obese humans can activate the pathways of conversion of hypoxanthine to urate (Saiki et al. 2001). When hypoxanthine is converted to urate, •O_2^- is formed. It has been concluded that sporadic strenuous physical activity may actually result in an acute increase in oxidant stress in obese persons (Saiki et al. 2001).

Nutritional and Physiological Antioxidant Deficiency

We have discussed in previous sections that diets containing higher amounts of fats can alter the oxygen metabolism rendering the lipid deposits vulnerable to oxidation. This can disrupt the delicate balance between ROS produced and antioxidant capacity of tissues leading to the obesity-associated complication such as atherosclerosis (Khan et al. 2006). These perturbations to the antioxidant defenses are frequently observed in obesity. There is a decrease in the activity of SOD and GPx, two major antioxidant enzymes in obesity (Beltowski et al. 2000). During the initial stages of obesity, there is an initial outburst in the antioxidant enzymes to counteract oxidative stress, whereas chronic obesity continually depletes the sources of antioxidant enzymes (Olusi 2002; Vincent et al. 2001). Furthermore, the changes in enzyme activities can also affect that degree of adiposity (Olusi 2002; Ozata et al. 2002).

Alternatively, replenishing tissues with dietary, enzymatic, and nonenzymatic antioxidant defenses are critical to maintain the antioxidant–prooxidant balance in tissues. Numerous studies have indicated that a direct relationship exists between low intakes of the protective antioxidants- and phytochemicals-rich foods (fruits, vegetables, whole grains, legumes, wine, olive oil, seeds, and nuts) in the diet and increased chances of obesity as measured by waist circumference, BMI, and plasma lipid peroxidation (Vincent et al. 2005c). Similar to these findings lower levels of other dietary antioxidants such as β-carotene, vitamins E and C and levels of trace minerals like zinc and selenium (cofactors for antioxidant enzymes) were found in the serum of obese subjects than their non-obese counterparts (Ozata et al. 2002; Ohrvall et al. 1993; Moor de Burgos et al. 1992; Decsi et al. 1997; Viroonudomphol et al. 2003). In addition to this cause–effect relationship, the antioxidant levels can also affect the degree of adiposity (Wallstrom et al. 2001; Reitman et al. 2002). For example, BMI is negatively correlated with plasma vitamin levels, with the obese having the lowest concentrations (Moor de Burgos et al. 1992; Myara et al. 2003; Strauss 1999). These studies lead to the conclusion that dietary antioxidants may be used in combating the excessive prooxidant processes during obesity.

In conclusion, it is the combination of inadequate dietary, enzymatic antioxidants and increased production of ROS formation that creates an imbalance favoring lipid and protein oxidation and oxidative stress seen in obesity.

Chronic Low-Grade Inflammation

Obesity is considered as low-intensity chronic inflammation (Saito et al. 2003; Kopp et al. 2003; Weyer et al. 2002) (Fig. 3.2) with concomitant increase in serum adipokines and fat mass, especially visceral fat. Also like other inflammatory conditions, it is also characterized by inflam matory cytokine expression, C-reactive protein (CRP) production, and increased white blood cell counts and white cell activity. Obese people have increase in white blood cell counts, especially increase in the monocyte subfraction, and trends toward elevation in the neutrophil subfraction (Kullo et al. 2002). These monocytes are source of $\bullet O_2^-$, H_2O_2, OH•, $ONOO^-$, hypochlorous acid, and myeloperoxidase. After developing into macrophages they produce interleukins and TNF-α. On the other hand, neutrophils generate $\bullet O_2^-$ via NADPH oxidase. ROS produced by these immune cells is associated with elevations in tyrosine cross-links, TBARS, oxidized linolenic acid, and oxidized serum proteins and lipoproteins discussed earlier (Garg et al. 2000). As obesity develops, there is a progressive infiltration of macrophages into the adipose tissue depots (Wellman and Friedberg 2002), due to leukocytes recruitment into adipose tissue by MDA and 4-HNE (by-products of fat-induced ROS generation) (Furukawa et al. 2004). This leads to changes in the adipocyte volume and overall fat pad size leading to increased secretion of TNF-α and leptin secretion, which again attracts more macrophages into fat tissue. This vicious cycle of events during obesity can potentially lead to obesity-associated oxidative damage and disease processes such as atherosclerosis (Wellen and Hotamisligil 2003). On the contrary nitric oxide (NO), an important antiatherogenic agent, inhibits platelet activation and aggregation, leukocyte chemotaxis, and endothelial adhesion (Chakraborty et al. 2003). In overweight and obesity conditions, the endothelium-dependent vasodilation of NO is impaired also seen in hypercholesterolemia (De Souza et al. 2005). Furthermore, increase in the production of superoxide and increased peroxynitrite in persons with obesity and high blood pressure diminish the availability of NO and cause vasoconstriction in the vasculature of the liver leading to other complications (Dobrian et al. 2003).

At subcellular level the inflammatory molecules TNF-α, IL-6, and CRP concentrations are positively associated with the level of adiposity (Olusi 2002; Saito et al. 2003; Weyer et al. 2002; Higdon and Frei 2003). These inflammatory molecules have been associated with CAD, infarction, stroke, thrombosis, and peripheral arterial disease (Saito et al. 2003; Kopp et al. 2003; Weyer et al.

2002). Fat expresses proinflammatory cytokines such as interleukin-6 (IL-6) and tumor necrosis factor alpha (TNF-α), and the increased IL-6 and TNF-α levels due to expansion of the adipose tissue during obesity may increase the production of CRP (Kopp et al. 2003; Maachi et al. 2004). These events can stimulate the expression of atherogenic endothelial adhesion molecules and promote the attachment and migration of monocytes into vessel walls, causing conversion of monocytes to macrophages (Lyon et al. 2003). Specifically, TNF-α is very critical because it not only increases the expression of adhesion molecules (Lyon et al. 2003) but also suppresses insulin signal transduction and expression of the insulin receptor in isolated adipocytes, leading indirectly to glucose dysregulation, hyperglycemia, and eventually pancreatic β-cell destruction (Weyer et al. 2002; Hotamisligil et al. 1994).

CRP levels on the other hand as indicators of vascular inflammation have been associated with F2-isoprostane levels and prothrombotic markers in obese persons (Block et al. 2002; Davi et al. 2002), suggesting that obesity is related to a chronic state of oxidant stress and platelet activation (Ridker et al. 2000). An inverse relationship exists between CRP levels and adiponectin levels (Ouchi et al. 1999). Adiponectin not only improves insulin sensitivity but also inhibits vascular inflammation (Lyon et al. 2003). These findings are in corroboration with studies showing increased plasma adiponectin levels with decreased CRP, TNF-α, and IL-6 levels (Kopp et al. 2003; Tsunekawa et al. 2003), weight loss, and vice versa (Yang et al. 2001). These findings suggest a shift in antioxidant–prooxidant balance in obesity.

Aging

Aging is characterized by decline in the biological functions with time leading to increased susceptibility to multiple forms of stress and diseases. According to Harman aging is the "progressive accumulation of the diverse deleterious changes in cells and tissue associated with a progressive increase in the chance of morbidity and mortality." Despite the well characterization of aging, the underlying biological mechanisms are not fully understood owing to its complex, integrative, and multi-factorial nature. Over the years numerous theories have been proposed to explain aging, such as oxidative theory of aging, mitochondrial theory, molecular inflammatory theory, cellular senescence theory, and so on. Interestingly, involvement of ROS in these theories is a common phenomenon and can also be correlated to the unifying mechanisms and downstream effects of aging in the form of different diseases. Therefore, the "free radical/oxidative stress theory" of aging is one of the most prevalent and widely accepted. According to this theory, the free radical-mediated damage to cellular macromolecules accompanies aging, inhibition of normal functions which coincides with pathological conditions, and even death.

Over the years a plethora of literature has been published that provides extensive information regarding the oxidative stress theory of aging (Ashok and Ali 1999; Balaban et al. 2005; Beckman and Ames 1998; Bokov et al. 2004; Droge 2002; Harman 2003; Li and Holbrook 2003). As we age the defense mechanisms preventing oxidation may decline in specific tissues, and accelerated oxidative damage could, therefore, trigger deterioration in physiological function. Since, oxidative stress is caused by the imbalance in oxidants (excess) and defects in antioxidants (impaired), it is important to address the molecular mechanisms of aging considering both these components of oxidative stress. Furthermore, based on the different theories of aging there are different factors that drive aging such as oxidation of molecules, cell death, inflammation, and so on (Fig. 3.3). Although it is not yet known if these factors are cause or effects of aging, it is clear that the cross talk between these factors characterizes aging. It is important to understand these factors because they are interconnected and thus make the process of aging multifaceted.

Oxidation of Biomolecules

Oxidation of biomolecules has been evidently related to the increased susceptibility to diseases, such as cancer and heart disease, as well as with

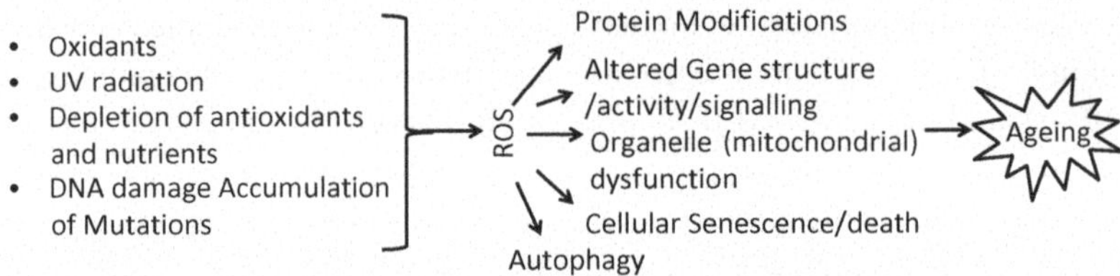

Fig. 3.3 ROS play an important role in the oxidative damage induced process of aging

the process of aging. Investigators have clearly shown a strong correlation between aging and an increase in oxidative damage to tissues throughout the body in species ranging from *C. elegans* to humans (Beckman and Ames 1998; Bokov et al. 2004; Navarro and Boveris 2004; Navarro et al. 2002; Sohal and Weindruch 1996). These studies have been focused on oxidative modification of intracellular macromolecules, primarily lipids, proteins, and DNA (Agarwal and Sohal 1994; Ames et al. 1993; Berlett and Stadtman 1997; Zhang et al. 2004, 2003) and showed increased oxidative stress in all major tissues in aged organisms, including mice (Hamilton et al. 2001), rats (Hamilton et al. 2001; Valls et al. 2005), hamsters (Takabayashi et al. 2004), and humans (Gianni et al. 2004; Short et al. 2005).

Since, the reactive metabolites are very short lived and difficult to detect directly in vivo, it is hard to delineate the exact pathways for oxidative cellular damage. Some of the most important and commonly used markers used in human aging studies include thiobarbituric acid-reactive substances (TBARS), alkenals, 8-isoprostane, 8-hydroxideoxyguanosine, and protein carbonyls. The hydroxyl radicals (HO•) formed from reactions catalyzed by free transition metals such as Fe^{2+}, $•O_2-$, and H_2O_2 are extremely reactive and are the major cause of damage to proteins, lipids, and DNA. Oxidative damage to these biomolecules seems to depend on hydrogen peroxide and a reduced transition metal. Therefore, molecules that contain transition metals, such as aconitase (a Krebs cycle enzyme), are likely to undergo oxidative damage (Berlett and Stadtman 1997; Stadtman 1992; Castro et al. 1994; Gardner et al. 1994; Hausladen and Fridovich 1994).

Substantially higher levels of lipid peroxidation products (e.g., MDA, 4-HNE, and F2-isoprostanes) have been observed in aged compared with young organisms in tissues, such as plasma (Mezzetti et al. 1996), kidney (Oxenkrug and Requintina 2003; Ward et al. 2005), brain (Poon et al. 2004; Rodrigues Siqueira et al. 2005; Wozniak et al. 2004), liver (Oxenkrug and Requintina 2003; Ward et al. 2005), lung (Lee et al. 1999; Wozniak et al. 2004), and muscle (Pansarasa et al. 1999; Judge et al. 2005). Additionally, age-related oxidative modifications to a large variety of proteins, including structural proteins (Grune et al. 2005), enzymes, and proteins, important in signal transduction pathways have been reported (Poon et al. 2004). Measurements of ROS-dependent total protein carbonyl are considered as indicators of protein oxidation (Levine et al. 1994; Reznick and Packer 1994). Fibroblasts obtained from the patients with diseases of accelerated aging (Progeria or Werner's syndrome) have dramatically higher levels of protein carbonyls (Oliver et al. 1987) compared to that of the age-matched controls. Numerous other studies have clearly shown a consistent increase in the protein-bound carbonyls with advanced age such as in dermal fibroblasts (Oliver et al. 1987), human brain (Smith et al. 1991), rat hepatocytes (Starke-Reed and Oliver 1989), and in several other degenerative diseases of aging.

In addition to HO• other radical species such as NO• have been implicated in oxidative protein damage and aging due to its high production to compete SOD for superoxide radicals consequently reacting with the superoxide to form $ONOO^-$. Therefore, NO• is a potential candidate relevant in aging. Protein nitration has been detected in progeria disease and in several other human disease states. Studies related to effect of NO• metabolites and aging are indicative of the fact that some proteins are selectively oxidized in vivo during aging due to the slow reactive nature of $ONOO^-$ (Beckman 1996; Beckman et al. 1990, 1994). For example, with age there is almost fourfold increase in nitration of the SERCA2a isoform of calcium ATPase in sarcoplasmic reticulum vesicles of rat skeletal muscle (Viner et al. 1996a). On the contrary in a closely related form of the protein (Viner et al. 1996b), no nitration was detected which strongly suggests that certain calcium ATPases are selectively modified by reactive nitrogen species, and these proteins accumulate during aging.

In the neurodegenerative aging-associated diseases, increased peroxynitrite and carbonyl have been reported. In subjects with Parkinson's disease, significant increases in protein-bound carbonyl levels (Alam et al. 1997) and free 3-nitrotyrosine (Bruijn et al. 1997) were found in post-mortem brain tissue of aged groups than the age-matched controls, whereas increased carbonyls but no 3-nitrotyrosines were detected or seen in Alzheimer's disease patients that their age-matched controls.

In addition to the increased ROS production, this increase in the oxidized protein is also a consequence of poor protein turnover (Farout and Friguet 2006; Friguet 2006) due to impairment of the proteolytic pathways, including proteasome proteases, lysosome proteases, and mitochondrial proteases (Shiba et al. 1993; Marshall et al. 1991). Age-associated impairment has generally been reported in the function of all these proteolytic pathways (Friguet 2002; Carrard et al. 2002) and the extent of this age-associated increase in oxidative damage to macromolecules varies greatly among different tissues, species, and detection methods. For example, some studies suggest no change in protein oxidation, such as o-tyrosine (a marker for hydroxyl radicals) and 3-nitrotyrosine (a marker for reactive nitrogen species in skeletal muscle, heart, and liver of aging rats). These effects were proposed to be attributed to the proteolytic degradation of intracellular proteins for these markers in these specific tissues (Leeuwenburgh et al. 1998).

Also, there are other potential age-associated damages that might be mutagenic to nuclear or mitochondrial DNA. These effects are discussed in detail in later sections of this chapter.

Although, the extent of age-associated increase in oxidative insult to macromolecules varies between different tissues, species, and detection methods, this increase can be generalized to the process of aging. These findings strongly suggest that reactive oxygen and nitrogen species such as peroxynitrite are involved in the age-related oxidative damage of macromolecules and could be active players in aging and the degenerative diseases of aging.

Depletion of Nutrients and Antioxidants

The relationship between aging and oxidative stress has been mainly studied in context of the antioxidant component rather than its oxidation potential. The research in this field is addressed in two ways. First approach involves the evaluation of changes in the antioxidant profiles of older compared with young organisms. Most of the studies in humans found an inverse relationship between aging and levels of the molecular antioxidants. Numerous studies have led credence to the fact that longevity is associated with higher levels of antioxidants such as Vitamin E, carotenoids, vitamin C, vitamin A, Se, and so on (Cutler 1991; Mecocci et al. 2000; Paolisso et al. 1998). Age-matched controls showed better immunological profile; endocrinological, metabolic characteristics; and nutritional status than the aged subjects (Sansoni et al. 1993; Mariotti et al. 1993; Paolisso et al. 1996). Another study by Berr et al. (2000) showed that higher plasma TBARS and concomitant decrease in the antioxidants such as vitamin E, selenium (Se), and carotenoids levels in elderly are associated with increased risk of cognitive decline. These studies indicate that good nutrition in the elderly, both of macro- and micronutrients, leads to healthy aging (Gonzalez-Cross et al. 2001; Barnett 1994).

The second approach is based on the nutritional intervention based amelioration of oxidative damage to slow down the rate of aging. Research impetus in the second approach started with the proposal of free radical theory of aging (Blackett and Hall 1980; Thomas 2004). These studies were aimed to increase the longevity by enhancing intracellular antioxidant defenses, either by dietary supplementation of the antioxidants or by overexpressing genes encoding antioxidant enzymes (e.g., SOD, catalase). Using synthetic antioxidant enzyme mimetics of SOD and catalase, such as EUK-8 and EUK-134, respectively, in a *C. elegans* model showed that treated organisms had significantly longer life spans than the untreated nematodes (Melov et al. 2000). Similar studies have been conducted using transgenic models overexpressing Cu/ZnSOD, MnSOD, catalase, and glutamate-cysteine ligase, a rate-limiting enzyme for de novo GSH biosynthesis in *Drosophila*. It was found that the overexpression of Cu/ZnSOD and MnSOD by an inducible promoter in adult *Drosophila* was associated with extended life span (Parkes et al. 1998; Sun et al. 2002; Sun and Tower 1999), while overexpression of catalase in *Drosophila* had no effect on longevity (Orr and Sohal 1992; Mockett et al. 2003). On the contrary, transgenic mice constitutively overexpressing human Cu/ZnSOD did not live longer than control animals (Huang et al. 2000), while heterozygous mice with reduced MnSOD activity have similar life expectancy to wild-type mice, coinciding with increased oxidative damage to DNA in these animals (Van Remmen et al. 2003). In contrast to these negative results, a recent study with transgenic mice overexpressing human catalase in mitochondria and human thioredoxin showed increase in life span (Mitsui et al. 2002) accompanied by the attenuated H_2O_2 production and H_2O_2-sensitive aconitase inactivation in heart and skeletal muscles (Schriner et al. 2005). These mixed results are difficult to reconcile and thus are inconclusive and raises the question whether the extrapolation of these findings in lower species such as *Drosophila* to more complex mammalian species is justified or not.

In terms of relation of human aging with concentrations and activity of enzymes involved in the antioxidant defense system, little is known. Although there is lack of agreement, reports suggest that the levels of superoxide dismutase, catalase, and glutathione peroxidase of a series of tissues (liver, brain, kidney, heart, etc.) are decreased in the senescent animals (Salminen et al. 1988; Cand and Verdetti 1989; Navarro et al. 2004). The levels of SOD, GSH-Px, and GR were found to be declined in older subjects (Guemouri et al. 1991; Andersen et al. 1997; Arthur et al. 1992). At the variance, Mecocci et al. (2000) demonstrated that plasma and red blood cell superoxide dismutase activities and plasma glutathione peroxidase activity increase with increasing age owing to an adaptive response to decline in nutritional antioxidants and increased level of oxidation products.

Though there are contrasting reports and studies are not conclusive, it can be postulated that the increase in oxidative stress and damage to cellular constituents associated with aging could be due to a decline in antioxidant defense systems.

Cellular Senescence and Death

Although the relevance of cellular senescence to aging and their commonality in their molecular mechanisms is unclear and controversial, recent studies have indicated that the cellular markers of senescence, such as telomere shortening, are exponentially increased with age in skin fibroblasts of primates (Herbig et al. 2006). This telomere hypothesis of aging which is an important human model dictates that many age-associated phenotypes are caused by the cellular senescence in response to one or more crucially shortened telomeres (Harley et al. 1992). Also increased levels of ROS were correlated with the signs of premature aging and cellular premature senescence (Barlow et al. 1999; Ito et al. 2004) in the hematopoietic stem cells obtained from mice that develop ataxia telangiectasia syndrome. However, the extent to which telomere shortening contributes to these events in vivo and mechanisms involved therein, remain unknown. Therefore, a better understanding of the molecular mechanisms of the cellular senescence may provide some insight into the biology of aging and potential sites for therapeutic interventions involving senescence pathways.

Previously we have discussed that accumulation of macromolecular damage may be the underlying cause of aging at a fundamental level.

In this context, it has been well recognized that cancer and aging might share similar molecular pathways. These pathways constitute accumulation of the oxidative DNA damage and p53 activation resulting in cell senescence and death. For example, the gene expression analysis in DNA repair-deficient mice demonstrated that nuclear DNA damage might invoke a longevity response by influencing the insulin-like growth factor 1 (IGF1) and growth hormone (GH), which are known to have important roles in determining the longevity from studies in model organisms (Kaeberlein 2007).

Since oxidative stress regulates the multiple steps of senescence signaling pathways, these responses to the DNA damage by oxidative stress may in turn mediated through p53 activation leading to cell apoptosis and senescence cascades (Grishko et al. 2003; Kil et al. 2006). Although, there are controversies in the literature regarding the role of apoptosis in aging, age-associated increases in apoptosis have been observed in several physiological systems, including the human immune system, human hair follicle, and rat skeletal muscle (Aggarwal and Gupta 1999; Arck et al. 2006; Song et al. 2006).

It has been clearly established that the ROS and ROS-modulated molecules can activate both intrinsic and extrinsic apoptotic pathways (Matsuzawa and Ichijo 2005) through mitochondrial damage and redox activation of MAPK cascade, respectively. Recently, a member of signal transduction adapters Shc protein family called p66Shc protein has been implicated as a potential link between the oxidative stress-mediated apoptosis and biological aging (Migliaccio et al. 2006). Evidence has suggested that p66Shc is an atypical signal transducer that can be regulated by the oxidative stress and also plays a role in H_2O_2 generation (Migliaccio et al. 1999; Giorgio et al. 2005). While mice lacking p66Shc (p66Shc-/-) live 30 % longer than the control animals, p66Shc-/- cells from knockout mice are resistant to ROS-induced apoptosis (Migliaccio et al. 1999; Napoli et al. 2003).

In addition to the apoptotic and senescence pathways, redox modification of transcriptional factors can regulate cellular proliferation, differentiation, senescence, death, and aging. For instance, an increase in the DNA binding activities of NF-κB and AP-1 in livers of old animals (Zhang et al. 2004) has been reported. In predominant animal models of longevity, *C. elegans* and *Drosophila*, activation of members of the Forkhead transcription factor family (Wang and Tissenbaum 2006; Hwangbo et al. 2004), has been associated with the extended life spans, whereas transgenic mice models of "superactive" form of p53 have a shortened life span and show signs of accelerated aging in some strains (Garcia-Cao et al. 2002; Tyner et al. 2002).

Inflammation

Since aging is accompanied by an increased incidence of various chronic diseases, understanding the basic mechanisms of chronic diseases is important to address their implications in the longevity and quality of life in humans. The free radical theory is the most acceptable theory of aging that not only successfully explains the aging mechanisms but also the pathogenesis of numerous chronic diseases such as atherosclerosis, dementia, arthritis, and osteoporosis commonly associated with aging (Beckman and Ames 1998; Cutler 2005). Studies have indicated that increased ROS is involved in various age-related pathologies, and these effects are possibly mediated through the state of the chronic inflammation (Chung et al. 2000, 2006; Lavrovsky et al. 2000; Sarkar and Fisher 2006). According to the "molecular inflammation hypothesis of aging" (Chung et al. 2001) aging and its related pathologies are mechanistically linked. Any imbalance in the redox balance during aging can activate the redox-sensitive transcription factors leading to the generation of numerous proinflammatory mediators (e.g., cytokines, chemokines, inducible nitric oxide (NO) synthase). These proinflammatory molecules, in turn, can lead to the generation of both reactive oxygen and reactive nitrogen species (Mariani et al. 2005; Oberley and Oberley 1986; Pryor et al. 2006; Stocker and Keaney 2005; Willcox et al. 2004) indicating there is a feedback loop causing cellular and tissue damage.

Although inflammation and ROS are inseparable, it is still not clear which of these is the initial stimulus. However, it is evident that accumulation of the reactive species is a major contributor to the pathogenesis of age-related diseases (Chung et al. 2006; Sarkar and Fisher 2006).

Additionally, there are substantial evidences that support the existence of a strong link between aging and inflammation as indicated by the increased infiltration of macrophages during age-associated diseases such as neurodegenerative diseases (McGeer and McGeer 2004, 2002) and atherosclerotic plaques (Giorgio et al. 2005). These activated macrophages are responsible to generate reactive species that pose the oxidative and nitrosative insult in specific tissues. Other than this prototypical marker of inflammation, increased levels of inflammatory cytokines like IL-1β, IL-6, and TNF-α (Baggio et al. 1998; Bruunsgaard et al. 2001; Dobbs et al. 1999; Ershler and Keller 2000; Forsey et al. 2003; Pedersen et al. 2000) and acute phase proteins such as C-reactive protein have been found in systemic circulation with advancing age. These studies indicate the role of oxidative stress-mediated chronic inflammation in the pathogenesis of specific age-related diseases. However, there is still lack of evidence to conclusively determine that inflammation is one of the causal agents of biological aging and issues of longevity in mammalian species.

Organelle Dysfunction (Mitochondria)

Literature is suggestive of the fact that oxidative damage to mitochondrial DNA (mtDNA) in animal tissues is concurrent with increased age (Miquel et al. 1980; Miquel 1991). Linnane et al. (1989) in "mitochondrial theory of aging" hypothesized that enhanced production of ROS and accumulation of the somatic mutations in mtDNA is a major contributor to the human aging and degenerative diseases. Under normal physiological conditions, mitochondria converts oxygen to superoxide anions, hydrogen peroxide, hydroxyl radicals, and other ROS that are taken care by the body's antioxidant defense system discussed earlier. During aging there is an increase in the production of superoxide anions and hydrogen peroxide in mitochondria (Sohal and Sohal 1991; Sohal et al. 1994; Perez-Campo et al. 1998). This causes oxidative damage to the mtDNA and membrane lipids of mitochondria (Sohal and Dubey 1994) leading to changes in the permeability of mitochondrial membranes culminating in cytochrome c release and other apoptogenic factor-mediated cell apoptosis (Tatton and Olanow 1999). Additionally, other studies suggest that there is decrease in the mitochondrial membrane potential concomitant with an increase in mitochondrial hydrogen peroxide in the older hepatocytes. Similarly, experiments in intact muscle mitochondria from house flies have shown that the rate of H_2O_2 generation progressively increases twofold as the house fly ages (Sohal and Sohal 1991). These studies suggest that increases in ROS with age, in mitochondria, could become deleterious to mitochondrial respiratory enzymes which can be used as "fingerprints" of oxidative damage in tissue proteins of aging rats (Leeuwenburgh et al. 1998; Crowley et al. 1998).

The critical role of mitochondria in aging is also due to the fact that mtDNA is more susceptible to oxidative insult than nuclear DNA damage (Richter et al. 1988), and this oxidative damage is concurrent with increased oxidation of mitochondrial glutathione with age (Stadtman 1992; Sohal et al. 1993; GarciadelaAsuncion et al. 1996). At the variance, administration of antioxidants protected animals from this damage. The oxidation of mtDNA in various tissues with aging leads to large-scale deletions, point mutations, and tandem duplications (Wei 1998; Lee and Wei 1997; Wallace 1994) and provide an extra source of oxidants (Lee and Wei 1997; Cortopassi and Arnheim 1990; Hattori et al. 1991; Katayama et al. 1991; Yen et al. 1991, 1992; Torii et al. 1992; Zhang et al. 1992; Lee et al. 1994a, b; Yang et al. 1994, 1995; Fahn et al. 1996). This age-dependent defects in the mitochondria have been associated with decline of respiratory function during the aging process. Studies by Ishii et al. (1998) with mutation in cytochrome *b* gene of *Caenorhabditis elegans* resulting in enhanced oxidative stress and shortened life span validate the fact that defects in mitochondrial respiratory system can lead to an increase of oxidative damage and premature aging or death. These studies though indicate that the mitochondrial role in aging is a "vicious cycle" (Wei 1998; Wei et al. 1996; Lee and Wei 2001); it remains unclear if this decline in mitochondrial function during aging mainly results from oxidative stress alone or is a cumulative effect of many other factors discussed earlier.

Autophagy

Autophagy as a housekeeping system for cells is responsible for the degradation and recycling of damaged cellular components to ensure cell survival under stress conditions, such as nutrition deprivation (Mortimore and Poso 1987). However, recently it has been suggested that autophagy can also induce cell death suggesting these dual roles of autophagy are important during aging process. In a *Caenorhabditis elegans* model, cellular autophagy has been found to be essential in the life span extension of the nematode (Melendez et al. 2003). On one hand, reports have shown that autophagic function declines with age in in vivo and in vitro settings (Del Roso et al. 2003; Massey et al. 2006; Terman 1995). Conversely, excessive activation of autophagy, leading to cell death, was observed in neurons with increased protein aggregation, suggesting that autophagy may play an important role in aging-related neurodegenerative diseases (Yu et al. 2005; Fortun et al. 2006). Although the regulation of autophagy is not yet completely understood, ROS have been implicated in the process. There is some evidence suggesting that aging-related increases in ROS production can result in elevated oxidative damage to proteins, including lysosomal proteins and proteins in autophagic pathways (Butler and Bahr 2006). However, more direct evidence for ROS involvement in autophagy comes from a recent report showing that cellular autophagy induced by caspase inhibition can lead to catalase degradation, resulting in ROS accumulation, lipid peroxidation, and loss of membrane integrity (Yu et al. 2006).

References

Agarwal S, Sohal RS (1994) Aging and protein oxidative damage. Mech Ageing Dev 75:11–19

Aggarwal S, Gupta S (1999) Increased activity of caspase 3 and caspase 8 in anti-Fas-induced apoptosis in lymphocytes from ageing humans. Clin Exp Immunol 117:285–290

Alam ZI, Jenner A, Daniel SE, Lees AJ, Cairns N, Marsden CD, Jenner P, Halliwell B (1997) Oxidative DNA damage in the parkinsonian brain: an apparent selective increase in 8-hydroxyguanine levels in substantia nigra. J Neurochem 69(3):1196–1203

Ames B, Shinenaga MK, Hagen TM (1993) Oxidants, antioxidants, and the degenerative diseases of aging. Proc Natl Acad Sci U S A 90:7915–7922

Andersen HR, Nielsen JB, Nielsen F, Gransdjean P (1997) Antioxidative enzyme activities in human erythrocytes. Clin Chem 43:562–568

Arck PC, Overall R, Spatz K, Liezman C, Handjiski B, Klapp BF, Birch-Machin MA, Peters EM (2006) Towards a "free radical theory of graying": melanocyte apoptosis in the aging human hair follicle is an indicator of oxidative stress induced tissue damage. FASEB J 20:1567–1569

Arthur Y, Herberth B, Guemouri L, Leconte E, Jeandel C, Siest G (1992) Age-related variations of enzymatic defenses against free radical and peroxides. In: Emerit I, Chance B (eds) Free radicals and aging. Birkhauser, Basel, pp 359–367

Ashok BT, Ali R (1999) The aging paradox: free radical theory of aging. Exp Gerontol 34:293–303

Baggio G, Donazzan S, Monti D, Mari D, Martini S, Gabelli C, Dalla Vestra M, Previato L, Guido M, Pigozzo S, Cortella I, Crepaldi G, Franceschi C (1998) Lipoprotein(a) and lipoprotein profile in healthy centenarians: a reappraisal of vascular risk factors. FASEB J 12:433–437

Bakker SJ, IJzerman RG, Teerlink T, Westerhoff HV, Gans RO, Heine RJ (2000a) Cytosolic triglycerides and oxidative stress in central obesity: the missing link between excessive atherosclerosis, endothelial dysfunction, and beta-cell failure? Atherosclerosis 148:17–21

Bakker SJ, IJzermen RG, Teerlink T, Westerhoff HV, Gans RO, Heine RJ (2000b) Cytosolic triglycerides and oxidative stress in central obesity: the missing link between excessive atherosclerosis, endothelial dysfunction, and beta-cell failure? Atherosclerosis 148:17–21

Balaban RS, Nemoto S, Finkel T (2005) Mitochondria, oxidants, aging. Cell 120:483–495

Barlow C, Dennery PA, Shigenaga MK, Smith MA, Morrow JD, Roberts LJ 2nd, Wynshaw-Boris A, Levine RL (1999) Loss of the ataxia-telangiectasia gene product causes oxidative damage in target organs. Proc Natl Acad Sci U S A 96:9915–9919

Barnes PJ, Karin M (1997) Nuclear factor-kappaB: a pivotal transcription factor in chronic inflammatory diseases. N Engl J Med 336:1066–1071

Barnett YA (1994) Nutrition and the ageing process. Br J Biomed Sci 51:278–287

Basu S, Riserus U, Turpeinen A, Vessby B (2000) Conjugated linoleic acid induces lipid peroxidation in men with abdominal obesity. Clin Sci 99:511–516

Baynes JW (1991) Role of oxidative stress in development of complications in diabetes. Diabetes 40:405–412

Baynes JW, Thorpe SR (1999) Role of oxidative stress in diabetic complications: a new perspective on an old paradigm. Diabetes 48:1–9

Beckman JS (1996) The physiological and pathological chemistry of nitric oxide. In: Lancaster JR Jr (ed) Nitric oxide: principles and actions. Academic, San Diego, pp 1–82

Beckman KB, Ames BN (1998) The free radical theory of aging matures. Physiol Rev 78:547–581

Beckman JS, Beckman TW, Chen J, Marshall PA, Freeman BA (1990) Apparent hydroxyl radical production by peroxynitrite: implications for endothelial injury from nitric oxide and superoxide. Proc Natl Acad Sci U S A 87:1620–1624

Beckman JS, Chen J, Ischiropoulos H, Crow JP (1994) Oxidative chemistry of peroxynitrite. Methods Enzymol 233:229–240

Begum N, Ragolia L (2000) High glucose and insulin inhibit VSMC MKP-1 expression by blocking iNOS via p38 MAPK activation. Am J Physiol 278: C81–C91

Beltowski J, Wojcicka G, Gorny D, Marciniak A (2000) The effect of dietary-induced obesity on lipid peroxidation, antioxidant enzymes and total plasma antioxidant capacity. J Physiol Pharmacol 51(Part 2):883–896

Benhamou PY, Moriscot C, Richard MJ, Beatrix O, Badet L, Pattou F, Kerr-Conte J, Chroboczek J, Lemarchand P, Halimi S (1998) Adenovirus-mediated catalase gene transfer reduces oxidant stress in human, porcine and rat pancreatic islets. Diabetologia 41:1093–1100

Berlett BS, Stadtman ER (1997) Protein oxidation in aging, disease, and oxidative stress. J Biol Chem 272:20313–20316

Berr C, Balansard B, Arnaud J, Roussel A, Alperovitch A (2000) Cognitive decline is associated with systemic oxidative stress: the EVA study. J Am Geriatr Soc 48:1285–1291

Bierhaus A, Schiekofer S, Schwaninger M, Andrassy M, Humpert PM, Chen J, Hong M, Luther T, Henle T, Kloting I, Morcos M, Hofmann M, Tritschler H, Weigle B, Kasper M, Smith M, Perry G, Schmidt AM, Stern DM, Haring HU, Schleicher E, Nawroth PP (2001) Diabetes-associated sustained activation of the transcription factor nuclear factor-κB. Diabetes 50: 2792–2808

Birnbaum MJ (2001) Turning down insulin signaling. J Clin Invest 108:655–659

Blackett AD, Hall DA (1980) The action of vitamin E on the ageing of connective tissues in the mouse. Mech Ageing Dev 14:305–316

Block G, Dietrich M, Norkus EP, Morrow JD, Hudes M, Caan B, Packer L (2002) Factors associated with oxidative stress in human populations. Am J Epidemiol 156:274–285

Boden G (1997) Role of fatty acids in the pathogenesis of insulin resistance and NIDDM. Diabetes 46(1): 3–10

Boden G, Chen X, Ruiz J, White JV, Rossetti L (1994) Mechanisms of fatty acid-induced inhibition of glucose uptake. J Clin Invest 93:2438–2446

Boden G, Ruiz J, Kim CJ, Chen X (1996) Effects of prolonged glucose infusion on insulin secretion, clearance, and action in normal subjects. Am J Physiol 270:E251–E258

Bokov A, Chaudhuri A, Richardson A (2004) The role of oxidative damage and stress in aging. Mech Ageing Dev 125:811–826

Bray GA (2000) Overweight, mortality and morbidity. In: Bouchard C (ed) Physical activity and obesity. Human Kinetics Publishers Inc., Champaign, pp 31–54

Brownlee M (1995) Advanced protein glycosylation in diabetes and aging. Annu Rev Med 46:223–234

Brownlee M (2001) Biochemistry and molecular cell biology of diabetic complications. Nature 414:813–820

Bruijn LI, Beal MF, Becher MW, Schulz JB, Wong PC, Price DL, Cleveland DW (1997) Elevated free nitrotyrosine levels, but not protein-bound nitrotyrosine or hydroxyl radicals, throughout amyotrophic lateral sclerosis (ALS)-like disease implicate tyrosine nitration as an aberrant in vivo property of one familial ALS-linked superoxide dismutase 1 mutant. Proc Natl Acad Sci U S A 94:7606–7611

Bruunsgaard H, Pedersen M, Pedersen BK (2001) Aging and proinflammatory cytokines. Curr Opin Hematol 8:131–136

Butler D, Bahr BA (2006) Oxidative stress and lysosomes: CNS-related consequences and implications for lysosomal enhancement strategies and induction of autophagy. Antioxid Redox Signal 8:185–196

Cand F, Verdetti J (1989) Superoxide dismutase, glutathione peroxidase, catalase, and lipid peroxidation in the major organs of the aging rats. Free Radic Biol Med 7:59–63

Candido R, Forbes JM, Thomas MC, Thallas V, Dean RG, Burns WC, Tikellis C, Ritchie RH, Twigg SM, Cooper ME, Burrell LM (2003) A breaker of advanced glycation end products attenuates diabetes-induced myocardial structural changes. Circ Res 92:785–792

Carlsson C, Borg LA, Welsh N (1999) Sodium palmitate induces partial mitochondrial uncoupling and reactive oxygen species in rat pancreatic islets in vitro. Endocrinology 140:3422–3428

Carrard G, Bulteau AL, Petropoulos I, Friguet B (2002) Impairment of proteasome structure and function in aging. Int J Biochem Cell Biol 34:1461–1474

Castro L, Rodriguez M, Radi R (1994) Aconitase is readily inactivated by peroxynitrite, but not by its precursor, nitric oxide. J Biol Chem 269(47):29409–29415

Ceriello A (2000) Oxidative stress and glycemic regulation. Metabolism 49:27–29

Chakraborty K, Khan GA, Banerjee P, Ray U, Sinha AK (2003) Inhibition of human blood platelet aggregation and the stimulation of nitric oxide synthesis by aspirin. Platelets 14:421–427

Chen YQ, Su M, Walia RR, Hao Q, Covington JW, Vaughan DE (1998) Sp1 sites mediate activation of the plasminogen activator inhibitor-1 promoter by glucose in vascular smooth muscle cells. J Biol Chem 273:8225–8231

Chrysohoou C, Panagiotakos DB, Pitsavos C, Skoumas I, Papademetriou L, Economou M, Stefanadis C (2007) The implication of obesity on total antioxidant capacity apparently healthy men and women: the ATTICA study. Nutr Metab Cardiovasc Dis 17:590–597

Chung HY, Kim HJ, Jung KJ, Yoon JS, Yoo MA, Kim KW, Yu BP (2000) The inflammatory process in aging. Rev Clin Gerontol 10:207–222

Chung HY, Kim HJ, Kim JW, Yu BP (2001) The inflammation hypothesis of aging: molecular modulation by calorie restriction. Ann N Y Acad Sci 928:327–335

Chung SS, Ho EC, Lam KS, Chung SK (2003) Contribution of polyol pathway to diabetes-induced oxidative stress. J Am Soc Nephrol 14:S233–S236

Chung HY, Sung B, Jung KJ, Zou Y, Yu BP (2006) The molecular inflammatory process in aging. Antioxid Redox Signal 8:572–581

Clark RJ, McDonough PM, Swanson E, Trost SU, Suzuki M, Fukuda M, Dillmann WH (2003) Diabetes and the accompanying hyperglycemia impairs cardiomyocyte calcium cycling through increased nuclear O-GlcNAcylation. J Biol Chem 278:44230–44237

Cohen B, Novick D, Rubinstein M (1996) Modulation of insulin activities by leptin. Science 274:1185–1188

Cortopassi GA, Arnheim N (1990) Detection of a specific mitochondrial DNA deletion in tissues of older humans. Nucleic Acids Res 18:6927–6933

Craven PA, Davidson CM, DeRubertis FR (1990) Increase in diacylglycerol mass in isolated glomeruli by glucose from de novo synthesis of glycerolipids. Diabetes 39:667–674

Craven PA, Studer RK, Felder J, Phillips S, DeRubertis FR (1997) Nitric oxide inhibition of transforming growth factor-beta and collagen synthesis in mesangial cells. Diabetes 46:671–681

Crowley JR, Yarasheski K, Leeuwenburgh C, Turk J, Heinecke JW (1998) Isotope dilution mass spectrometric quantification of 3-nitrotyrosine in proteins and tissues is facilitated by reduction to 3-aminotyrosine. Anal Biochem 259(1):127–135

Cutler RG (1991) Antioxidants and aging. Am Clin Nutr 53:373S–379S

Cutler RG (2005) Oxidative stress profiling. Part I. Its potential importance in the optimization of human health. Ann N Y Acad Sci 1055:93–135

Davi G, Guagnano MT, Ciabattoni G, Basili S, Falco A, Marinopiccoli M et al (2002) Platelet activation in obese women: role of inflammation and oxidant stress. JAMA 288:2008–2014

Davi G, Chiarelli F, Santilli F, Pomilio M, Vigneri S, Falco A et al (2003) Enhanced lipid peroxidation and platelet activation in the early phase of type 1 diabetes mellitus: role of interleukin-6 and disease duration. Circulation 107:3199–3203

De Souza C, Van Guilder G, Greiner J, Smith D, Hoetzer G, Stauffer B (2005) Basal endothelial nitric oxide release is preserved in overweight and obese adults. Obes Res 13:1303–1306

Decsi T, Molnar D, Koletzko B (1997) Reduced plasma concentrations of alpha-tocopherol and beta-carotene in obese boys. J Pediatr 130:653–655

DeFronzo RA (1997) Pathogenesis of type 2 diabetes mellitus: metabolic and molecular implications for identifying diabetes genes. Diabetes Rev 5:117–119

Del Roso A, Vittorini S, Cavallini G, Donati A, Gori Z, Masini M, Pollera M, Bergamini E (2003) Ageing-related changes in the in vivo function of rat liver macroautophagy and proteolysis. Exp Gerontol 38:519–527

Derubertis FR, Craven PA (1994) Activation of protein kinase C in glomerular cells in diabetes: mechanism and potential links to the pathogenesis of diabetic glomerulopathy. Diabetes 43:1–8

Dewanjee S, Bose SK, Sahu R, Mandal SC (2008) Antidiabetic effect of matured fruits of Diospyros peregrine in alloxan-induced diabetic rats. Int J Green Pharm 2:95–99

Dichtl W, Nilsson L, Goncalves I, Ares MP, Banfi C, Calara F, Hamsten A, Eriksson P, Nilsson J (1999) Very low-density lipoprotein activates nuclear factor-kappa B in endothelial cells. Circ Res 84:1085–1094

Dobbs RJ, Charlett A, Purkiss AG, Dobbs SM, Weller C, Peterson DW (1999) Association of circulating TNF-alpha and IL-6 with ageing and parkinsonism. Acta Neurol Scand 100:34–41

Dobrian AD, Davies MJ, Prewitt RL, Lauterio TJ (2000) Development of hypertension in a rat model of diet-induced obesity. Hypertension 35:1009–1015

Dobrian A, Schriver S, Lynch T, Prewitt R (2003) Effect of salt on hypertension and oxidative stress in a rat model of diet-induced obesity. Am J Physiol Ren Physiol 285:619–628

Droge W (2002) Free radicals in the physiological control of cell function. Physiol Rev 82:47–95

Du XL, Edelstein D, Rossetti L, Fantus IG, Goldberg H, Ziyadeh F, Wu J, Brownlee M (2000) Hyperglycemia-induced mitochondrial superoxide overproduction activates the hexosamine pathway and induces plasminogen activator inhibitor-1 expression by increasing Sp1 glycosylation. Proc Natl Acad Sci U S A 97:12222–12226

Duplus E, Glorian M, Forest C (2000) Fatty acid regulation of gene transcription. J Biol Chem 275:30749–30752

Duvnjak M, Lerotic I, Barsic N, Tomasic V, Jukic L, Velagic V (2007) Pathogenesis and management issues for non-alcoholic fatty liver disease. World J Gastroenterol 13:4539–4550

Ershler WB, Keller ET (2000) Age-associated increased interleukin-6 gene expression, late-life diseases, and frailty. Annu Rev Med 51:245–270

Fahn HJ, Wang LS, Hsieh RH, Chang SC, Kao SH, Huang MH, Wei YH (1996) Age-related 4,977 bp deletion in human lung mitochondrial DNA. Am J Respir Crit Care Med 154:1141–1145

Farout L, Friguet B (2006) Proteasome function in aging and oxidative stress: implications in protein maintenance failure. Antioxid Redox Signal 8:205–216

Feener EP, Xia P, Inoguchi T, Shiba T, Kunisaki M, King GL (1996) Role of protein kinase C in glucose- and angiotensin II-induced plasminogen activator inhibitor expression. Contrib Nephrol 118:180–187

Fonseca-Alaniz MH, Takada J, Alonso-Vale MI, Lima FB (2007) Adipose tissue as an endocrine organ: from theory to practice. J Pediatr 83(Suppl 5):S192–S203

Forsey RJ, Thompson JM, Ernerudh J, Hurst TL, Strindhall J, Johansson B, Nilsson BO, Wikby A (2003) Plasma cytokine profiles in elderly humans. Mech Ageing Dev 124:487–493

Fortun J, Go JC, Li J, Amici SA, Dunn WA Jr, Notterpek L (2006) Alterations in degradative pathways and protein aggregation in a neuropathy model based on PMP22 overexpression. Neurobiol Dis 22:153–164

Friguet B (2002) Protein repair and degradation during aging. Sci World J 2:248–254

Friguet B (2006) Oxidized protein degradation and repair in ageing and oxidative stress. FEBS Lett 580:2910–2916

Furukawa S, Fujita T, Shimabukuro M, Iwaki M, Yamada Y, Nakajima Y et al (2004) Increased oxidative stress in obesity and its impact on metabolic syndrome. J Clin Invest 114:1752–1761

Gallagher KA, Liu ZJ, Xiao M, Chen H, Goldstein LJ, Buerk DG, Nedeau A, Thom SR, Velazquez OC (2007) Diabetic impairments in NO-mediated endothelial progenitor cell mobilization and homing are reversed by hyperoxia and SDF-1 alpha. J Clin Invest 117:1249–1259

Ganz MB, Seftel A (2000) Glucose-induced changes in protein kinase C and nitric oxide are prevented by vitamin E. Am J Physiol Endocrinol Metab 278:E146–E152

Garcia-Cao I, Garcia-Cao M, Martin-Caballero J, Criado LM, Klatt P, Flores JM, Weill JC, Blasco MA, Serrano M (2002) "Super p53" mice exhibit enhanced DNA damage response, are tumor resistant and age normally. EMBO J 21:6225–6235

Garcia de la Asuncion J, Millán A, Pla R, Bruseghini L, Esteras A, Pallardó FV, Sastre J, Viña J (1996) Mitochondrial glutathione oxidation correlates with age-associated oxidative damage to mitochondrial DNA. FASEB J 10:333–338

Gardner PR, Nguyen DD, White CW (1994) Aconitase is a sensitive and critical target of oxygen poisoning in cultured mammalian cells and in rat lungs. Proc Natl Acad Sci U S A 91(25):12248–12252

Garg R, Kumbkarni Y, Aljada A, Mohanty P, Ghanim H, Hamouda W et al (2000) Troglitazone reduces reactive oxygen species generation by leukocytes and lipid peroxidation and improves flow-mediated vasodilatation in obese subjects. Hypertension 36:430–435

Geraldes P, King GL (2010) Activation of protein kinase C isoforms and its impact on diabetic complications. Circ Res 106:1319–1331

Geraldes P, Hiraoka-Yamamoto J, Matsumoto M, Clermont A, Leitges M, Marette A, Aiello LP, Kern TS, King GL (2009) Activation of PKC-delta and SHP-1 by hyperglycemia causes vascular cell apoptosis and diabetic retinopathy. Nat Med 15:1298–1306

Gianni P, Jan KJ, Douglas MJ, Stuart PM, Tarnopolsky MA (2004) Oxidative stress and the mitochondrial theory of aging in human skeletal muscle. Exp Gerontol 39:1391–1400

Giorgio M, Migliaccio E, Orsini F, Paolucci D, Moroni M, Contursi C, Pelliccia G, Luzi L, Minucci S, Marcaccio M, Pinton P, Rizzuto R, Bernardi P, Paolucci F, Pelicci PG (2005) Electron transfer between cytochrome *c* and p66Shc generates reactive oxygen species that trigger mitochondrial apoptosis. Cell 122:221–233

Goldin A, Beckman JA, Schmidt AM, Creager MA (2006) Advanced glycation end products: sparking the development of diabetic vascular injury. Circulation 114:597–605

Gonzalez-Cross M, Marcos A, Pietrzik K (2001) Nutrition and cognitive impairment in the elderly. Br J Nutr 86:313–321

Griffin ME, Marcucci MJ, Cline GW, Bell K, Barucci N, Lee D, Goodyear LJ, Kraegen EW, White MF, Shulman GI (1999) Free fatty acid-induced insulin resistance is associated with activation of protein kinase C θ and alterations in the insulin signaling cascade. Diabetes 48:1270–1274

Grishko V, Pastukh V, Solodushko V, Gillespie M, Azuma J, Schaffer S (2003) Apoptotic cascade initiated by angiotensin II in neonatal cardiomyocytes: role of DNA damage. Am J Physiol Heart Circ Physiol 285:H2364–H2372

Grodsky GM (2000) Kinetics of insulin secretion: underlying metabolic events in diabetes mellitus. In: Le Roith D, Taylor SI, Olefsky JM (eds) Diabetes mellitus: a fundamental and clinical text. Lippincott Williams & Wilkins, Philadelphia, pp 2–11

Grune T, Merker K, Jung T, Sitte N, Davies KJA (2005) Protein oxidation and degradation during postmitotic senescence. Free Radic Biol Med 39:1208–1215

Guemouri L, Artur Y, Herberth B, Jeandel C, Cuny G, Siest G (1991) Biological variability of superoxide dismutase, glutathione peroxidase, and catalase in blood. Clin Chem 37:1932–1937

Ha H, Lee HB (2000) Reactive oxygen species as glucose signaling molecules in mesangial cells cultured under high glucose. Kidney Int Suppl 77:S19–S25

Halliwell B, Gutteridge JM (1990) Role of free radicals and catalytic metal ions in human disease: an overview. Methods Enzymol 186:1–85

Hamilton ML, Van Remmen H, Drake JA, Yang H, Guo ZM, Kewitt K, Walter CA, Richardson A (2001) Does oxidative damage to DNA increase with age? Proc Natl Acad Sci U S A 98:10469–10474

Harley CB, Vaziri H, Counter CM, Allsopp RC (1992) The telomere hypothesis of cellular aging. Exp Gerontol 27:375–382

Harman D (2003) The free radical theory of aging. Antioxid Redox Signal 5:557–561

Harmon JS, Gleason CE, Tanaka Y, Poitout V, Robertson RP (n.d.) Antecedent hyperglycemia, not hyperlipidemia, is associated with increased islet triacylglycerol content and decreased insulin gene mRNA level in Zucker diabetic fatty rats. Diabetes 50(11):2482–2486

Hart GW (1997) Dynamic O-linked glycosylation of nuclear and cytoskeletal proteins. Annu Rev Biochem 66:315–335

Hartwich J, Goralska J, Siedlecka D, Gruca A, Trzos M, Dembinska-Kiec A (2007) Effect of supplementation with vitamin E and C on plasma hsCPR level and cobalt-albumin binding score as markers of plasma oxidative stress in obesity. Genes Nutr 2:151–154

Hattori K, Tanaka M, Sugiyama S, Obayashi T, Ito T, Satake T, Hanaki Y, Asai J, Nagano M, Ozawa T (1991) Age-dependent increase in deleted mitochondrial DNA in the human heart: possible contributory factor to presbycardia. Am Heart J 121:1735–1742

Hausladen A, Fridovich I (1994) Superoxide and peroxynitrite inactivate aconitases, but nitric oxide does not. J Biol Chem 269(47):29405–29408

Hennig B, Meerarani P, Toborek M, McClain CJ (1999) Antioxidant-like properties of zinc in activated endothelial cells. J Am Coll Nutr 18:152–158

Hennig B, Meerarani P, Ramadass P, Watkins BA, Toborek M (2000) Fatty acid-mediated activation of vascular endothelial cells. Metabolism 49:1006–1013

Herbig U, Ferreira M, Condel L, Carey D, Sedivy JM (2006) Cellular senescence in aging primates. Science 311:1257

Higdon JV, Frei B (2003) Obesity and oxidative stress: a direct link to CVD? Arterioscler Thromb Vasc Biol 23:365–367

Ho E, Bray TM (1999) Antioxidants, NF-kappaB activation, and diabetogenesis. Proc Soc Exp Biol Med 222:205–213

Ho E, Chen G, Bray TM (1999) Supplementation of N-acetylcysteine inhibits NF-kappaB activation and protects against alloxan-induced diabetes in CD-1 mice. FASEB J 13:1845–1854

Ho FM, Liu SH, Liau CS, Huang PJ, Lin-Shiau SY (2000a) High glucose-induced apoptosis in human endothelial cells is mediated by sequential activations of c-Jun NH(2)-terminal kinase and caspase-3. Circulation 101:2618–2624

Ho E, Chen G, Bray TM (2000b) Alpha-phenyl-tert-butylnitrone (PBN) inhibits NFkappaB activation offering protection against chemically induced diabetes. Free Radic Biol Med 28:604–614

Hogg N, Kalyanaraman B, Joseph J, Struck A, Parthasarathy S (1993) Inhibition of low-density lipoprotein oxidation by nitric oxide. Potential role in atherogenesis. FEBS Lett 334(2):170–174

Hori O, Yan SD, Ogawa S, Kuwabara K, Matsumoto M, Stern D, Schmidt AM (1996) The receptor for advanced glycation end-products has a central role in mediating the effects of advanced glycation end-products on the development of vascular disease in diabetes mellitus. Nephrol Dial Transplant 11(Suppl 5):13–16

Horie K, Miyata T, Maeda K, Miyata S, Sugiyama S, Sakai H, van Ypersole de Strihou C, Monnier VM, Witztum JL, Kurokawa K (1997) Immunohistochemical colocalization of glycoxidation products and lipid peroxidation products in diabetic renal glomerular lesions: implication for glycoxidative stress in the pathogenesis of diabetic nephropathy. J Clin Invest 100:2995–3004

Hotamisligil GS, Spiegelman BM (1994) Tumor necrosis factor α: a key component of the obesity-diabetes link. Diabetes 43:1271–1278

Hotamisligil GS, Murray DL, Choy LN, Spiegelman BM (1994) Tumor necrosis factor-a inhibits signaling from the insulin receptor. Proc Natl Acad Sci U S A 91:4854–4858

Huang T, Carlson E, Gillepsie A, Shi Y, Epstein C (2000) Ubiquitous overexpression of CuZn superoxide dismutase does not extend life span in mice. J Gerontol B Psychol Sci Soc Sci 55:B5–B9

Hwangbo DS, Gersham B, Tu MP, Palmer M, Tatar M (2004) Drosophila dFOXO controls lifespan and regulates insulin signalling in brain and fat body. Nature 429:562–566

Igarashi M, Wakasaki H, Takahara N, Ishii H, Jiang ZY, Yamauchi T, Kuboki K, Meier M, Rhodes CJ, King GL (1999) Glucose or diabetes activates p38 mitogen-activated protein kinase via different pathways. J Clin Invest 103:185–195

Ii S, Ohta M, Kudo E, Yamaoka T, Tachikawa T, Moritani M, Itakura M, Yoshimoto K (2004) Redox state-dependent and sorbitol accumulation-independent diabetic albuminuria in mice with transgene-derived human aldose reductase and sorbitol dehydrogenase deficiency. Diabetologia 47:541–548

Inoguchi T, Battan R, Handler E, Sportsman JR, Heath W, King GL (1992) Preferential elevation of protein kinase C isoform beta II and diacylglycerol levels in the aorta and heart of diabetic rats: differential reversibility to glycemic control by islet cell transplantation. Proc Natl Acad Sci U S A 89:11059–11063

Inoguchi T, Li P, Umeda F, Yu HY, Kakimoto M, Imamura M et al (2000) High glucose level and free fatty acid stimulate reactive oxygen species production through protein kinase C-dependent activation of NAD(P)H oxidase in cultured vascular cells. Diabetes 49:1939–1945

Ishii N, Fujii M, Hartman PS, Tsuda M, Yasuda K, Senoo-Matsuda N, Yanase S, Ayusawa D, Suzuki K (1998) A

mutation in succinate dehydrogenase cytochrome *b* causes oxidative stress and ageing in nematodes. Nature 394:695–697
Ito K, Hirao A, Arai F, Matsuoka S, Takubo K, Hamaguchi I, Nomiyama K, Hosokawa K, Sakurada K, Nakagata N, Ikeda Y, Mak TW, Suda T (2004) Regulation of oxidative stress by ATM is required for self-renewal of haematopoietic stem cells. Nature 431:997–1002
Jacqueminet S, Briaud I, Rouault C, Reach G, Poitout V (2000) Inhibition of insulin gene expression by long-term exposure of pancreatic beta cells to palmitate is dependent on the presence of a stimulatory glucose concentration. Metabolism 49:532–536
Ji LL (1995) Exercise and oxidative stress: role of the cellular antioxidant systems. Exerc Sport Sci Rev 23:135–166
Ji LL (1996) Exercise, oxidative stress, and antioxidants. Am J Sports Med 24(Suppl 6):S20–S24
Jiang ZY, Woollard AC, Wolff SP (1990) Hydrogen peroxide production during experimental protein glycation. FEBS Lett 268(1):69–71
Judge S, Jang YM, Smith A, Hagen T, Leeuwenburgh C (2005) Age associated increases in oxidative stress and antioxidant enzyme activities in cardiac interfibrillar mitochondria: implications for the mitochondrial theory of aging. FASEB J 19:419–421
Kaeberlein M (2007) Molecular basis of ageing. EMBO Rep 8:907–911. doi:10.1038/sj.embor.7401066
Kahn CR (1994) Insulin action, diabetogenes, and the cause of type 2 diabetes. Diabetes 43:1066–1084
Kaneto H, Xu G, Song KH, Suzuma K, Bonner-Weir S, Sharma A, Weir GC (2001) Activation of the hexosamine pathway leads to deterioration of pancreatic beta-cell function through the induction of oxidative stress. J Biol Chem 276:31099–31104
Katayama M, Tanaka M, Yamamoto H, Ohbayashi T, Nimura Y, Ozawa T (1991) Deleted mitochondrial DNA in the skeletal muscle of aged individuals. Biochem Int 25:47–56
Kawamura M, Heinecke JW, Chait A (1994) Pathophysiological concentrations of glucose promote oxidative modification of low density lipoprotein by a superoxide-dependent pathway. J Clin Invest 94(2):771–778
Khan N, Naz L, Yasmeen G (2006) Obesity: an independent risk factor systemic oxidative stress. Pak J Pharm Sci 19:62–69
Kikkawa R, Haneda M, Uzu T, Koya D, Sugimoto T, Shigeta Y (1994) Translocation of protein kinase C alpha and zeta in rat glomerular mesangial cells cultured under high glucose conditions. Diabetologia 37:838–841
Kil IS, Huh TL, Lee YS, Lee YM, Park JW (2006) Regulation of replicative senescence by NADP_-dependent isocitrate dehydrogenase. Free Radic Biol Med 40:110–119
Kolm-Litty V, Sauer U, Nerlich A, Lehmann R, Schleicher ED (1998) High glucose-induced transforming growth factor beta1 production is mediated by hexosamine pathway in porcine glomerular mesangial cells. J Clin Invest 101:160–169
Kopp HP, Kopp CW, Festa A, Krzyzanowska K, Kriwanek S, Minar E et al (2003) Impact of weight loss on inflammatory proteins and their association with the insulin resistance syndrome in morbidly obese patients. Arterioscler Thromb Vasc Biol 23:1042–1047
Koya D, King GL (1998) Protein kinase C activation and the development of diabetic complications. Diabetes 47:859–866
Kuboki K, Jiang ZY, Takahara N, Ha SW, Igarashi M, Yamauchi T, Feener EP, Herbert TP, Rhodes CJ, King GL (2000) Regulation of endothelial constitutive nitric oxide synthase gene expression in endothelial cells and in vivo: a specific vascular action of insulin. Circulation 101:676–681
Kullo IJ, Hensrud DD, Allison TG (2002) Comparison of numbers of circulating blood monocytes in men grouped by body mass index (o25, 25 to o30, X30). Am J Cardiol 89:1441–1443
Kyriakis JM, Avruch J (1996) Sounding the alarm: protein kinase cascades activated by stress and inflammation. J Biol Chem 271:24313–24316
Lameloise N, Muzzin P, Prentki M, Assimacopoulos-Jeannet F (2001) Uncoupling protein 2: a possible link between fatty acid excess and impaired glucose-induced insulin secretion? Diabetes 50:803–809
Lavrovsky Y, Chatterjee B, Clark RA, Roy AK (2000) Role of redox regulated transcription factors in inflammation, aging and age-related diseases. Exp Gerontol 35:521–532
Lee AY, Chung SS (1999) Contributions of polyol pathway to oxidative stress in diabetic cataract. FASEB J 13:23–30
Lee HC, Wei YH (1997) Mutation and oxidative damage of mitochondrial DNA and defective turnover of mitochondria in human aging. J Formos Med Assoc 96:770–778
Lee HC, Wei YH (2001) Mitochondrial alterations, cellular response to oxidative stress and defective degradation of proteins in aging. Biogerontology 2:231–244
Lee HC, Pang CY, Hsu HS, Wei YH (1994a) Differential accumulations of 4,977 bp deletion in mitochondrial DNA of various tissues in human ageing. Biochim Biophys Acta 1226:37–43
Lee HC, Pang CY, Hsu HS, Wei YH (1994b) Ageing-associated tandem duplications in the D-loop of mitochondrial DNA of human muscle. FEBS Lett 354:79–83
Lee HC, Lim MLR, Lu CY, Liu VWS, Fahn HJ, Zhang C, Nagley P, Wei YH (1999) Concurrent increase of oxidative DNA damage and lipid peroxidation together with mitochondrial DNA mutation in human lung tissues during aging–smoking enhances oxidative stress on the aged tissues. Arch Biochem Biophys 362:309–316
Lee JY, Sohn KH, Rhee SH, Hwang D (2001) Saturated fatty acids, but not unsaturated fatty acids, induce the

expression of cyclooxygenase-2 mediated through toll-like receptor 4. J Biol Chem 276:16683–16689
Leeuwenburgh C, Hansen P, Shaish A, Holloszy JO, Heinecke JW (1998) Markers of protein oxidation by hydroxyl radical and reactive nitrogen species in tissues of aging rats. Am J Physiol 274:R453–R461
Levine RL, Williams JA, Stadtman ER, Shacter E (1994) Carbonyl assays for determination of oxidatively modified proteins. Methods Enzymol 233:346–357
Li J, Holbrook NJ (2003) Common mechanisms for declines in oxidative stress tolerance and proliferation with aging. Exp Gerontol 35:292–299
Linnane AW, Marzuki S, Ozawa T, Tanaka M (1989) Mitochondrial DNA mutations as an important contributor to ageing and degenerative disease. Lancet i:642–645
Lyon CJ, Law RE, Hsueh WA (2003) Minireview: adiposity, inflammation, and atherogenesis. Endocrinology 144:2195–2200
Maachi M, Pieroni L, Bruckert E, Jardel C, Fellahi S, Hainque B et al (2004) Systemic low-grade inflammation is related to both circulating and adipose tissue TNFalpha, leptin and IL-6 levels in obese women. Int J Obes Relat Metab Disord 28:993–997
Maddux BA, See W, Lawrence JC Jr, Goldfine AL, Goldfine ID, Evans JL (2001) Protection against oxidative stress-induced insulin resistance in rat L6 muscle cells by micromolar concentrations of α-lipoic acid. Diabetes 50:404–410
Maechler P, Jornot L, Wollheim CB (1999) Hydrogen peroxide alters mitochondrial activation and insulin secretion in pancreatic beta cells. J Biol Chem 274:27905–27913
Mariani E, Polidori MC, Cherubini A, Mecocci P (2005) Oxidative stress in brain aging, neurodegenerative and vascular diseases: an overview. J Chromatogr B Biomed Appl 827:65–75
Mariotti S, Barbesino G, Caturegli P, Bartalena L, Sansoni P, Fagnoni F, Monti D, Fagiolo U, Franceschi C, Pinchera A (1993) Complex alteration of thyroid function in healthy centenarians. J Clin Endocrinol Metab 77:1130–1134
Marshall S, Garvey WT, Traxinger RR (1991) New insights into the metabolic regulation of insulin action and insulin resistance: role of glucose and amino acids. FASEB J 5:3031–3036
Massey AC, Kiffin R, Cuervo AM (2006) Autophagic defects in aging: looking for an "emergency exit"? Cell Cycle 5:1292–1296
Matsuzawa A, Ichijo H (2005) Stress-responsive protein kinases in redox regulated apoptosis signaling. Antioxid Redox Signal 7:472–481
McCarthy AD, Etcheverry SB, Cortizo AM (2001) Effect of advanced glycation endproducts on the secretion of insulin-like growth factor-I and its binding proteins: role in osteoblast development. Acta Diabetol 38(3):113–122
McGarry JD (2002) Banting lecture 2001: dysregulation of fatty acid metabolism in the etiology of type 2 diabetes. Diabetes 51(1):7–18
McGeer PL, McGeer EG (2002) Inflammatory processes in amyotrophic lateral sclerosis. Muscle Nerve 26:459–470
McGeer PL, McGeer EG (2004) Inflammation and neurodegeneration in Parkinson's disease. Parkinsonism Relat Disord 10(Suppl 1):S3–S7
Mecocci P, Polidori MC, Troiano L, Cherubini A, Cecchetti R, Pini G, Straatman M, Monti D, Stahl W, Sies H, Franceschi C, Senin U (2000) Plasma antioxidants and longevity: a study on healthy centenarians. Free Radic Biol Med 28:1243–1248
Meglasson MD, Matschinsky FM (1986) Pancreatic islet glucose metabolism and regulation of insulin secretion. Diabetes Metab Rev 2:163–214
Melendez A, Talloczy Z, Seaman M, Eskelinen EL, Hall DH, Levine B (2003) Autophagy genes are essential for dauer development and life-span extension in *C. elegans*. Science 301:1387–1391
Melov S, Ravenscroft J, Malik S, Gill MS, Walker DW, Clayton PE, Wallace DC, Malfroy B, Doctrow SR, Lithgow GJ (2000) Extension of life-span with superoxide dismutase/catalase mimetics. Science 289:1567–1569
Mezzetti A, Lafenna D, Romano F, Costantini F, Pierdomenico SD, Cesare DD, Cuccurullo F, Riario-Sforza G, Zuliani G, Fellini R (1996) Systemic oxidative stress and its relationship with age and illness. J Am Geriatr Soc 44:823–827
Michael JC, James MC, Robbins VK (2000) Pathologic basis of disease, "the pancreas", 6th edn. Harcourt Publishers, pp 902–929
Migliaccio E, Giorgio M, Mele S, Pelicci G, Reboldi P, Pandolfi PP, Lanfrancone L, Pelicci PG (1999) The p66shc adaptor protein controls oxidative stress response and life span in mammals. Nature 402:309–313
Migliaccio E, Giorgio M, Pelicci PG (2006) Apoptosis and aging: role of p66Shc redox protein. Antioxid Redox Signal 8:600–608
Miquel J (1991) An integrated theory of aging as the result of mitochondrial DNA mutation in differentiated cells. Arch Gerontol Geriatr 12:99–117
Miquel J, Economos JE, Johnson JE Jr (1980) Mitochondrial role in cell ageing. Exp Gerontol 15:575–591
Mitsui A, Hamuro J, Nakamura H, Kondo N, Hirabayashi Y, Ishizaki- Koizumi S, Hirakawa T, Inoue T, Yodoi J (2002) Overexpression of human thioredoxin in transgenic mice controls oxidative stress and life span. Antioxid Redox Signal 4:693–696
Miwa I, Ichimura N, Sugiura M, Hamada Y, Taniguchi S (2000) Inhibition of glucose-induced insulin secretion by 4-hydroxy-2-nonenal and other lipid peroxidation products. Endocrinology 141:2767–2772
Mockett RJ, Bayne ACV, Kwong LK, Orr WC, Sohal RS (2003) Ectopic expression of catalase in *Drosophila* mitochondria increases stress resistance but not longevity. Free Radic Biol Med 34:207–217
Mohamed AK, Bierhaus A, Schiekofer S, Tritschler H, Ziegler R, Nawroth PP (1999a) The role of oxidative

stress and NF-kB activation in late diabetic complications. Biofactors 10(2/3):157–167
Mohamed AK, Bierhaus A, Schiekofer S, Tritschler H, Ziegler R, Nawroth PP (1999b) The role of oxidative stress and NF-κB activation in late diabetic complications. Biofactors 10:157–167
Moor de Burgos A, Wartanowicz M, Ziemlanski S (1992) Blood vitamin and lipid levels in overweight and obese women. Eur J Clin Nutr 46:803–808
Morrow J (2003) Is a oxidative stress a connection between obesity and atherosclerosis. Arterioscler Thromb Vasc Biol 23:368–370
Mortimore GE, Poso AR (1987) Intracellular protein catabolism and its control during nutrient deprivation and supply. Annu Rev Nutr 7:539–564
Mullarkey CJ, Edelstein D, Brownlee M (1990) Free radical generation by early glycation products: a mechanism for accelerated atherogenesis in diabetes. Biochem Biophys Res Commun 173(3):932–939
Musicki B, Kramer MF, Becker RE, Burnett AL (2005) Inactivation of phosphorylated endothelial nitric oxide synthase (Ser-1177) by O-GlcNAc in diabetes-associated erectile dysfunction. Proc Natl Acad Sci U S A 102:11870–11875
Myara I, Alamowitch C, Michel O, Heudes D, Bariety J, Guy-Grand B et al (2003) Lipoprotein oxidation and plasma vitamin E in nondiabetic normotensive obese patients. Obes Res 11:112–120
Napoli C, Martin-Padura I, de Nigris F, Giorgio M, Mansueto G, Somma P, Condorelli M, Sica G, De Rosa G, Pelicci P (2003) Deletion of the p66Shc longevity gene reduces systemic and tissue oxidative stress, vascular cell apoptosis, and early atherogenesis in mice fed a high-fat diet. Proc Natl Acad Sci U S A 100:2112–2116
Natarajan R, Scott S, Bai W, Yerneni KKV, Nadler J (1999) Angiotensin II signaling in vascular smooth muscle cells under high glucose conditions. Hypertension 33:378–384
Navarro A, Boveris A (2004) Rat brain and liver mitochondria develop oxidative stress and lose enzymatic activities on aging. Am J Physiol Regul Integr Comp Physiol 287:R1244–R1249
Navarro A, Sanchez Del Pino MJ, Gomez C, Peralta JL, Boveris A (2002) Behavioral dysfunction, brain oxidative stress, and impaired mitochondrial electron transfer in aging mice. Am J Physiol Regul Integr Comp Physiol 282:R985–R992
Navarro A, Gomez C, Lopez-Cepero JM, Boveris A (2004) Beneficial effects of moderate exercise on mice aging: survival, behavior, oxidative stress, and mitochondrial electron transfer. Am J Physiol Regul Integr Comp Physiol 286:R505–R511
Nishikawa T, Edelstein D, Du XL, Yamagishi S, Matsumura T, Kaneda Y, Yorek MA, Beebe D, Oates PJ, Hammes HP, Giardino I, Brownlee M (2000a) Normalizing mitochondrial superoxide production blocks three pathways of hyperglycaemic damage. Nature 404:787–790
Nishikawa T, Edelstein D, Brownlee M (2000b) The missing link: a single unifying mechanism for diabetic complications. Kidney Int 58:26–30
Nishino T, Horii Y, Shiiki H, Yamamoto H, Makita Z, Bucala R, Dohi K (1995) Immunohistochemical detection of advanced glycosylation end products within the vascular lesions and glomeruli in diabetic nephropathy. Hum Pathol 26:308–313
Niwa T, Katsuzaki T, Miyazaki S, Miyazaki T, Ishizaki Y, Hayase F, Tatemichi N, Takei Y (1997) Immunohistochemical detection of imidazolone, a novel advanced glycation end product, in kidneys and aortas of diabetic patients. J Clin Invest 99:1272–1280
Oberley LW, Oberley DT (1986) Free radicals, cancer and aging. In: Johnson JEJ, Walford R, Harmon D, Miquel J (eds) Free radical, aging, and degenerative diseases. Alan R. Liss, New York
Ohrvall M, Tengblad S, Vessby B (1993) Lower tocopherol serum levels in subjects with abdominal adiposity. J Intern Med 234:53–60
Oliver CN, Ahn B, Moerman EJ, Goldstein S, Stadtman ER (1987) Age-related changes in oxidized proteins. J Biol Chem 262:5488–5491
Olusi SO (2002) Obesity is an independent risk factor for plasma lipid peroxidation and depletion of erythrocyte cytoprotective enzymes in humans. Int J Obes Relat Metab Disord 26:1159–1164
Orr WC, Sohal RS (1992) The effects of catalase gene overexpression on life span and resistance to oxidative stress in transgenic *Drosophila melanogaster*. Arch Biochem Biophys 297:35–41
Ouchi N, Kihara S, Arita Y, Maeda K, Kuriyama H, Okamoto Y et al (1999) Novel modulator for endothelial adhesion molecules: adipocyte-derived plasma protein adiponectin. Circulation 100:2473–2476
Oxenkrug GF, Requintina PJ (2003) Mating attenuates aging-associated increase of lipid peroxidation activity in C57BL/6J mice. Ann N Y Acad Sci 993:161–167
Ozata M, Mergen M, Oktenli C, Aydin A, Sanisoglu SY, Bolu E et al (2002) Increased oxidative stress and hypozincemia in male obesity. Clin Biochem 35:627–631
Packer L, Rosen P, Tritschler H, King GL, Azzi A (eds) (2000) Antioxidants and diabetes management. Marcel Dekker, New York
Pang Y, Bounelis P, Chatham JC, Marchase RB (2004) Hexosamine pathway is responsible for inhibition by diabetes of phenylephrine-induced inotropy. Diabetes 53:1074–1081
Pansarasa O, Bertorelli L, Vecchiet J, Felzani G, Marzatico F (1999) Age-dependent changes of antioxidant activities and markers of free radical damage in human skeletal muscle. Free Radic Biol Med 27:617–622
Paolisso G, Giugliano D (1996) Oxidative stress and insulin action. Is there a relationship? Diabetologia 39:357–363
Paolisso G, Gambardella A, Ammendola S, D'Amore A, Balbi V, Varricchio M, D'Onofrio F (1996) Glucose

tolerance and insulin action in healthy centenarians. Am J Physiol 270:E890–E894

Paolisso G, Tagliamonte MR, Rizzo MR, Manzella D, Gambardella A, Varricchio M (1998) Oxidative stress and advancing age: results in healthy centenarians. J Am Geriatr Soc 46:833–838

Parkes TL, Elia AJ, Dickinson D, Hilliker AJ, Phillips JP, Boulianne GL (1998) Extension of *Drosophila* life span by overexpression of human SOD1 in motor neurons. Nat Genet 19:171–174

Patel C, Ghanim H, Ravishankar S, Sia CL, Viswanathan P, Mohantym P, Dandona P (2007) Prolonged reactive oxygen species generation and Nuclear Factor- kB activation after a high-fat, high-carbohydrate meal in the obese. J Clin Endocrinol Metab 92:4476–4479

Paz K, Hemi R, LeRoith D, Karasik A, Elhanany E, Kanety H, Zick Y (1997) A molecular basis for insulin resistance: elevated serine/threonine phosphorylation of IRS-1 and IRS-2 inhibits their binding to the juxtamembrane region of the insulin receptor and impairs their ability to undergo insulin-induced tyrosine phosphorylation. J Biol Chem 272:29911–29918

Pedersen BK, Bruunsgaard H, Ostrowski K, Krabbe K, Hansen H, Krzywkowski K, Toft A, Sondergaard SR, Petersen EW, Ibfelt T, Schjerling P (2000) Cytokines in aging and exercise. Int J Sports Med 21(Suppl 1):S4–S9

Perez-Campo R, Lopez-Torres M, Cadenas S, Rojas C, Barja G (1998) The rate of free radical production as a determinant of the rate of aging: evidence from the comparative approach. J Comp Physiol 168:149–158

Pieper GM, Riaz-ul-Haq (1997) Activation of nuclear factor-kappaB in cultured endothelial cells by increased glucose concentration: prevention by calphostin C. J Cardiovasc Pharmacol 30:528–532

Pihl E, Zilmer K, Kullisaar T, Kairane C, Magi A, Zilmer M (2006) Atherogenic inflammatory and oxidative stress markers in relation to overweight values in male former athletes. Int J Obes 30:141–146

Poitout V, Robertson RP (2002) Minireview: secondary beta-cell failure in type 2 diabetes–a convergence of glucotoxicity and lipotoxicity. Endocrinology 143(2):339–342

Poon HF, Calabrese V, Scapagnini G, Butterfield DA (2004) Free radicals and brain aging. Clin Geriatr Med 20:329–359

Pryor WA, Houk KN, Foote CS, Fukuto JM, Ignarro LJ, Squadrito GL, Davies KJ (2006) Free radical biology and medicine: it's a gas, man! Am J Physiol Regul Integr Comp Physiol 291:R491–R511

Pugliese G, Pricci F, Pugliese F, Mene P, Lenti L, Andreani D, Galli G, Casini A, Bianchi S, Rotella CM (1994) Mechanisms of glucose-enhanced extracellular matrix accumulation in rat glomerular mesangial cells. Diabetes 43:478–490

Purves T, Middlemas A, Agthong S, Jude EB, Boulton AJ, Fernyhough P, Tomlinson DR (2001) A role for mitogen-activated protein kinases in the etiology of diabetic neuropathy. FASEB J 15:2508–2514

Ramasamy R, Goldberg IJ (2010) Aldose reductase and cardiovascular diseases, creating human-like diabetic complications in an experimental model. Circ Res 106:1449–1458

Randle PJ, Kerbey AL, Espinal J (1988) Mechanisms decreasing glucose oxidation in diabetes and starvation: role of lipid fuels and hormones. Diabetes Metab Rev 623:638

Reaven GM (2000) Insulin resistance and its consequences: type 2 diabetes mellitus and coronary heart disease. In: LeRoith D, Taylor SI, Olefsky JM (eds) Diabetes mellitus: a fundamental and clinical text. Lippincott Williams & Wilkins, Philadelphia, pp 604–615

Reitman A, Friedrich I, Ben-Amotz A, Levy Y (2002) Low plasma antioxidants and normal plasma B vitamins and homocysteine in patients with severe obesity. Israel Med Assoc J 4:590–593

Reznick AZ, Packer L (1994) Oxidative damage to proteins: spectrophotometric method for carbonyl assay. Methods Enzymol 233:357–363

Richter C, Park JW, Ames BN (1988) Normal oxidative damage to mitochondrial and nuclear DNA is extensive. Proc Natl Acad Sci U S A 85:6465–6467

Ridker PM, Hennekens CH, Buring JE, Rifai N (2000) C-reactive protein and other markers of inflammation in the prediction of cardiovascular disease in women. N Engl J Med 342:836–843

Rivard A, Silver M, Chen D, Kearney M, Magner M, Annex B, Peters K, Isner JM (1999) Rescue of diabetes-related impairment of angiogenesis by intramuscular gene therapy with adeno-VEGF. Am J Pathol 154:355–363

Robertson RP, Harmon JS, Tanaka Y, Sacchi G, Tran PO, Gleason CE, Poitout V (2000) Glucose toxicity of the β-cell: cellular and molecular mechanisms. In: Le Roith D, Taylor SI, Olefsky JM (eds) Diabetes mellitus: a fundamental and clinical text. Lippincott Williams & Wilkins, Philadelphia, pp 125–132

Rodrigues Siqueira I, Fochesatto C, da Silva Torres IL, Dalmaz C, Netto Alexandre C (2005) Aging affects oxidative state in hippocampus, hypothalamus and adrenal glands of Wistar rats. Life Sci 78:271–278

Rodriguez-Porcel M, Lerman LO, Holmes DR Jr, Richardson D, Napoli C, Lerman A (2002) Chronic antioxidant supplementation attenuates nuclear factor-kappa B activation and preserves endothelial function in hypercholesterolemic pigs. Cardiovasc Res 53:1010–1018

Rolo AP, Palmeira CM (2006) Diabetes and mitochondrial function: role of hyperglycemia and oxidative stress. Toxicol Appl Pharmacol 212:167–178

Rösen P, Nawroth PP, King G, Möller W, Tritschler HJ, Packer L (2001) The role of oxidative stress in the onset and progression of diabetes and its complications: a summary of a Congress Series sponsored by UNESCO-MCBN, the American Diabetes Association and the German Diabetes Society. Diabetes Metab Res Rev 17(3):189–212

Rudich A, Kozlovsky N, Potashnik R, Bashan N (1997) Oxidant stress reduces insulin responsiveness in 3T3–L1 adipocytes. Am J Physiol 35:E935–E940
Rudich A, Tirosh A, Potashnik R, Khamaisi M, Bashan N (1999) Lipoic acid protects against oxidative stress induced impairment in insulin stimulation of protein kinase B and glucose transport in 3T3–L1 adipocytes. Diabetologia 42:949–957
Russell AP, Gastaldi G, Bobbioni-Harsch E, Arboit P, Gobelet C, Deriaz O et al (2003) Lipid peroxidation in skeletal muscle of obese as compared to endurance-trained humans: a case of good vs bad lipids? FEBS Lett 551:104–106
Saiki S, Sato T, Kohzuki M, Kamimoto M, Yosida T (2001) Changes in serum hypoxanthine levels by exercise in obese subjects. Metab Clin Exp 50:627–630
Saito I, Yonemasu K, Inami F (2003) Association of body mass index, body fat, and weight gain with inflammation markers among rural residents in Japan. Circ J 67:323–329
Salminen A, Saari P, Kihlstrom M (1988) Age- and sex-related difference in lipid peroxidation of mouse cardiac and skeletal muscle. Comp Biochem Physiol 89B:689–695
Salvadori A, Fanari P, Fontana M, Buontempi L, Saezza A, Baudo S et al (1999) Oxygen uptake and cardiac performance in obese and normal subjects during exercise. Respiration 66:25–33
Sansoni P, Cossarizza A, Brianti V, Fagnoni F, Snelli G, Monti D, Marcato A, Passeri G, Ortolani C, Forti E (1993) Lymphocyte subsets and natural killer cell activity in healthy old people and centenarians. Blood 82:2767–2773
Sarkar D, Fisher PB (2006) Molecular mechanisms of aging-associated inflammation. Cancer Lett 236:13–23
Sayeski PP, Kudlow JE (1996) Glucose metabolism to glucosamine is necessary for glucose stimulation of transforming growth factor-alpha gene transcription. J Biol Chem 271:15237–15243
Schatteman GC, Hanlon HD, Jiao C, Dodds SG, Christy BA (2000) Blood derived angioblasts accelerate blood-flow restoration in diabetic mice. J Clin Invest 106:571–578
Schleicher ED, Weigert C (2000) Role of the hexosamine biosynthetic pathway in diabetic nephropathy. Kidney Int 58(Suppl 77):S13–S18
Schriner SE, Linford NJ, Martin GM, Treuting P, Ogburn CE, Emond M, Coskun PE, Ladiges W, Wolf N, Van Remmen H, Wallace DC, Rabinovitch PS (2005) Extension of murine life span by overexpression of catalase targeted to mitochondria. Science 308:1909–1911
Scivittaro V, Ganz MB, Weiss MF (2000) AGEs induce oxidative stress and activate protein kinase C-beta (II) in neonatal mesangial cells. Am J Physiol 278:F676–F683
Segall L, Lameloise N, Assimacopoulos-Jeannet F, Roche E, Corkey P, Thumelin S, Corkey BE, Prentki M (1999) Lipid rather than glucose metabolism is implicated in altered insulin secretion caused by oleate in INS-1 cells. Am J Physiol 277:E521–E528
Shiba T, Inoguchi T, Sportsman JR, Heath WF, Bursell S, King GL (1993) Correlation of diacylglycerol level and protein kinase C activity in rat retina to retinal circulation. Am J Physiol 265:E783–E793
Short KR, Bigelow ML, Kahl J, Singh R, Coenen-Schimke J, Raghavakaimal S, Nair KS (2005) Decline in skeletal muscle mitochondrial function with aging in humans. Proc Natl Acad Sci U S A 102:5618–5623
Shulman GI (2000) Cellular mechanisms of insulin resistance. J Clin Invest 106:171–176
Smith CD, Carney JM, Starke-Reed PE, Oliver CN, Stadtman ER, Floyd RA, Markesbery WR (1991) Excess brain protein oxidation and enzyme dysfunction in normal aging and in Alzheimer disease. Proc Natl Acad Sci U S A 88:10540–10543
Sohal RS, Dubey A (1994) Mitochondrial oxidative damage, hydrogen peroxide release, and aging. Free Radic Biol Med 16:621–626
Sohal RS, Sohal BH (1991) Hydrogen peroxide release by mitochondria increases during aging. Mech Ageing Dev 57:187–202
Sohal RS, Weindruch R (1996) Oxidative stress caloric restriction, aging. Science 273:59–63
Sohal RS, Agarwal S, Dubey A, Orr WC (1993) Protein oxidative damage is associated with life expectancy of houseflies. Proc Natl Acad Sci U S A 90:7255–7259
Sohal RS, Ku HH, Agarwal S, Forster MJ, Lal H (1994) Oxidative damage, mitochondrial oxidant generation, and antioxidant defenses during aging and in response to food restriction in the mouse. Mech Ageing Dev 74:121–133
Song W, Kwak HB, Lawler JM (2006) Exercise training attenuates age induced changes in apoptotic signaling in rat skeletal muscle. Antioxid Redox Signal 8:517–528
Stadtman ER (1992) Protein oxidation and aging. Science 257:1220–1224
Starke-Reed PE, Oliver CN (1989) Protein oxidation and proteolysis during aging and oxidative stress. Arch Biochem Biophys 275:559–567. doi:10.1016/0003-9861(89)90402-5
Steppan CM, Bailey ST, Bhat S, Brown EJ, Banerjee RR, Wright CM, Patel HR, Ahima RS, Lazar MA (2001) The hormone resistin links obesity to diabetes. Nature 409:307–312
Stevens MJ, Obrosova I, Feldman EL, Greene DA (2000) The sorbitol-osmotic and sorbitol-redox hypothesis. In: LeRoith D, Taylor SI, Olefsky JM (eds) Diabetes mellitus: a fundamental and clinical text. Lippincott Williams & Wilkins, Philadelphia, pp 972–983
Stitt AW, Li YM, Gardiner TA, Bucala R, Archer DB, Vlassara H (1997) Advanced glycation end products (AGEs) co-localize with AGE receptors in the retinal vasculature of diabetic and of AGE-infused rats. Am J Pathol 150:523–531
Stitt AW, Moore JE, Sharkey JA, Murphy G, Simpson DA, Bucala R, Vlassara H, Archer DB (1998) Advanced glycation end products in vitreous: structural and

functional implications for diabetic vitreopathy. Invest Ophthalmol Vis Sci 39:2517–2523
Stocker R, Keaney JF (2005) New insights on oxidative stress in the artery wall. J Thromb Haemost 3:1825–1834
Strauss RS (1999) Comparison of serum concentrations of alphatocopherol and beta-carotene in a cross-sectional sample of obese and nonobese children (NHANES III). National Health and Nutrition Examination Survey. J Pediatr 134:160–165
Sun J, Tower J (1999) FLP recombinase-mediated induction of Cu/Zn-superoxide dismutase transgene expression can extend the life span of adult *Drosophila melanogaster* flies. Mol Cell Biol 19:216–228
Sun J, Folk D, Bradley TJ, Tower J (2002) Induced overexpression of mitochondrial Mn-superoxide dismutase extends the life span of adult *Drosophila melanogaster*. Genetics 161:661–672
Tajiri Y, Moller C, Grill V (1997) Long-term effects of aminoguanidine on insulin release and biosynthesis: evidence that the formation of advanced glycosylation end products inhibits β-cell function. Endocrinology 138:273–280
Takabayashi F, Tahara S, Kaneko T, Miyoshi Y, Harada N (2004) Accumulation of 8-oxo-2 -deoxyguanosine (as a biomarker of oxidative DNA damage) in the tissues of aged hamsters and change in antioxidant enzyme activities after single administration of *N*-nitrosobis(2-oxopropyl) amine. Gerontology 50:57–63
Tanaka Y, Gleason CE, Tran PO, Harmon JS, Robertson RP (1999) Prevention of glucose toxicity in HIT-T15 cells and Zucker diabetic fatty rats by antioxidants. Proc Natl Acad Sci U S A 96:10857–10862
Tatton WG, Olanow CW (1999) Apoptosis in neurodegenerative diseases: the role of mitochondria. Biochim Biophys Acta 1410:195–213
Terman A (1995) The effect of age on formation and elimination of autophagic vacuoles in mouse hepatocytes. Gerontology 41(Suppl 2):319–326
Thangarajah H, Yao D, Chang EI, Shi Y, Jazayeri L, Vial IN, Galiano RD, Du XL, Grogan R, Galvez MG, Januszyk M, Brownlee M, Gurtner GC (2009) The molecular basis for impaired hypoxia-induced VEGF expression in diabetic tissues. Proc Natl Acad Sci U S A 106:13505–13510
The Diabetes Control and Complications Trial Research Group (1993) The Effect of intensive treatment of diabetes on the development and progression of long-term complications in insulin-dependent diabetes mellitus. N Engl J Med 329:977–986
Thomas DR (2004) Vitamins in health and aging. Clin Geriatr Med 20:259–274
Tiedge M, Lortz S, Drinkgern J, Lenzen S (1997) Relation between antioxidant enzyme gene expression and antioxidative defense status of insulin-producing cells. Diabetes 46:1733–1742
Tiedge M, Lortz S, Munday R, Lenzen S (1998) Complementary action of antioxidant enzymes in the protection of bioengineered insulin-producing RINm5F cells against the toxicity of reactive oxygen species. Diabetes 47:1578–1585
Toborek M, Hennig B (1994) Fatty acid-mediated effects on the glutathione redox cycle in cultured endothelial cells. Am J Clin Nutr 59:60–65
Torii K, Sugiyama S, Tanaka M, Takagi K, Hanaki Y, Iida K, Matsuyama M, Hirabayashi N, Uno H, Ozawa T (1992) Ageing-associated deletions of human diaphragmatic mitochondrial DNA. Am J Respir Cell Mol Biol 6:543–549
Tsai EC, Hirsch IB, Brunzell JD, Chait A (1994) Reduced plasma peroxyl radical trapping capacity and increased susceptibility of LDL to oxidation in poorly controlled IDDM. Diabetes 43(8):1010–1014
Tsunekawa T, Hayashi T, Suzuki Y, Matsui-Hirai H, Kano H, Fukatsu A et al (2003) Plasma adiponectin plays an important role in improving insulin resistance with glimepiride in elderly type 2 diabetic subjects. Diabetes Care 26:285–289
Tyner SD, Venkatachalam S, Choi J, Jones S, Ghebranious N, Igelmann H, Lu X, Soron G, Cooper B, Brayton C, Hee Park S, Thompson T, Karsenty G, Bradley A, Donehower LA (2002) p53 mutant mice that display early ageing-associated phenotypes. Nature 415:45–53
Unger RH, Zhou YT (2001) Lipotoxicity of β-cells in obesity and in other causes of fatty acid spillover. Diabetes 50(Suppl 1):S118–S121
Valls V, Peiro C, Muniz P, Saez GT (2005) Age-related changes in antioxidant status and oxidative damage to lipids and DNA in mitochondria of rat liver. Process Biochem 40:903–908
Van Gaal LF, Vertommen J, De Leeuw IH (1998) The in vitro oxidizability of lipoprotein particles in obese and non-obese subjects. Atherosclerosis 137:S39–S44
Van Remmen H, Ikeno Y, Hamilton M, Pahlavani M, Wolf N, Thorpe SR, Alderson NL, Baynes JW, Epstein CJ, Huang TT, Nelson J, Strong R, Richardson A (2003) Life-long reduction in MnSOD activity results in increased DNA damage and higher incidence of cancer but does not accelerate aging. Physiol Genomics 16:29–37
Vikramadithyan RK, Hu Y, Noh HL, Liang CP, Hallam K, Tall AR, Ramasamy R, Goldberg IJ (2005) Human aldose reductase expression accelerates diabetic atherosclerosis in transgenic mice. J Clin Invest 115:2434–2443
Vincent HK, Powers SK, Dirks AJ, Scarpace PJ (2001) Mechanism for obesity-induced increase in myocardial lipid peroxidation. Int J Obes Relat Metab Disord 25:378–388
Vincent HK, Morgan JW, Vincent KR (2004) Obesity exacerbates oxidative stress levels after acute exercise. Med Sci Sports Exerc 36:772–779
Vincent HK, Bourguignon C, Frick KI, Rutkowksi JR, Vincent KR, Weltman AL et al (2005a) Contributing factors to post-exercise oxidative stress in obesity. Obes Res; in review
Vincent HK, Vincent KR, Bourguignon C, Braith RW (2005b) Obesity and post-exercise oxidative stress in older women. Med Sci Sports Exerc 37:213–219

Vincent HK, Bourguignon C, Taylor AG (2005c) Relationship between the newly proposed dietary 'Phytochemical index', obesity and oxidative stress in young healthy adults. Int J Obes Relat Metab Disord; in review

Viner RI, Ferrington DA, Huehmer AFR, Bigelow DJ, Schowneich C (1996a) Accumulation of nitrotyrosine on the SERCA2a isoform of SR Ca-ATPase of rat skeletal muscle during aging: a peroxynitrite-mediated process? FEBS Lett 379:286–290

Viner RI, Huhmer AF, Bigelow DJ, Schoneich C (1996b) The oxidative inactivation of sarcoplasmic reticulum Ca(2+)-ATPase by peroxynitrite. Free Radic Res 24:243–259

Viroonudomphol D, Pongpaew P, Tungtrongchitr R, Changbumrung S, Tungtrongchitr A, Phonrat B et al (2003) The relationships between anthropometric measurements, serum vitamin A and E concentrations and lipid profiles in overweight and obese subjects. Asia Pac J Clin Nutr 12:73–79

Wallace DC (1994) Mitochondrial DNA, sequence variation in human evolution and disease. Proc Natl Acad Sci U S A 91:8739–8746

Wallstrom P, Wirfalt E, Lahmann PH, Gullberg B, Janzon L, Berglund G (2001) Serum concentrations of beta-carotene and alphatocopherol are associated with diet, smoking, and general and central adiposity. Am J Clin Nutr 73:777–785

Wang Y, Tissenbaum HA (2006) Overlapping and distinct functions for a *Caenorhabditis elegans* SIR2 and DAF-16/FOXO. Mech Ageing Dev 127:48–56

Ward WF, Qi W, Remmen HV, Zackert WE, Roberts LJ II, Richardson A (2005) Effects of age and caloric restriction on lipid peroxidation: measurement of oxidative stress by F2-isoprostane levels. J Gerontol A Biol Sci Med Sci 60:847–851

Wautier JL, Schmidt AM (2004) Protein glycation: a firm link to endothelial cell dysfunction. Circ Res 95:233–238

Wei YH (1998) Oxidative stress and mitochondrial DNA mutations in human aging. Proc Soc Exp Biol Med 217:53–63

Wei YH, Kao SH, Lee HC (1996) Simultaneous increase of mitochondrial DNA deletions and lipid peroxidation in human aging. Ann N Y Acad Sci 786:24–43

Wellen KE, Hotamisligil GS (2003) Obesity-induced inflammatory changes in adipose tissue. J Clin Invest 112:1785–1788

Wellman NS, Friedberg B (2002) Causes and consequences of adult obesity: health, social and economic impacts in the United States. Asia Pac J Clin Nutr 11:S705–S709

West IC (2000) Radicals and oxidative stress in diabetes. Diabet Med 17:171–180

Weyer C, Yudkin JS, Stehouwer CD, Schalkwijk CG, Pratley RE, Tataranni PA (2002) Humoral markers of inflammation and endothelial dysfunction in relation to adiposity and in vivo insulin action in Pima Indians. Atherosclerosis 161:233–242

Willcox JK, Ash SL, Catignani GL (2004) Antioxidants and prevention of chronic disease. Crit Rev Food Sci Nutr 44:275–295

Williams B, Gallacher B, Patel H, Orme C (1997) Glucose-induced protein kinase C activation regulates vascular permeability factor mRNA expression and peptide production by human vascular smooth muscle cells in vitro. Diabetes 46:1497–1503

Wojtczak L, Schonfeld P (1993) Effect of fatty acids on energy coupling processes in mitochondria. Biochim Biophys Acta 1183:41–57

Wolff SP, Dean RT (1987) Glucose autoxidation and protein modification. The potential role of 'autoxidative glycosylation' in diabetes. Biochem J 245(1):243–250

Wozniak A, Drewa G, Wozniak B, Schachtschabel DO (2004) Activity of antioxidant enzymes and concentration of lipid peroxidation products in selected tissues of mice of different ages, both healthy and melanoma-bearing. Z Gerontol Geriatr 37:184–189

Yamagishi SI, Edelstein D, Du XL, Brownlee M (2001) Hyperglycemia potentiates collagen-induced platelet activation through mitochondrial superoxide overproduction. Diabetes 50:1491–1494

Yang JH, Lee HC, Lin KJ, Wei YH (1994) A specific 4,977 bp deletion of mitochondrial DNA in human ageing skin. Arch Dermatol Res 286:386–390

Yang JH, Lee HC, Wei YH (1995) Photoageing-associated mitochondrial DNA length mutations in human skin. Arch Dermatol Res 287:641–648

Yang WS, Lee WJ, Funahashi T, Tanaka S, Matsuzawa Y, Chao CL et al (2001) Weight reduction increases plasma levels of an adipose-derived anti-inflammatory protein, adiponectin. J Clin Endocrinol Metab 86:3815–3819

Yao D, Taguchi T, Matsumura T, Pestell R, Edelstein D, Giardino I, Suske G, Rabbani N, Thornalley PJ, Sarthy VP, Hammes HP, Brownlee M (2007) High glucose increases angiopoietin-2 transcription in microvascular endothelial cells through methylglyoxal modification of mSin3A. J Biol Chem 282:31038–31045

Yaworsky K, Somwar R, Klip A (2000) Interrelationship between oxidative stress and insulin resistance. In: Packer L, Rösen P, Tritschler HJ, King GL (eds) Antioxidants in diabetes management. Marcel Dekker, New York, pp 275–302

Yen TC, Su JH, King KL, Wei YH (1991) Ageing-associated 5 kb deletion in human liver mitochondrial DNA. Biochem Biophys Res Commun 178:124–131

Yen TC, Pang CY, Hsieh JH, Su CH, King KL, Wei YH (1992) Age dependent 6 kb deletion in human liver mitochondrial DNA. Biochem Int 26:457–468

Yerneni KK, Bai W, Khan BV, Medford RM, Natarajan R (1999) Hyperglycemia-induced activation of nuclear transcription factor kappaB in vascular smooth muscle cells. Diabetes 48:855–864

Yu WH, Cuervo AM, Kumar A, Peterhoff CM, Schmidt SD, Lee JH, Mohan PS, Mercken M, Farmery MR, Tjernberg LO, Jiang Y, Duff K, Uchiyama Y, Naslund J, Mathews PM, Cataldo AM, Nixon RA (2005)

Macroautophagy–a novel beta-amyloid peptide-generating pathway activated in Alzheimer's disease. J Cell Biol 171:87–98

Yu L, Wan F, Dutta S, Welsh S, Liu Z, Freundt E, Baehrecke EH, Lenardo M (2006) Autophagic programmed cell death by selective catalase degradation. Proc Natl Acad Sci U S A 103:4952–4957

Yuan M, Konstantopoulos N, Lee J, Hansen L, Li ZW, Karin M, Shoelson SE (2001) Reversal of obesity- and diet-induced insulin resistance with salicylates or targeted disruption of IKKβ. Science 293: 1673–1677

Zhang C, Baumer A, Maxwell RJ, Linnane AW, Nagley P (1992) Multiple mitochondrial DNA deletions in an elderly human individual. FEBS Lett 297:34–38

Zhang Z, Apse K, Pang J, Stanton RC (2000) High glucose inhibits glucose- 6-phosphate dehydrogenase via cAMP in aortic endothelial cells. J Biol Chem 275:40042–40047

Zhang HJ, Xu L, Drake VJ, Xie L, Oberley LW, Kregel KC (2003) Heat-induced liver injury in old rats is associated with exaggerated oxidative stress and altered transcription factor activation. FASEB J 17:2293–2295

Zhang HJ, Doctrow SR, Xu L, Oberley LW, Beecher B, Morrison J, Oberley TD, Kregel KC (2004) Redox modulation of the liver with chronic antioxidant enzyme mimetic treatment prevents age-related oxidative damage associated with environmental stress. FASEB J 18:1547–1549

4 Oxidative Stress and Carcinogenesis

Apart from the external environment, cell's internal environment also plays a big role in the carcinogenesis. Cell proliferation, a hallmark of cancerous growth, involves cell-cycle events and internal environment. Involvement of the ROS in cell-cycle machinery, in cancerous growth, development of the cell characteristics in cancer progression as metastasis, and various other factors especially p53 play an important role in cancer regulation. It is aimed to discuss these parameters under the influence of the oxidative stress during carcinogenesis. Several review articles have been published on the current topic of interest (Boonstra and Post 2004; Nishikawa 2008; Sarsour et al. 2009; Suzuki and Malsubara 2011; Montero and Jassem 2011: Verbon et al. 2012; and others).

Cell-Cycle Regulation by ROS

The mammalian cells duplicate by a process known as cell cycle. All cells starting from the quiescence stage (G_0) pass through the proliferative phases in sequence: the G1, S (DNA synthesis), G_2, and M (mitosis). The progression through these phases is regulated by the cell's intrinsic molecular mechanisms. Popularly in response to the mitogenic stimuli, quiescent cells enter the proliferative cycle and may transit back to the quiescence state. Quiescence stage is essential to prevent aberrant proliferation as well as to protect the cellular life span. Unlike differentiation and cellular senescence, the quiescence is a reversible process that protects the proliferative capacity of cells essential for the cell and tissue renewal. Correct progression through the cell cycle and thus its precise regulation are essential to an organism's survival. Three main components have been identified that play a role in the cell cycle: (a) cyclin-dependent kinases (CDK)-cyclin complexes and related kinases, (b) metabolic enzymes and related metabolites, and (c) ROS/cellular redox status. Errors in any of these processes may lead to a variety of problems, including the development of cancer. Cell-cycle regulation is exerted by the endogenously produced cyclins and cyclin-dependent kinases (CDKs). The production and activity of these proteins/enzymes are affected by a wide variety of external factors, such as growth factors and hormones. These factors specifically activate or inhibit various signaling pathways that ultimately result in the cell-cycle progression or arrest, differentiation, or apoptosis (Romeo et al. 2012). ROS are essential components of these signaling pathways involving some key factors, for example, p21 or MAPK (Boonstra and Post 2004).

ROS influence the redox status in cells and can, according to their concentration, cause either a positive response (cell proliferation) or a negative cell response (growth arrest or cell death) and thus play a very important physiological role as second messengers (Suzuki et al. 1997). Changes in the cellular redox environment by using antioxidant like N-acetyl-L-cysteine (NAC, a glutathione precursor) inhibited the prolifera-

M. Bansal and N. Kaushal, *Oxidative Stress Mechanisms and their Modulation*,
DOI 10.1007/978-81-322-2032-9_4, © Springer India 2014

tion in mouse embryonic fibroblasts, hepatic stellate cells, and vascular smooth muscle cells (Menon et al. 2003). Also dietary antioxidants help in controlling the tumor onset and progression by preventing DNA damage and by acting on the cell-cycle checkpoints (Caputo et al. 2012). Altered cellular redox status and the redox-sensitive thiols through the activity of redox enzyme thioredoxin reductase contribute toward resistance to the cellular response to ionizing radiation (Selenius et al. 2012). However, changes towards a more oxidizing environment are required for the entry into S phase. Inhibition with antioxidant was associated with an increase in MnSOD activity and a decrease in cyclin D1 protein levels (Menon et al. 2007). MnSOD activity has been shown to regulate the mitochondrial "ROS-switch," in which a superoxide signal promotes proliferation, and a H_2O_2 signal supports quiescence. The concept of the cellular redox environment regulating cell-cycle progression is further supported by the observations of SOD activity influencing the oxidative stress-induced activation of G_2-checkpoint pathway in human oral squamous cancer, pancreatic cancer, and glioma cells (Kalen et al. 2006). Phytochemical induction of the cell-cycle arrest by glutathione oxidation and reversal by NAC in human colon carcinoma cells was observed (Odom et al. 2009). In cultured hamster fibroblasts, sublethal doses of ROS added exogenously stimulated proliferation, and H_2O_2 in nanomolar concentration generated from growth factor receptor-ligand binding also is known to facilitate the cell proliferation (Burch and Heintz 2005). NADPH oxidases, such as NOX1 and NOX4, are required for the growth factor-mediated production of H_2O_2, which subsequently activates multiple signaling pathways (Rhee et al. 2000, 2003). However, in the neuroblastoma cells, in vitro-stimulated environment with CO_2-induced oxidative stress and cell damage lead to the p53 upregulation and S-phase arrest with reduction in the cyclin B1 expression (Montalto et al. 2013).

A point in the G1 phase, called restriction point, commits the cell to enter S phase, may also be due to the necessary redox status (Pardee 1974), and is

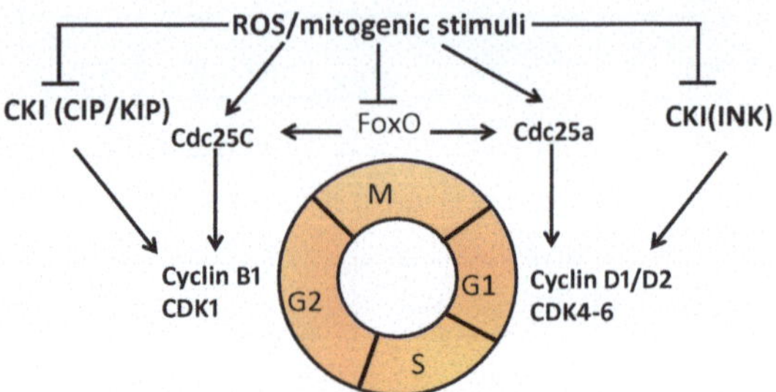

Fig. 4.1 Cell-cycle regulation by ROS

regulated by the cyclin D1/D2 in association with CDK4–6 (Grana and Reddy 1995). CDK4–6 kinase activity in the early G1 is low primarily because of the lower levels of cyclin D1. After the mitogenic stimulation, cyclin D1 increases sharply in the mid-to-late G_1 which coincides with the higher levels of CDK4–6 kinase activity. Further, cyclin D1 has been shown to be transcriptionally downregulated by the FoxO3a (Forkhead box class O transcription factor), which subsequently inhibits cell-cycle progression (Schmidt et al. 2002) (Fig. 4.1). Also the activity of FoxO in human cells is directly regulated by the cellular redox state. The FoxO-family of transcription factors are known to be phosphorylated by the mitogenic-signaling pathway, phosphatidylinositol-3 kinase (PI3K)/protein kinase B (AKT). Phosphorylated FoxO is excluded from the nucleus, thereby relieving FoxO-mediated gene repression (Brownwell et al. 2001). ROS induce the formation of cysteine–thiol–disulfide-dependent complexes of the FoxO for the modulation of its activity (Dansen et al. 2009).

ROS have been demonstrated to be important in cell-cycle regulation. The cell-cycle proteins like p21, Rb, cyclin D1, CDK4–6 kinase, and CDC25 phosphatase have been shown to be redox regulated in NAC-treated mouse and human fibroblasts (Liu et al. 1999; Savitsky and Finkel 2002). These changes especially in cyclin D1 (nuclear protein) are inversely correlated with MnSOD (mitochondrial) activity. One possible mechanism of this could be due to the FoxO-mediated transcriptional control of cyclin D1 and MnSOD expression. FoxO3a is known to activate MnSOD transcription, while inhibiting cyclin D1 transcription (Schmidt et al. 2002).

FoxO3a-mediated induction in MnSOD transcription is associated with the quiescence state. During the quiescence state, FoxO3a transcriptionally upregulate p27 expression (Medma et al. 2000). Inhibition of the FoxO3a activity is anticipated to relieve the cyclin D1 from transcriptional repression, which in turn is believed to support the cellular proliferation. Alternatively, the redox sensitivity in cyclin D1 expression could also be regulated by posttranslational mechanisms. NIH3T3 mouse fibroblasts carrying the Thr286A cyclin D1 mutation suppressed NAC-induced cyclin D1 degradation. This suggests that redox-sensitive phosphorylation of Thr286 could influence cyclin D1 protein levels (Menon et al. 2007). Furthermore, the redox sensitivity in cyclin accumulation could also be regulated by the thiol-redox reactions of critical cysteine residues. In contrast to the effect of NAC on cyclin D1, H_2O_2 inhibits cyclin D1 protein degradation in fibroblasts, resulting in cyclin D1 accumulation (Martinez et al. 2001).

Further, CDKs, serine/threonine kinases are specific for the G1, G1/S, S, or G2/M phase, and during each of these phases, a specific cyclin forms a complex with a specific CDK: cyclin D with CDK4 or CDK6 depending on the cell type, cyclin E with CDK2, and cyclins A and B with CDK1 (Satyanarayana and Kaldis 2009). Cyclin binding is essential to start activation process of CDKs, which consists of several steps of reversible phosphorylation. For example, CDK1 activation requires phosphorylation on Thr-161 by CDK-activating kinase (CAK) (Wu et al. 1994) and dephosphorylation on Thr-14 and Thr-15 by Cdc25. Phosphorylation of these sites by Wee1 (nuclear kinase) inhibits CDK1 activation (Domingo-Sananes et al. 2011). In addition, the presence of many types of cyclins and CDKs has been described in mammalian cells, but not all CDKs are required for the cell-cycle progression. The cyclin A and CDK2 kinase complex regulates the progression through S and G_2 phases. The cyclin B1 and CDK1 kinase complex, along with CDC25C phosphatase, regulates the progression from G_2 to M phase. Earlier it was believed that the functions of individual cyclins and CDKs are specific to a specific cell-cycle phase. However, recent reports demonstrate the redundancy in these cell-cycle regulatory protein functions as explained through the knockout experiments. However, knockout of cyclin A and knockout cyclin B have also been shown to be lethal in mice (Aleem and Kaldis 2006; Hochegger et al. 2008). Cyclins are the positive regulators of cell-cycle progression, and cyclin-dependent kinase inhibitors (CKIs) are the negative regulators. The INK family of CKIs specifically inhibits cyclin D/CDK4-6 kinase complexes. The inhibitory effect of p21 is ubiquitous, and it can inhibit all cyclin and CDK kinase activities. p21 and p27 also are known to facilitate the assembly of cyclin and CDK complexes (LaBaer et al. 1997).

In addition to the cell-cycle regulation by the expression of cyclins and CDKs, it can also be influenced by modulating CDK activity. Kinases, such as Wee1, of the cyclin CDK complexes and phosphatases, such as Cdc25, regulate the reversible phosphorylation state of cyclin/CDK complexes, which are essential for CDK activation. Different homologs of Cdc25 phosphatases stimulate cell-cycle progression. Cdc25c homolog is required for entry into the M phase because it activates the CDK1/cyclin B complex (Hoffmann et al. 1993) by dephosphorylating its Thr-15 (Yamaura et al. 2009). Increased expression, transportation activity into the nucleus, and phosphorylation/dephosphorylation activity at the specific sites determine the activity of Cdc25 in different situations (Boutros et al. 2006). CDK activity is also regulated by the CDK inhibitors (CKIs). These play an inhibitory role by inhibiting the CDK activation by binding to cyclin/CDK complexes and also play role by promoting cell-cycle progression by facilitating entry of cyclins and CDKs into the nucleus (Cheng et al. 1999) (Fig. 4.1). The CKI family consists of two subfamilies: the inhibitor of CDKs (INK4) family and the CDK-interacting protein (CIP) family. The genes encoding these proteins are located in regions that are often mutated in cancer cell lines (Molenaar et al. 2012), thus indicating their importance in cell-cycle regulation. The CIP family inhibits all cyclin-bound CDKs (Lu and Hunter 2010). CKI expression and activity are mediated by the ubiquitin-mediated degradation (Boonstra and Post 2004). Ubiquitination is a process involving the activation, conjugation, and ligation through respective enzymes named E1, E2, and E3 (Lu and Hunter 2010). The ubiquitinated proteins are recognized by the proteasome and degraded. The amount of CKI present is mediated by at least seven E3s. The different E3s are specific for different CKIs and cyclin/CDK complexes and are active during different cell-cycle phases (Starostina and Kipreos 2011). Regulation of cell-cycle progression depends primarily on the cyclins and their selective kinases (CDKs) produced by a cell. Different combinations of cyclins and CDKs are active during different phases of the cell cycle by which activation and inhibition of different phases occur. At any event when a cyclin has ensured progression through its associated phase, it is targeted for the ubiquitination and degradation. Based on all these activities and events, the cell cycle is considered a reversible process (Boonstra and Post 2004).

Physiological levels of cyclin D1 decreased aerobic glycolysis and mitochondrial activities and function in cells. Also the mitochondria activity was enhanced by the genetic deletion of cyclin D1. Subsequent study showed that cyclin D1 and CDK4-6 complex phosphorylates nuclear respiratory factor 1 (NRF1) at Ser47, suppressing its transcriptional activation of nuclear-encoded mitochondrial gene expression (Wang et al. 2006), and on the other hand, dephosphorylation of NRF1 in the absence of cyclin D1 promotes expression of nuclear-encoded mitochondrial genes. These events provide strong evidence for cyclin D1-coordinating cellular metabolism and cell-cycle progression. CDC25, a cell-cycle-regulatory protein, is a phosphatase and exhibits redox sensitivity in its function. CDC25 phosphatases are a family of dual specific phosphatases that dephosphorylate pThr14 and pTyr15 on CDKs and activate the cyclin/CDK kinase complex activity (Sebastian et al. 1993). In an in vitro study, the phosphatase activity of CDC25 was inhibited by using N-ethylmaleimide, a thiol-alkylating agent, and also by mutating a single conserved cysteine residue in it, demonstrating its redox activity (Dunphy and Kumagai 1991). Further, it was shown that the H_2O_2 treatment of HeLa cells induces an intramolecular disulfide bond between Cys377 and Cys330 of CDC25 and this thiol–disulphide redox reaction is associated with an inhibition in CDC25 phosphatase activity (Savitsky and Finkel 2002). H_2O_2 could influence the redox state of protein thiols by two-electron reactions. The reduced form of cysteine in protein (RSH) can be oxidized to sulfenic acid (RSOH), which can be further oxidized to sulfinic (RSO_2H) and sulfonic (RSO_3H) acids. The sulfinic and sulfonic forms of proteins are believed to be targeted for the degradation. The sulfenic form can react with another RSH to form the disulfide, RSSR, which can then be reduced back to RSH by cellular antioxidant machinery. This thiol–disulfide exchange reaction can regulate many of the cell-cycle-regulatory protein functions during the redox regulation of the cell cycle. In addition, superoxides can initiate one-electron reactions that can alter the redox state of metal cofactors (e.g., Fe^{3+} and Zn^{2+}) present in many kinases and phosphatases, thereby affecting their activities. Thus, both one- and two-electron reactions can participate in the redox regulation of cell-cycle proteins during progression of cell-cycle proteins during progression from one cell-cycle phase to the next.

ROS have been well implicated in cell proliferation, inhibition of apoptosis, or mutagenesis, and its high level contributes to the development of cancer. On the other hand, antioxidants have shown inhibitory effects on the tumor formation (Zhang et al. 2002]. An effect of ROS on cell proliferation was shown in a study on platelet-derived growth factor (PDGF) stimulation. The proliferation induced by PDGF was considered to be mediated by the ROS production by NOX1. In another study, ROS produced by NOX4 and NOX5 were also shown to be involved in the regulation of cell proliferation (Bedard and Krause 2007). Low amounts of ROS result in correct cell-cycle progression whereas high amounts can lead to uncontrolled cell proliferation, i.e., cancerous growth (Freinbichler et al. 2011). ROS had been shown to influence proliferation in the form of second messengers in many pathways regulating the cell proliferation, such as those involving p21 and MAPK (Boonstra and Post 2004). In one of the recent studies, a multifunctional factor APE1/Ref-1 prevented the oxidative inactivation of ERK for cyclin D1 expression and G1-to-S progression following lead acetate exposure (Wang et al. 2013). Further, ROS were shown to be important modulators of ubiquitination and phosphorylation of the cell-cycle-associated enzymes (Boonstra and Post 2004), which was considered important to the regulation of cell-cycle progression by ROS. Also ROS was found to activate growth factor receptors in the absence of the growth factor receptor ligands thereby controlling proliferation (Inoguchi et al. 2003).

Epidermal growth factor receptor (EGFR, a receptor tyrosine kinase, RTK) is involved in the regulation of cell proliferation, survival, migration, and differentiation. Its expression is upregulated in the cancerous tissues like breast, head and neck, ovarian, and esophageal cancer (Nicholson et al. 2001). The EGFR consists of an extracellular ligand-binding domain and cytoplasmic domain with enzymatic activity. Upon ligand binding, this cytoplasmic domain gets activated through signaling cascades such as the Ras/MAPK and PI3K/Akt pathways (Huang et al. 2009) which further lead to the gene transcription and proliferation (Yarden 2001). ROS are known to activate EGFR signaling. ROS was found to directly activate the EGFR in breast tissue and in turn overexpress CYP2E1 (Cytochrome P450 mixed function oxidase). Increasing ROS production is observed after the addition of ethanol in the cell culture experiments, which on metabolizing by CYP2E1 result in ROS production. This increase in ROS production has been shown to result in EGFR activation (Leon-Buitimea et al. 2012). Proof of EGFR activation by ROS was also provided in lung cells subjected to hypoxia, by the effect of diphenyleneiodonium (DPI), an inhibitor of ROS production both by NOX and mitochondria (Yunbo and Trush 1998). Usually hyperoxia leads to EGFR activation, resulting in tissue injury and possibly tissue death. The addition of DPI and the concomitant decrease in ROS inhibit EGFR activation in these cells (Papaiahgari et al. 2006). However, this effect might also be partially due to the concomitant inhibition of ATP production by DPI (Wind et al. 2010). Both the ethanol and hyperoxia-induced EGFR activation by ROS might be mediated by the transactivation, a process in which a receptor is activated through the activation of another receptor while its own ligand is absent.

Overexpression of Cdc25 in many cancer tissues stimulates cell-cycle progression (Turowski et al. 2003; Vijayakumar et al. 2011), and therefore, Cdc25 inhibitors are used in cancer therapy. The effect of Cdc25 inhibitors, such as the ortho-quinonoid compound LGH00031, generally depends on the presence of ROS. LGH00031 inhibits the activity of Cdc25b which in turn inhibit the G2/M phase transition and further decrease the tumor cell proliferation. The inhibitory effect of LGH00031 is abolished when the ROS content of a cell is diminished prior to the drug administration by the addition of antioxidant NAC, a glutathione precursor. These findings strongly suggest a direct link between ROS and the activity status of Cdc25b. ROS exert their effect on Cdc25 activity via enhancing phosphorylation of Cdc25 or alternatively through inactivation of Cdc25 by sulfonation of the cysteine at the active site. However, inhibition of ROS production by NOX4 leads to the hyperphosphorylation and inactivation of Cdc25c in melanoma cells (Yamaura et al. 2009). Thus, ROS production induced by LGH00031 and ROS production in melanoma cells have opposite effects on the activity of Cdc25. This demonstrates that the effects of ROS vary from case to case and are influenced by the local environmental differences in different cells.

Further, as explained earlier in this chapter, under various cell-cycle events, cyclin ubiquitination and subsequent degradation are important mechanisms contributing to the irreversibility of the cell cycle. In addition, ubiquitination is essential for the regulation of the expression of CKIs

(Lu and Hunter 2010), such as p21, p27, and p57 (Starostina and Kipreos 2011). ROS influence the ubiquitination by inhibiting E1 and E2 (ubiquitination-associated enzymes) activity via an increase in the ratio of oxidized versus reduced glutathione (redox ratio), leading to the S-thiolation of the active sites of E1 and E2. ROS also decreases ubiquitin-mediated degradation (next step in ubiquitination) by inhibiting the proteasome (Boonstra and Post 2004). Thus, it suggests clearly that ROS regulate the cell-cycle progression via ubiquitination.

Studies on the human foreskin fibroblasts also further prove the regulatory role of ROS via ubiquitination. Treatment with antioxidants prevents necessary accumulation of cyclin A at the end of G1 phase and thus results in G1-phase arrest. It is the differences in degradation, not the translation of cyclin A, that are responsible for the absence of accumulation, implying that ROS influence cell-cycle progression via ubiquitination. However, in this particular case, ROS inactivate anaphase-promoting complex (APC) via phosphorylation; thus, cyclin A is not degraded, hence leading to its accumulation (Havens et al. 2006). Interestingly, ubiquitination via phosphorylation is also induced by ROS that are produced as by-products, thereby allowing them to influence the cell-cycle progression. For example, doxorubicin (DOX), a chemotherapeutic agent, leads to the apoptosis via the production of mitochondrial superoxide in keratinocytes. Superoxides decrease ERK1/2 phosphorylation, thereby enhancing the ubiquitination of the antiapoptotic enzyme Bcl2 (Luanpitpong et al. 2012). ROS thus influences ubiquitination in two ways: by an effect on the redox state of the cell and subsequent S-thiolation of the ligase (component of the ubiquitination-associated enzyme, D3) and also by the induction phosphorylation and thereby effecting activity of the ubiquitination machinery.

Thus, effect of ROS is determined by the multiple factors, including the specific type and amount of ROS produced (Boonstra and Post 2004), the localization of the ROS (Li et al. 2007), and the presence or absence of other enzymes, such as a specific Cdc25 isoform.

Role of ROS and Oxidative Stress in Initiation, Promotion, Progression, and Metastasis of Cancers

Environmental agents such as cigarette smoke, xenobiotics, lifestyle, diet, chronic ultraviolet exposure, and sustained cellular injuries provide sources of endogenous ROS production, which can function as chemical effectors in tumorigenesis (Holiday 2005; Kundu and Surh 2008). Cancer usually manifests late in life and could be due to an increase in ROS production or decrease in ROS removal or both (Van Remmen et al. 2003). DNA damage leading to activation of oncogenes and/or non-expression of tumor suppressor proteins is one of the plausible mechanisms by which ROS can promote carcinogenesis. Mutations in the oncogenes like ras and p53 (also a tumor suppressor gene) have been observed in many types of human cancers (Rajalingam et al. 2007; Starano et al. 2007).

Carcinogenesis can be divided into three distinct stages: initiation, promotion, and progression. Initiation can occur because of the mutations in one or more genes, which result in loss or gain of function. Promotion is the functional enhancement and alteration of the pathway induced by the initiation. Progression is the continuing change of unstable karyotype, often leading to its aberrant change and further leading to the aberrant proliferation. More precisely, tumor formation classically includes DNA damage and mutagenesis, causing transformation of the normal cells into preneoplastic cells (initiation), followed by selective clonal expansion (promotion), and further some malignant cells to acquire more aggressive characteristics (progression). In the tumor development, progression is a stage of uncontrolled tumor expansion due to the inability of the organism to recognize cancer cells as abnormal, to persistence of the causal factor, or to disturbance in the mechanisms that repair oxidized DNA. Oxidative stress contributes to this uncontrolled development. For instance, ROS at physiological concentrations control the activity of transcription factor

AP-1 (which regulates the expression of the cell growth mediators), but if ROS are produced in excess, AP-1 is overproduced and this leads to the abnormal cell proliferation. The $p21^{Ras}$ oncogene (C-ras) is involved in various processes, including cell proliferation and differentiation and the organization of the cytoskeleton. In various tumors, Ras overexpression has been evidenced. In vitro work has demonstrated the involvement of ROS as mediator of a Ras-induced cell propagation cycle (Irani et al. 1997).

Cancer cells in general exhibit lower levels of antioxidant enzyme activities compared with their respective normal cells, in particular the activity of MnSOD enzyme (Oberley and Buettner 1979). Studies suggest that oxidative stress could significantly contribute to the cancer progression, possibly by perturbing the redox control of the cell cycle (Pennington et al. 2005; Halliwell 2007). Redox potential in normal cells correlates with Rb (retinoblastoma protein) phosphorylation status during the cell cycle, suggesting that perturbations in the cellular redox potential could significantly affect the function of a tumor suppressor gene (Hoffman et al. 2008). Furthermore, it is hypothesized that the metabolic redox-signaling pathways could initiate as well as promote carcinogenesis. This hypothesis is based on the numerous studies demonstrating a regulatory role of MnSOD activity in cancer cell growth in both cell culture and tumor xenograft animal model systems (Zhang et al. 2002; Weydert et al. 2003). These results suggest that reestablishing the redox control of the cell cycle by manipulating the expression of ROS-removal enzymes (e.g., MnSOD) could suppress or inhibit (or both) carcinogenesis.

Further, majority of the human cancers are carcinomas originating from the epithelial cells, and several studies in aged normal human breast and prostate fibroblasts support these epithelial malignancies (Dong-LeBourhis et al. 1997; Barcellos-Hoff and Ravani 2000). Fibroblasts are the primary component of the stroma (extracellular supporting structure of tissues) that supports these epithelial tissues. Senescent human fibroblasts enhanced cellular proliferation in premalignant and malignant epithelial cells in vitro and tumor growth and metastasis in mice in vivo (Campisi 2005). Prostate epithelial cells from tissue with aged stroma can become tumorigenic when cocultured with tumor-burden fibroblasts. Likewise, exposure of the mammary gland stroma to irradiation or carcinogens stimulates nonmalignant epithelial cell proliferation and promotes tumor formation (Barcellos-Hoff and Ravani 2000). The secreted growth factors, cytokines, and extracellular matrix proteins from the aged fibroblasts are believed to enhance the premalignant and malignant epithelial cell proliferation. Because many of the growth factors and cytokines are known to generate ROS, it is hypothesized that ROS signaling derived from the aged fibroblasts could provide mitogenic stimuli to premalignant and malignant epithelial cells.

In practice, although cancer is a genetic disease, mutagenic transformations are not sufficient to acquire metastatic competence, and also many oncogene-driven mouse models of cancer are still not able to automatically establish metastases (Minna et al. 2003). Although seeding can occur in multiple organs, in many cases, however, metastatic tumors grow only in a few (Husemann et al. 2008). This shows that a receptive microenvironment is required for the development of metastasis (Psaila and Lyden 2009). Some cancer cells remain in a dormancy state even for many years (Ngugen et al. 2009) or remain in a balanced state of the proliferation and apoptosis (Ngugen et al. 2009). Microenvironment may suppress the malignancy of potentially metastatic cells, but their reactivation also occurs through the perturbation in the microenvironment. Tumor invasion and metastasis, responsible for most cancer deaths in humans, involve transferring of the cells from a primary tumor to the blood in circulation, then infiltrating into a new organ, initiation and maintenance of the new growth, and vascularization of the metastatic tumor. All these events are regulated by multiple factors and must be successfully completed to permit the outgrowth of metastatic tumors in the new microenvironment (Chambers et al. 2002). ROS plays an important role in the initiation of activities of the migration

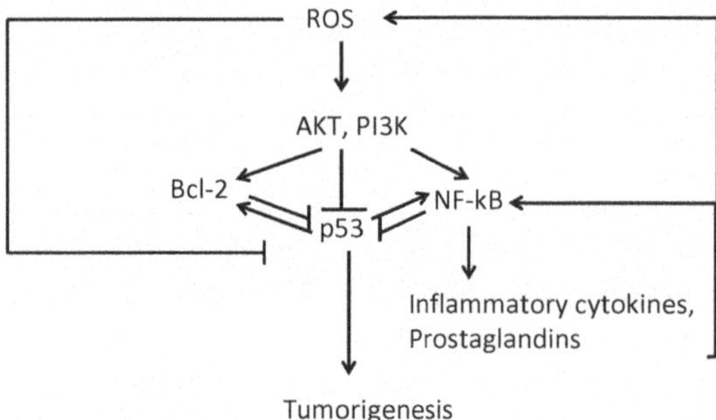

Fig. 4.2 ROS plays a central role in carcinogenesis by modulating key molecular pathways

and invasion of cancer cells in tumor metastasis. Various molecular mechanisms in ROS and metastasis have been recently well reviewed (Lee and Kang 2013). As indicated earlier also, oxidative stress often prevails in cancer cells, and its different levels induce different outcomes in cancer cells. Mild oxidative stress activates cell signaling mechanisms such as proliferation, migration, and invasion, but high oxidative stress can induce cell death (Nishikawa 2008). Cell migration in the pathological conditions is an important process (Hurd et al. 2012), and it involves various types of the cellular changes involving alterations in cell structure by the regulation of cytoskeleton dynamics and expression of adhesion molecules. ROS are known to actively participate in these events (Fig. 4.2).

ROS regulate many integrin (transmembrane receptor)-mediated cellular responses, such as adhesion, cytoskeleton organization, migration, proliferation, and differentiation. The integrin activation triggers a transient ROS production either independently or in cooperation with growth factor receptors. Integrin interacts with extracellular matrix (ECM) proteins and produces cellular ROS by promoting changes in the mitochondrial metabolic function (Taddei et al. 2007) and activation of several ROS generating oxidases, such as NADPH oxidases (NOX), lipoxygenase (LOX), and cyclooxygenase (COX) (Broom et al. 2006). Rac1 (a GTPase) acts upstream of both the NOX (Ushio-Fukai 2009) and arachidonic acid (AA)-metabolizing enzymes (LOX and COX). Arachidonic acid metabolism regulates NOX and mitochondrial ROS production, suggesting that Rac1 can activate complex networks of regulation for the ROS production (Lee et al. 2006). ROS also play a major role in the stabilization or destabilization of the cell–cell junction of the vascular endothelial (VE)-cadherin. VE-cadherin is a cell–cell adhesion glycoprotein needed for the vascular development. Rac1-induced ROS function as a signaling molecule to disrupt VE-cadherin-based cell–cell adhesion, leading to the endothelial barrier dysfunction and permeability changes as well as endothelial migration and proliferation involved in angiogenesis.

In addition, Src and Pyk2 kinases involved in the phosphorylation of the adherent junction proteins such as β-catenin and p120-catenin result in loss of cell–cell adhesion. ROS has been implicated in the intercellular dissociation (Nigam et al. 1998; Inumaru et al. 2009). The p120-catenin family members are important cell–cell functional proteins that stabilize catenins at the cell membrane (Anastasiadis et al. 2000). In an oxidative stress environment, tyrosine phosphorylation of p120-catenin results in its translocation to the cytoplasm (Inumaru et al. 2009), thus leading to intercellular dissociation accompanied by the cytoskeletal remodeling. In addition, ROS scavengers inhibit H_2O_2-induced intercellular dissociation, demonstrating that ROS levels regulate cell adhesion. Loss of cell–cell adhesion was shown to be mediated by the Rac1–ROS signaling pathway (Inumaru et al. 2009). Further antioxidants have been shown to protect the VE-cadherin at cell–cell contacts (Lin et al. 2003). This was achieved by inhibiting the VEGF-induced angiogenesis through the disruption of ROS-dependent Src kinases activation which otherwise phosphorylate at tyrosine in the VE-cadherin tyrosine. Another protein IQGAP (IQ motif containing GTPase-activating protein, component of Rac1–ROS signaling pathway) is also required for the establishment of VE-cadherin-based cell–cell contacts and co-localized with VE-cadherin at cell–cell contact sites (Yamaoka-Tojo et al. 2006). This has been shown to play an important role in the regulation of cadherin-adhesive functions and act as a scaffold protein involved in the cellular mobility and morphogenesis. Exposure of several cancer cell lines to the inflammation or chemically induced ROS boost their migratory and invasive behavior (Payne et al. 2005). This is linked to a process named EMT (epithelial–mesenchymal transition) by which epithelial cells lose their polarity and cell–cell adhesion and gain migratory and invasive properties. EMT is a biological process allowing epithelial cells to undergo several biochemical alterations that permit the achievement of a mesenchymal phenotype. More precisely, this is accompanied by loss of epithelial markers (VE-cadherin, laminin-1, desmoplakin, cytokeratin, and collagen), induction of mesenchymal markers (N-cadherin, smooth muscle actin, and vimentin), and upregulation of transcription factors. This not only involves a physiological mechanism for development and tissue remodeling but also a pathological mechanism associated with various diseases including inflammation, fibrosis, and cancer (Thiery 2002).

Malignant tumors are considered invasive and may reach distant sites through the circulatory system. A cancer cell is considered malignant with the following characteristics: local invasion, intravasation, survival in the blood and lymphatic system, extravasation, and colonization (Ngugen et al. 2009). A number of signaling pathways involved in EMT are known to be activated during tumor metastasis (Giannoni et al. 2012). EMT is essential for the migration and invasion of many cancer cells (Baum et al. 2008; Barnett et al. 2011). Studies have suggested the role of ROS in the induction of EMT in cancer cells through activation of Snail (Dong et al. 2007). Snail overexpression in prostate cancer increases intracellular ROS levels (Barnett et al. 2011) and further results in the induction of EMT. During EMT, cell–cell molecular adhesion decreases and cell–extracellular matrix adhesion increases; these events favor cell migration and invasion. ROS play a pivotal role in the cell–cell dissociation process, since they can regulate the activity of Src kinase. Src is known to be activated in several cancers, and there is convincing evidence that increased Src activity is associated with a more invasive and aggressive phenotype (Frame 2002). A number of studies suggest that Src also plays an important role in the cellular response to ROS, because Src-specific inhibitors and dominant-negative Src mutants strongly attenuate cellular response to ROS (Holliday 2005). Furthermore, ROS induce cell–cell dissociation by endocytosis of N-cadherin mediated by Src kinase phosphorylation and internalization of p120-catenin, resulting in loss of epithelial integrity and transient Rho/Rho kinase pathway activation. The H_2O_2-induced Src activation also induces activation of NF-κB leading to MnSOD expression, which reduces oxidative stress. This indicates that oxidative stress-induced cell–cell dissociation might be required for the initial step of EMT (Inumaru et al. 2009). The expression of genes implicated in hypoxia-induced EMT and angiogenesis may also be regulated by ROS. HIF-1α is induced by hypoxia and then activates the downstream transcription of EMT-related genes. Under hypoxic conditions, ROS are produced due to the aberrant function of mitochondria complex III (Giannoni et al. 2012).

Further, after the release from a primary tumor, cancer cells interact with various components, including blood and immune cells, and form embolism in blood vessels, and these processes are known to generate ROS around cancer cells in metastasis. ROS-generating and detoxifying systems determine the ROS level in tumor growth and metastasis. Changes in three types of genes, namely, oncogenes, tumor suppressor genes, and stability genes, are usually involved in the carcinogenesis with a concomitant reduced level of antioxidant defense in cancer cells. Low activities of Cu/Zn-SOD, Mn-SOD, catalase, and glutathione peroxidase are reported in a variety of transformed cell lines, and some tumor cells were reported to have reduced SOD activities compared with their normal counterparts (Sykes et al. 1978). SOD and catalase activities were much lower in mouse colon carcinoma CT26 and mouse liver hepatoma Hepa 1–6 than in mouse fibroblast NIH 3T3. Attenuation of catalase in the malignantly transformed cell lines was mainly responsible for the elevated ROS levels in these cells. It was reported that some tumor cells produce large amounts of ROS compared with normal cells (Szatrowski and Nathan 1991). These findings strongly suggest that ROS, especially hydrogen peroxide, are not efficiently removed in most tumor tissues. When the blood flow to tumor tissues is insufficient, which is the case for most solid tumors, the ROS level in tumors increases compared with that in surrounding normal tissues. Macrophages and monocytes activated by contact with tumor cells produce very high-level ROS (Mytar et al. 1999). Solid tumors consist of not only cells but also stroma cells such as fibroblasts and endothelial cells and various infiltrating types of leukocytes, including granulocytes, monocytes, activated macrophages, lymphocytes, and dendritic cells (Condeelis and Pollard 2006). Therefore, interaction between cancer cells and other cells continuously takes place in tumor tissues. Upon interaction, ROS and other molecules, including cytokines, growth factors, protease, and angiogenic factors, are released, and most of such responses can accelerate tumor metastasis. Cytokines produced by immune responses enhance ROS production, as observed in the human fibroblasts stimulated with TNF-α or interleukin-1 (Meier et al. 1989). In liver, Kupffer cells on binding to the cancer cells destroy them by various mechanisms, including phagocytosis and the release of ROS and proteases (Decker 1990). Inhibition of Kupffer cell function prior to tumor cell challenge increases metastatic growth in the liver (Bayon et al. 1996), suggesting that Kupffer cells act as a defense against hepatic metastasis. Thus, immune cells, including monocytes, neutrophils, activated macrophages, and Kupffer cells, can destroy cancer cells through various immune reactions.

Embolization of cancer cells in blood vessels of downstream organs binds with coagulation factors and platelets and facilitates the arrest of cancer cells in capillaries (Nash et al. 2002). Some specific adhesive interactions also take place, which determine the tissue specificity of the tumor metastasis. Majority of the melanoma B16-BL6 cells injected into the tail vein of mice are initially trapped by the lung. However, subsequently, less than 10 % of cells trapped were found in the organ 24 h after injection, suggesting that arrested cancer cells are efficiently cleared. Mechanical stress as well as immune cytosis is involved in the clearance of the arrested cancer cells. Irrespective of the mechanisms involved, the clearance of arrested cancer cells reconstructs the obstructed blood flow. Reperfusion following ischemia is associated with the production of large quantities of the proinflammatory cytokines and ROS, which are the causes of ischemia/reperfusion injuries of various organs and tissues. An ischemia followed by reperfusion triggers the infiltration of neutrophils, and activated neutrophils release ROS in the vasculature (Jaeschke 2003). In addition, NADPH oxidase also contributes to the ROS production after reperfusion. It has been reported that most cancer cells are killed by ROS and other reactive species produced in the process of ischemia/reperfusion (Jessup et al. 1999).

Various anticancer agents in clinical use cause DNA damage and ROS production (Sawa et al. 2000; Pelicano et al. 2004; Valerie et al. 2007). Cancer cells encounter high-level ROS at the site of carcinogenesis, en route to distant organs, and in metastatic sites. Anticancer treatment further increases the level of ROS around cancer cells, which kill most but not all, and therefore, changes in the surroundings of cancer cells will affect various aspects of the cancer cells including the invasion and adhesion processes. The importance of ROS in the killing of tumor cells has been demonstrated in cell-free systems generating superoxide anions and hydrogen peroxide (Ioannidis and deGroot 1993). Low-level ROS produce mitogenic effects to stimulate the proliferation of several types of cells (Burdon 1995). NIH-3T3 fibroblasts transfected with a ras oncogene showed an excessive production of the superoxide radical by the action of an NADPH oxidase enzyme complex, and this increased level of ROS promoted abnormal proliferation (Fruehauf and Meyskens 2007). Suppression of apoptosis is also involved in the oxidative stress – induced increase in proliferation (Fruehauf and Meyskens 2007). Overexpression of the catalase in cells was closely related to the reduced growth rates of rat aortic smooth muscle cells and fibroblasts (Arnold et al. 2001).

Involvement of ROS in the signaling pathways and transcription factors as explained in other chapter in this write-up is linked to the ROS-mediated changes in the expression of a number of genes (Allen and Tresini 2000). Those regulated include molecules closely related to the various stages of tumor metastasis, including matrix metalloproteinases (MMPs), adhesion molecules, EGF, EGF receptor (EGFR), and vascular endothelial growth factor. In most cases, ROS increase the expression and/or activate these proteins, leading to the aggravation of tumor metastasis. In metastasis, before forming metastatic colonies in distant organs, the cancer cells enter into the systemic circulation. Cancer cells migrate close to the blood (or lymphatic) vessels and enter or intravasate them. In these processes, active MMPs are required to destroy the extracellular matrix and basement membranes for the migration of cancer cells. MMPs are secreted in the latent forms from the cancer cells and various surrounding cells and activated by ROS. Hydrogen peroxide causes increased expression of MMPs responsible for the degradation of the basement membranes in endothelial cells (Belkhiri et al. 1997). Further, an essential step in the tumor metastasis is breach of the extracellular matrix (ECM) and invasion of the surrounding stroma. The ECM-degrading enzymes allow the cancer cells to migrate to the new sites. These proteases include the matrix metalloproteinases (MMPs), cathepsins, and urokinase plasminogen activator (uPA) (Meyer and Hart 1998; Brooks et al. 2010). ROS have been implicated in the abnormal activation of these proteases in cancer cells. Growth factor signaling induces the invasive ability in cancer cells by increasing the MMPs and uPA activities in an ROS-dependent manner (Kim et al. 2007). The activities of MMPs are regulated by the endogenous inhibitors, known as the tissue inhibitors of metalloproteinases (TIMPs). ROS-dependent regulation of MMPs is reported in special types of cancer cells including breast cancer, glioblastoma (Chiu et al. 2010), and pancreatic cancer (Binker et al. 2009). ROS have been shown to directly regulate the expression and loss of activity of MMPs or indirectly to regulate the blockade of TIMPs (Brenneisen et al. 2002). The NF-κB-induced and uPA expressions have also been shown to be regulated by ROS produced from NOX and mitochondria (Tobar et al. 2010). The uPA and uPAR have been shown to play a key role in many physiological processes, including embryogenesis, angiogenesis, wound healing, and metastasis. uPA is an extracellular protease that is activated upon binding to the receptor uPAR. When activated, uPA cleaves plasminogen, producing plasmin which degrades ECM components by proteolysis and also inducing the activation of MMPs. The activity of uPA is regulated by the plasminogen activator inhibitors (PAI-1 and PAI-2) and the protease nexin-1 (Meyer and Hart 1998; Brooks et al. 2010). Similar to MMPs, uPA and uPAR expression can be induced by ROS in a MAPK-dependent manner (Kim et al. 2007; Lee et al. 2009).

Pathways/Networks of p53 Inhibition and Cancer Progression

p53 is one of the most frequently mutated genes in human cancers, and approximately half of all tumors carry mutant p53. In large number of humans with tumors, the p53 pathway is partially abolished by the inactivation of other signaling components (Brown et al. 2009). p53 exhibits diverse and global biological functions, including regulation of the cell cycle, apoptosis, senescence, DNA metabolism, angiogenesis, cellular differentiation, and the immune response. Through these pathways, p53 facilitates the repair and survival of the damaged cells or eliminates severely injured cells from the respective pool to protect the organism.

The p53 tumor suppressor is commonly described as a sequence-specific transcription factor that is kept at low levels in healthy cells. Engagement of the p53 signaling pathways occurs in response to a broad range of stressors, intrinsic and extrinsic to the cell, which stabilize and affect the p53 by a series of posttranslational modifications. Phosphorylation is classically regarded as the first crucial step of the p53 stabilization, requiring a number of kinases, such as ATM (ataxia telangiectasia mutated) Chk1/2,

JNK, P38, and others, and followed by other modifications, which further allows it to adequately respond by modulating the expression of different subsets of target genes (Riley et al. 2008). Key mechanism by which p53 is regulated is through the control of protein stability, a mechanism primarily mediated by Mdm-2, a transcriptional target of p53. Mdm-2 protein can inhibit p53 by regulating its stability, cellular localization, and transactivation (Brooks et al. 2007).

The key role of p53 as a tumor suppressor is to block cell-cycle progression and/or to induce apoptosis, in response to the cellular stresses such as DNA damage. Impaired p53 activity promotes the accumulation of the DNA damage in cells, which leads to a cancer phenotype. Most TP53 (p53 encoding gene) mutations in human cancers result in mutations within the DNA-binding domain, thus preventing p53 from transcribing its target genes. Mutant p53 has led to a loss of normal function of the wild-type protein and also abilities to promote cancer. p53 plays an important role in detecting and repair of the damaged genome. When p53 responds to DNA damage, it elicits cell-cycle arrest or apoptosis (Hanahan and Weinberg 2000), and induction of wild-type p53 can induce apoptosis in leukemia cells (Yonish-Rouach et al. 1991). Mice that have a specific p53 mutant lack the ability to induce cell-cycle arrest but retain the ability to induce apoptosis, allowing them to efficiently suppress oncogene-induced tumors (Toledo et al. 2006), thus suggesting that the proapoptotic function of p53 may play a more important role in its antitumor effects than in its induction of cell-cycle arrest.

p53 mainly functions as transcription factor and induces apoptosis. In many cells, apoptosis occurs through one of the two major pathways: intrinsic mitochondrial or extrinsic death receptor pathway (Kroemer et al. 2007) (Fig. 4.3). In the mitochondrial pathway, death stimuli target mitochondria either directly or through transduction by proapoptotic members of the Bcl-2 family, such as Bax. Bax is a member of the Bcl-2 family, which forms heterodimers with Bcl-2, inhibiting its activity. The Bcl-2 protein family plays an important role in apoptosis and cancer (Yip and Reed 2008). For example, Bcl-2 controls the release of cytochrome c from the mitochondria, which activates the apoptotic pathway by activating caspase 9 which further activates executioner caspase 3. Both caspases play key roles in the apoptotic pathway. Several human cancers, including colon and stomach cancer, have altered expression of Bcl-2 (Krajewska et al. 1996). In the breast cancer, a study showed that a low level of expression of Bax is associated with a poor prognosis (Krajewski et al. 1995), whereas other reports have shown no correlation between the Bax expression level and prognosis (Sjostrom et al. 2002). In the death receptor pathway, the receptors located at the cellular membrane recruit adaptor proteins such as initiator caspase-8, triggering the activation of caspases to control apoptosis. CD95 (also called Fas and Apo-1) is a "death receptor" indicating its major role in apoptosis. The first report of CD95 showed that an anti-CD95 antibody reduced the growth of human B-cell xenograft tumors (Trauth et al. 1989). Several reports have indicated the CD95 pathway to play an important role in apoptosis induced by cytotoxic agents and that this system involves the activation of wild-type p53 (Muller et al. 1998a). Therefore, the p53 status may influence chemosensitivity via CD95 signaling. However, CD95 is reported to promote tumor growth (Chen et al. 2010). Programmed cell death is very complicated and depends on a variety of factors.

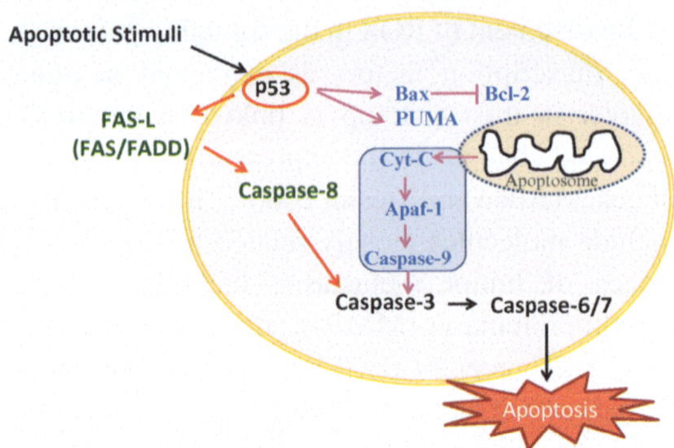

Fig. 4.3 p53-mediated extrinsic and intrinsic apoptotic pathways

A Bcl-2 subfamily contains only the BH3 domain. Several BH3 proteins have been identified, and p53 acts as a transcription factor for PUMA, p53-upregulated modulator of apoptosis (Yu et al. 2001), and NOXA (Oda et al. 2000). PUMA is also a key mediator of the apoptotic pathway induced by p53. When PUMA is disrupted in the colon cancer cells, p53-induced apoptosis is prevented (Yu et al. 2001). PUMA may play a pivotal role in determining the cell fate (programmed cell death versus cell-cycle arrest) in response to the p53 activation. A report about PUMA knockout mice (Jeffers et al. 2003) showed that knockout of PUMA involves the apoptotic deficiency observed in p53 knockout mice. PUMA is an essential mediator for the p53-dependent and p53-independent apoptosis in vivo (Yu and Zhang 2003). Since evading apoptosis is one of the hallmarks of cancer (Hanahan and Weinberg 2000), PUMA may play an important role during carcinogenesis. However, in certain situations, apoptosis can promote carcinogenesis (Michalak et al. 2010). PUMA-deficient hemopoietic stem cells are protected from the gamma-irradiation-induced cell death, which reduces compensatory proliferation and replication. On the other hand, wild-type mice experience massive cell death when they received gamma irradiation, which subsequently led to repopulation of the region by stem/progenitor cells. These reports indicate that the homeostasis stem/progenitor structure of tissue may suppress tumor formation.

The p53 protein suppresses tumor formation not only by inducing apoptosis but also by causing cell-cycle arrest. Depending on the type of cellular stress, p53 can induce G1 arrest through the activation of transcription of cyclin-dependent kinase inhibitor, p21 (Giono and Manfredi 2006). P53 also regulates the G2/M transition. For example, p53 can block the cell entry into mitosis by inhibition of Cdc2, whereas Cdc2 needs to bind to cyclin B1 in order to function. Repression of cyclin B1 by p53 also arrests cells in G2 (Taylor and Stark 2001). However, transient cell cycle may not lead to the tumor eradication, because a cell with oncogenic potential that cannot be repaired may resume proliferation (Vousden and Prives 2009).

The other mechanism, cellular senescence, may play an important role in p53-mediated tumor suppression. Cellular senescence is permanent cell-cycle arrest. There are many reports regarding the correlation between tumor development, p53, and senescence (Campisi and d'Adda di Fagagna 2007). Oncogenic ras expressed in human and rodent primary cells results in accumulation of p53/p16 and in turn cellular senescence. Inactivation of p53 or p16 prevents ras-induced senescence (Serrano et al. 1997). This report implies that the cellular senescence has an important role in suppressing tumor development. The inactivation of p53, as is present in most human cancers, allows cells to evade cellular senescence, thus resulting in tumor development. P53 also seems to prevent premalignant lesions from developing into malignant tumors by activating senescence programs (Vousden and Prives 2009). Cellular senescence induced by p53 is important not only for cancer prevention but also for the anticancer effect induced by any wild-type p53 introduced in established tumors.

Role of ROS and p53 in Apoptosis

Under normal conditions, the cell cycle is constantly active. If dividing cells are damaged for whatever reason, they are able to interrupt temporarily their cycle at stage G1, S, or G2 (checkpoints), repair the damage, and resume division (Shackelford et al. 2000). If the damage is too great, apoptosis eliminates the altered cells selectively. During apoptosis, which is a normal physiological process, cells initiate a programmed suicide mechanism leading to many morphological changes. During cell proliferation, protein p53 plays a primordial role: checking the integrity of the DNA. It triggers mechanisms that eliminate, for instance, the oxidized DNA bases that cause mutations. When cell damage is too great, p53 triggers cell death by apoptosis.

Uncontrolled apoptosis can be harmful to an organism, leading to a subtle regulatory system consisting of proapoptotic factors (e.g., p53) and antiapoptotic factors. Among the latter is the NF-κB transcription factor, whose activation is itself regulated by ROS. In its inactive form, NF-kB exists as a trimer composed of subunits p65, p50, and IkBα. In response to oxidative stress, the IkBα subunit dissociates from the trimer and the active form of NF-kB (p65/p50 dimer) can then migrate to the nucleus and activate genes coding for the expression of antiapoptotic molecules. However, different in vitro responses have been observed (activation or inhibition of the factor) according to the cell type and ROS concentration used. Apoptosis is crucial for the normal development and homeostasis of all multicellular organisms, triggered through the extrinsic (receptor-mediated) or the intrinsic (mitochondria-mediated) pathway. The intrinsic pathway can be triggered by many stimuli including ROS. Mitochondria are the major site of ROS production and accumulation of ROS may lead to the initiation of apoptosis (Meyer et al. 2005). Many cytotoxic agents induce ROS, which are involved in the induction of apoptotic cell death. H_2O_2 can cause the release of cytochrome c from mitochondria into the cytosol. Moreover, H_2O_2 may also activate nuclear transcription factors, such as NF-κB, AP-1, and p53 (Price et al. 1998), which may lead to upregulation of death proteins or production of inhibitors of survival proteins. Several studies imply that inhibition of apoptosis by Bcl-2 is associated with protection against ROS (Ribeiro and Oilario 1998). High oxidative stress level kills cells either by necrosis or by apoptosis (Adjuiket al. 2004; Efferth 2006). In various apoptosis models, changes in the redox status of the cells to a more oxidizing environment occur prior to the activation of the final phase of caspase activation (Schulze-Bergkamen and Krar 2004; Efferth 2006). This is further supported by the ability of various antioxidants such as NAC to block apoptosis in a similar way that caspase inhibitors do (Debatin and Krammer 2004). In addition, the antioxidant properties of Bcl-2, a potent inhibitor of apoptosis, further confirm this view (Gorman et al. 1997; Dell'Eva et al. 2004). Under normal conditions, aerobic cells are endowed with extensive antioxidant defense mechanisms to counteract the damaging effects of ROS (Muller et al. 1998a, b; Orrenius 2004). When prooxidants overwhelm antioxidant defense mechanisms, oxidative stress occurs. So apoptosis has been said as a fail-safe device to prevent cells from proliferating uncontrollably, in the face of a persistent oxidative stress (Efferth et al. 2007). Current chemotherapeutic agents such as anthracycline derivatives, which are frequently used as chemotherapeutics in the treatment of various types of cancers, target some of these apoptotic pathways. For example, Adriamycin is known to chelate iron and generate ROS that result in apoptosis of cancer cells (Eskelinen 2008). Another example of a chemotherapeutic agent that generates ROS for cancer treatment is artesunate (ART), which induces apoptosis in leukemic T cells mainly through the mitochondrial pathway via ROS generation (Galluzzi et al. 2007).

In conclusion, in this write-up initially, various cell-cycle events like involvement of particular cyclin/kinase selection normally and as influenced by the ROS have been discussed extensively. Secondly, the cancer progression leading to the metastasis has been discussed in detail. Involvement of the various proteins influencing cell migration and also involvement of ROS at each event have been discussed. In the end, the role of p53 in cancer development has been taken up with various environmental conditions.

References

Adjuik M, Babker A, Garner P, Olliato P, Taylor W, White N, International Artemisinin Study Group (2004) Artesunate combinations for treatment of malaria: meta-analysis. Lancet 363:9–17

Aleem E, Kaldis P (2006) Mouse models of cell cycle regulators: new paradigms. Results Probl Cell Differ 42:271–328

Allen RG, Tresini M (2000) Oxidative stress and gene regulation. Free Radic Biol Med 28:463–499

Anastasiadis PZ, Moon SY, Thoreson MA, Mariner DJ, Crawford HC, Zheng Y, Reynolds AB (2000) Inhibition of RhoA by p120 catenin. Nat Cell Biol 2:637–644

Arnold RS, Shi J, Murad E, Whalen AM, Sun CQ, Polavarapu R, Partlusarathy S, Petros JA, Lambeth JD (2001) Hydrogen peroxide mediates the cell growth and transformation caused by the mitogenic oxidase Nox1. Proc Natl Acad Sci USA 98:5550–5555

Bamett P, Arnold RS, Mezenencev R, Chung LW, Zayzaloon M, Odero-Marah V (2011) Snail-mediated regulation of reactive oxygen species in ARCaP human prostate cancer cells. Biochem Biophys Res Commun 404:34–39

Barcellos-Hoff MH, Ravani SA (2000) Irradiated mammary gland stroma promotes the expression of tumorigenic potential by unirradiated epithelial cells. Cancer Res 60:1254–1260

Baum B, Settleman J, Quinian MP (2008) Transitions between epithelial and mesenchymal states in development and disease. Semin Cell Dev Biol 19:294–308

Bayon LG, Izquierdo MA, Sirovich I, Van Rooijen N, Beelen RH, Meijer S (1996) Role of Kupffer cells in arresting circulating tumor cells and controlling metastatic growth in the liver. Hepatology 23: 1224–1231

Bedard K, Krause K (2007) The NOX family of ROS-generating NADPH oxidases: physiology and pathophysiology. Physiol Rev 87:245–313

Belkhiri A, Richards C, Whaley M, McQueen SA, Orr FW (1997) Increased expression of activated matrix metalloproteinase-2 by human endothelial cells after sublethal H_2O_2 exposure. Lab Invest 77:533–539

Binker MG, Binker-Cosen AA, Richards D, Oliver B, Cosen-Binker LI (2009) EGF promotes invasion by PANC-1 cells through Rac1/ROS-dependent secretion and activation of MMP-2. Biochem Biophys Res Commun 379:445–450

Boonstra J, Post JA (2004) Molecular events associated with reactive oxygen species and cell cycle progression in mammalian cells. Gene 337:1–13

Boutros R, Dozier C, Ducommun B (2006) The when and wheres of CDC25 phosphatases. Curr Opin Cell Biol 18:185–191

Brenneisen P, Sies H, Scharffetter-Kochanek K (2002) Ultraviolet-B irradiation and matrix metalloproteinases: from induction via signaling to initial events. Ann N Y Acad Sci 973:31–43

Brooks CL, Li M, Gu W (2007) Mechanistic studies of MDM2-mediated ubiquitination in p53 regulation. J Biol Chem 282:22804–22815

Brooks SA, Lomax-Browne HJ, Carter TM, Kinch CE, Hall DM (2010) Molecular interactions in cancer cell metastasis. Acta Histochem 112:3–25

Broom OJ, Massoumi R, Sjolander A (2006) Alpha2beta integrin signaling enhances cyclooxygenase-2 expression on intestinal epithelial cells. J Cell Physiol 209:950–958

Brown CJ, Lain S, Verma CS, Fersht AR, Lane DP (2009) Awakening guardian angels: drugging the p53 pathway. Nat Rev Cancer 9:862–873

Brownwell AM, Kops GJ, Macara IG, Burgering BM (2001) Inhibition of nuclear import by protein kinase B (Akt) regulates the subcellular distribution and activity of the forkhead transcription factor AFX. Mol Cell Biol 21:3534–3546

Burch PM, Heintz NH (2005) Redox regulation of cell-cycle re-entry: cyclin D1 as a primary target for the mitogenic effects of reactive oxygen and nitrogen species. Antioxid Redox Signal 7:741–751

Burdon RH (1995) Superoxide and hydrogen peroxide in relation to mammalian cell proliferation. Free Radic Boil Med 18:775–794

Campisi J (2005) Aging, tumor suppression and cancer: high wire-act! Mech Ageing Dev 126:51–58

Campisi J, d'Adda di Fagagna F (2007) Cellular senescence: when bad things happen to good cells. Nat Rev Mol Cell Biol 8:729–740

Caputo F, Vegliante R, Ghibelli L (2012) Redox modulation of DNA damage response. Biochem Pharmacol 84:1292–1306

Chambers AF, Groom AC, MacDonald IC (2002) Dissemination and growth of cancer cell in metastatic sites. Nat Rev Cancer 2:563–572

Chen L, Park SM, Tumanov AV et al (2010) CD95 promotes tumour growth. Nature 465:492–496

Cheng M et al (1999) The p21(Cip1) and p27(Kip1) CDK 'inhibitors' are essential activators of cyclin D-dependent kinases in murine fibroblasts. EMBO J 18:1571–1583

Chiu WT, Shen SC, Chow JM, LinCW SLT, Chen YC (2010) Contribution of reactive oxygen species to migration/invasion of human glioblastoma cells U87 via ERK –dependent COX-2PGE(2) activation. Neurobiol Dis 37:118–129

Condeelis J, Pollard JW (2006) Macrophages obligate partners for tumor cell migration, invasion and metastasis. Cell 124:263–266

Dansen TB, Smits LM, van Triest MH, de Keizer PL, van Leenen D, Koerkamp MG, Szypowska A, Meppelink A, Brenkman AB, Yodoi J, Holstege FC, Burgering BM (2009) Redox-sensitive cysteines bridge p300/CBP-mediated acetylation and FoxO4 activity. Net Chem Biol 5:664–672

Debatin KM, Krammer PH (2004) Death receptors in chemotherapy and cancer. Oncogene 23:2950–2966

Decker K (1990) Biologically active products of stimulated liver macrophages (Kupffer cells). Eur J Biochem 192:245–261

Dell'Eva R, Pfeffer U, Vene R, Anfosso L, Forlani A, Albini A, Efferth T (2004) Inhibition of angiogenesis in vivo and growth of Kaposi's sarcoma xenograft tumors by the anti-malarial artesunate. Biochem Pharmacol 68:2359–2366

Domingo-Sananes MR, Kapuy O, Hunt Y, Novak B (2011) Switches and latches: a biochemical tug-of-war between the kinases and phosphatases that control mitosis. Philos Trans R Soc B 366:3584–3594

Dong R, Wang Q, He XL, Chu YK, Lu JG, Mao J (2007) Role of nuclear factor kappa B and reactive oxygen species in the tumor factor – alpha-induced epithelial mesenchymal transition of MCF-7 cells. Braz J Med Biol Res 40:1071–1078

Dong-Le Bourhis X, Berthois Y, Millot G, Degeorges A, Sylvi M, Martin PM, Calvo F (1997) Effect of stromal and epithelial cells derived from normal and tumorous breast tissue on the proliferation of human breast cancer cell lines in co-culture. Int J Cancer 71:42–48

Dunphy WG, Kumagai A (1991) The cdc25 protein contains an intrinsic phosphatase activity. Cell 67:189–196

Efferth T (2006) Molecular pharmacology and pharmacogenomics of artemisinin and its derivatives in cancer cells. Curr Drug Targets 7:407–421

Efferth T, Giaisi M, Merling A, Krammer PH, Li-Weber M (2007) Artesunate induces ROS-mediated apoptosis in doxorubicin-resistant T leukemia cells. PLoS One 2:e693

Eskelinen EL (2008) New insights into the mechanisms of macroautophagy in mammalian cells. Inter Rev Cell Mol Biol 266:207–247

Frame MC (2002) Src in cancer: deregulation and consequences for cell behavior. Biochim Biophys Acta 1602:114–130

Freinbichler W et al (2011) Highly reactive oxygen species: detection, formation and possible functions. Cell Mol Life Sci 68:2067–2079

Fruehauf JP, Meyskens FL Jr (2007) Reactive oxygen species a breath of life or death? Clin Cancer Res 13:789–794

Galluzzi L, Maiuri MC, Vitale I, Zischka H, Castedo M, Zitvogel L, Kroemer G (2007) Cell death modalities: classification and pathophysiological implications. Cell Death Differ 14:1237–1243

Giannoni E, Parri M, Charugi P (2012) EMT and oxidative stress: a bidirectional interplay affecting tumor malignancy. Antioxid Redox Signal 16:1248–1263

Giono LE, Manfredi JJ (2006) The p53 tumor suppressor participates in multiple cell cycle checkpoints. J Cell Physiol 209:13–20

Gorman A, McGowan, Cotter TG (1997) Role of peroxide anion during tumour cell apoptosis. FEBS Lett 404:27–33

Grana X, Reddy EP (1995) Cell cycle control in mammalian cells: role of cyclins, cyclin dependent kinases (CDKs), growth suppressor genes and cyclin-dependent kinase inhibitors (CKIs). Oncogene 11:211–219

Halliwell B (2007) Oxidative stress and cancer: have we moved forward? Biogeosciences J401:1–11

Hanahan D, Weinberg RA (2000) The hallmarks of cancer. Cell 100:57–70

Havens CG, Alan H, Yoshioka N, Dowdy SF (2006) Regulation of late G1/S phase transition and APCCdh1 by reactive oxygen species. Mol Cell Biol 26(12):4701–4711

Hochegger H, Takeda S, Hunt T (2008) Cyclin-dependent kinases and cell-cycle transitions: does one fit all? Nat Rev Mol Cell Biol 9:910–916

Hoffman A, Spetner LM, Burke M (2008) Ramifications of a redox switch within a normal cell: its absence in a cancer cell. Free Radic Biol Med 45:265–268

Hoffmann I, Clarke PR, Marcote MK, Karsenti E, Draetta G (1993) Phosphorylation and activation of human Cdc25-C by cdc2—cyclin B and its involvement in the self-amplification of MPF at mitosis. EMBO J 12:53–63

Holliday GM (2005) Inflammation, gene mutation and photoimmunosuppression in response to UVR-induced oxidative damage contributes to photocarcinogenesis. Mutat Res 571:107–120

Huang S, Zhang A, Ding G, Chen R (2009) Aldosterone-induced mesangial cell proliferation is mediated by EGF receptor transactivation. Am J Physiol Renal Physiol 296:F1323–F1333

Hurd TR, DeGennaro M, Lehmann R (2012) Redox regulation of cell migration and adhesion. Trends Cell Biol 22:107–115

Husemann Y, Geigl JB, Schubert F, Musiani P, Meyer M, Burghart E, Forni G, Eils R, Fehn T, Riethmuller G, Klein CA (2008) Systemic spread is an early step in breast cancer. Cancer Cell 13:58–68

Inoguchi T et al (2003) Protein kinase C-dependent increase in reactive oxygen species (ROS) production in vascular tissues of diabetes: role of vascular NAD(P)H oxidase. J Am Soc Nephrol 14:s227–s232

Inumaru J, Nagano O, Takahashi E, Ishimoto T, Nakamura S, Suzuki Y, Niwa S, Umozawa K, Tenihara H, Saya H (2009) Molecular mechanisms regulating dissociation of cell-cell function of epithelial cells by oxidative stress. Genes Cells 14:703–716

Ioannidis I, de Groot H (1993) Cytotoxicity of nitric oxide in Fu5 rat hepatoma cells: evidence for co-operative action with hydrogen peroxide. Biochem J 296:341–345

Irani K, Xia Y, Zweler JI et al (1997) Mitogenic signaling mediated by oxidants in Ras – transformed fibroblasts. Science 275:1649–1651

Jaeschke H (2003) Molecular mechanisms of hepatic ischemia –reperfusion injury and preconditioning. Am J Physiol Gastrointest Liver Physiol 284:G15–G26

Jeffers JR, Parganas E, Lee Y et al (2003) Puma is an essential mediator of p53-dependent and -independent apoptotic pathways. Cancer Cell 4:321–328

Jessup JM, Battle P, Waller H, Edmiston KH, Stolz DB, Watkins SC, Locker J, Skena K (1999) Reactive nitrogen and oxygen radicals formed during hepatic ischemia – reperfusion kill weekly metastatic colorectal cancer cells. Cancer Res 59:1825–1829

Kalen AL, Sarsour EH, Venkataraman S, Goswami PC (2006) Mn-superoxide dismutase overexpression enhances G_2 accumulation and radioresistance in human oral squamous carcinoma cells. Antioxid Redox Signal 8:1273–1281

Kim MH, Yoo HS, Kim MY, Jang HJ, Back MK, Kim HR, Kim KK, Shin BA, Ahn BW, Jung YD (2007) Helicobacter pylori stimulates urokinase plasminogen activator receptor expression and cell invasiveness through reactive oxygen species and NF-kappaB signaling in human gastric carcinoma cells. Int J Mol Med 19:689–697

Krajewska M, Moss SF, Krajewski S, Song KI, Holt PR, Reed JC (1996) Elevated expression of Bcl-X and reduced Bak in primary colorectal adenocarcinomas. Cancer Res 56:2422–2427

Krajewski S, Blomqvist C, Franssila K et al (1995) Reduced expression of proapoptotic gene *bax* is associated with poor response rates to combination chemotherapy and shorter survival in women with metastatic breast adenocarcinoma. Cancer Res 55:4471–4478

Kroemer G, Galluzzi L, Brenner C (2007) Mitochondrial membrane permeabilization in cell death. Physiol Rev 87(1):99–163

Kundu JK, Surh YJ (2008) Inflammation: gearing the journey to cancer. Mutat Res 659:15–30

LaBaer J, Garrett MD, Stevenson LF, Slingerland JM, Sandhu C, Chou HS, Fattaey A, Harlow E (1997) New functional activities for the p21 family of CDK inhibitors. Genes Dev 11:847–862

Lee DJ, Kang SW (2013) Reactive oxygen species and tumor metastasis. Mol Cells 35:93–98

Lee SB, Bae LH, Bae YS, Um HD (2006) Link between mitochondria and NADPH oxidase 1 isozyme for the sustained production of reactive oxygen species and cell death. J Biol Chem 281:36228–36235

Lee KH, Kim S, Kim JR (2009) Reactive oxygen species regulate urokinase plasminogen activator expression and cell invasion via mitogen-activated protein kinase pathways after treatment with hepatocyte growth factor in stomach cancer cells. J Exp Clin Cancer Res 28:73

Léon-Buitimea A, Rodríguez-Fragoso L, Lauer FT, Bowles H, Thompson TA, Burchiel SW (2012) Ethanol-induced oxidative stress is associated with EGF receptor phosphorylation in MCF-10A cells overexpressing CYP2E1. Toxicol Lett 209:161–165

Li X et al (2007) Localization changes of endogenous hydrogen peroxide during cell division cycle of xanthomonas. Mol Cell Biochem 300:207—213

Lin MT, Yen ML, Lin CY, Kuo ML (2003) inhibition of vascular endothelial growth factor-induced angiogenesis by resveratrol through interruption of Src-dependent vascular endothelial cadherin tyrosine phosphorylation. Mol Pharmacol 64:1029–1036

Liu M, Wikonkal NM, Brash DE (1999) Induction of cyclin-dependent kinase inhibitors and G(1) prolongation by the chemopreventive agent N-acetylcysteine. Carcinogenesis 20:1869–1872

Lu Z, Hunter T (2010) Ubiquitylation and proteasomal degradation of the p21Cip1, p27Kip1 and p57Kip2 CDK inhibitors. Cell Cycle 9(2010):2342–2352

Luanpitpong S et al (2012) Mitochondrial superoxide mediates doxorubicin-induced keratinocyte apoptosis through oxidative modification of ERK and Bcl-2 ubiquitination. Biochem Pharmacol 15:1643–1654

Martinez Munoz C, Post JA, Verkleij AJ, Verrips CT, Boonstra J (2001) The effect of hydrogen peroxide on the cyclin D expression in fibroblasts. Cell Mol Life Sci 58:990–996

Medema RH, Kops GJ, Bos JL, Burgering BM (2000) AFX-like forkhead transcription factors mediate cell-cycle regulation by Ras and PKB through p27kip1. Nature 404:782–787

Meier B, Radeke HH, Selle S, Younes M, Sies H, Resch K, Habermeh I (1989) Human fibroblasts release reactive oxygen species in response to interleukin-1 or tumor necrosis factor-alpha. Biochem J 263: 539–545

Menon SG, Sarsour EH, Spitz DR, Higashikubo R, Sturm M, Zhang H, Goswami PC (2003) Redox regulation of the G_1 to S phase transition in the mouse embryo fibroblast cell cycle. Cancer Res 63:2109–2117

Menon SG, Sarsour EH, Kalen AL, Venkataraman S, Hitchler MJ, Domann FE, Oberley LW, Goswami PC (2007) Superoxide signaling mediates N-acetyl-L-cysteine-induced G_1 arrest: regulatory role of cyclin D1 and manganese superoxide dismutase. Cancer Res 67:6392–6399

Meyer T, Hart IR (1998) Mechanisms of tumour metastasis. Eur J Cancer 34:214–221

Meyer M, Schreck R, Baeuerle PA (2005) H_2O_2 and antioxidants have opposite effects on activation of NF-kappa B and AP-1 in intact cells: AP-1 as secondary anti-oxidant responsive factor. EMBO J 12:2005–2015

Michalak EM, Vandenberg CJ, Delbridge AR et al (2010) Apoptosis-promoted tumorigenesis: gamma-irradiation-induced thymic lymphomagenesis requires Puma-driven leukocyte death. Genes Dev 24:1608–1613

Minna JD, Kurie JM, Jacks T (2003) A big step in the study of small cell lung cancer. Cancer Cell 4:163–166

Molenaar JJ et al (2012) Copy number defects of G1-cell cycle genes in neuroblastoma are frequent and correlate with high expression of E2F target genes and a poor prognosis. Genes Chromosome Cancer 51(2012):10–19

Montalto AS, Curro M, Russo T, Visalli G, Impellizzeri P, Antonuccio P, Arena S, Borruto FA, Scaffari G, Ientile R, Romeo C (2013) In vitro CO_2-induced ROS production impairs cell cycle in SH-SY5Y neuroblastoma cells. Pediatr Surg Int 29:51–59

Montero AJ, Jassem J (2011) Cellular redox pathways as a therapeutic target in the treatment of cancer. Drugs 71:1385–1396

Muller I, Niethammer D, Bruchelt G (1998a) Anthracycline-derived chemotherapeutics in apoptosis and free radical cytotoxicity. Int J Mol Med 1:491–494

Muller M, Wilder S, Bannasch D et al (1998b) P53 activates the CD95 (APO-1/Fas) gene in response to DNA damage by anticancer drugs. J Exp Med 188:2033–2045

Mytar B, Siedlar M, Woloszyn M, Ruggiero IR, Pryjmu J, Zembala M (1999) Induction of reactive oxygen intermediates in human monocytes by tumor cells and their role in spontaneous monocytes toxicity. Br J Cancer 79:737–743

Nash GF, Turner LF, Scully MF, Kakkar AK (2002) Platelets and cancer. Lancet Oncol 3:425–430

Nguyen DX, Bos PD, Massague J (2009) Metastasis: from dissemination to organ-specific colonization. Nat Rev Cancer 9:274–284

Nicholson RI, Gee JMW, Harper ME (2001) EGFR and cancer prognosis. Eur J Cancer 37:9–15

Nigam S, Weston CE, Liu CH, Simon EE (1998) The actin cytoskeleton and integrin expression in the recovery of cell adhesion after oxidant stress to a proximal tubule cell line (JTC-12). J Am Soc Nephrol 9:1787–1797

Nishikawa M (2008) Reactive oxygen species in tumor metastasis. Cancer Lett 266:53–59
Oberley LW, Buettner GR (1979) Role of superoxide dismutase in cancer: a review. Cancer Res 39:1141–1149
Oda E, Ohki R, Murasawa H et al (2000) Noxa, a BH3-only member of the Bcl-2 family and candidate mediator of p53-induced apoptosis. Science 288:1053–1058
Odom RY, Dansby MY, Rollins-Hairston AM, Jackson KM, Kirlin WG (2009) Phytochemical induction of cell cycle arrest by glutathione oxidation and reversal by N-acetylcysteine in human colon carcinoma cells. Nutr Cancer 61:332–339
Orrenius S (2004) Mitochondrial regulation of apoptotic cell death. Toxicol Lett 149:19–23
Papaiahgari S, Zhang Q, Kleeberger SR, Cho H, Reddy SP (2006) Hyperoxia stimulates an Nrf2-ARE transcriptional response via ROS-EGFR-PI3K-Akt/ERK MAP kinase signaling in pulmonary epithelial cells. Antioxid Redox Signal 8(2006):43–52
Pardee AB (1974) A restriction point for control of normal animal cell proliferation. Proc Natl Acad Sci U S A 71:1286–1290
Payne SL, Fogelgren B, Hess AR, Seftor EA, Wiley EL, Fong SF, Csiszar K, Hendrix MJ, Kirschmann DA (2005) Lysyl oxidase regulates breast cancer cell migration and adhesion through a hydrogen peroxide – mediated mechanism. Cancer Res 65:11429–11436
Pelicano H, Carney D, Huang P (2004) ROS stress in cancer cells and therapeutic implications. Drug Resist Updat 7:97–110
Pennington JD, Wang TJ, Nguyen P, Sun L, Bisht K, Smart D, Gius D (2005) Redox-sensitive signaling factors as a novel molecular targets for cancer therapy. Drug Resist Update 8:322–330
Price R, Vugt MV, Nosten F, Luxemburger C, Brockman A, Phaipun L, Chongsuphajansiddhi T, White N (1998) Artesunate versus artemether for the treatment of recrudescent multidrug-resistant Plasmodium falciparum malaria. Am J Trop Med 59:883–888
Psaila B, Lyden D (2009) The metastatic niche: adapting the foreign soil. Nat Rev Cancer 9:285–293
Rajalingam K, Schreck R, Rapp UR, Albert S (2007) Ras oncogenes and their downstream targets. Biochim Biophys Acta 1773:1177–1195
Rhee SG, Bae YS, Lee SR, Kwon J (2000) Hydrogen peroxide: a key messenger that modulates protein phosphorylation through cysteine oxidation. Sci STKE 2000:PE1
Rhee SG, Chang TS, Bae YS, Lee SR, Kang SW (2003) Cellular regulation by hydrogen peroxide. J Am Soc Nephrol 14:S211–SW215
Ribeiro IR, Ollario P (1998) Safety of artemisinin and its derivatives, A review of published and unpublished clinical trials. Med Trop (Mars) 58:50–53
Riley T, Sontag E, Chen P, Levine A (2008) Transcriptional control of human p53-regulated genes. Nat Rev Mol Cell Biol 9:402–412
Romeo Y, Zhang X, Roux PP (2012) Regulation and function of the RSK family of protein kinases. Biochem J 441(2012):553–569
Sarsour EH, Kumar MG, Chaudhuri L, Kalen AL, Goswami PC (2009) Redox control of the cell cycle in health and disease. Antioxid Redox Signal 11:2985–3011
Satyanarayana A, Kaldis P (2009) Mammalian cell-cycle regulation: several CDKs, numerous cyclins and diverse compensatory mechanisms. Oncogene 28(2009):2925–2939
Savitsky PA, Finkel T (2002) Redox regulation of Cdc25C. J Biol Chem 277:20535–20540
Sawa T, Wu J, Akaike T, Maeda H (2000) Tumor-targeting chemotherapy by a xanthine oxidase – polymer conjugate that generate oxygen-free radicals in tumor tissue. Cancer Res 60:666–671
Schmidt M, Fernandez de Mattos S, van der Horst A, Klompmaker R, Kops GJ, Lam EW, Burgering BM, Medema RH (2002) Cell cycle inhibition by FoxO forkhead transcription factors involves downregulation of cyclin D. Mol Cell Biol 22:7842–7852
Schulze-Bergkamen H, Krar PH (2004) Apoptosis in cancer-implications for therapy. Semin Oncol 31:90–119
Sebastian B, Kakizuka A, Hunter T (1993) Cdc25M2 activation of cyclin-dependent kinases by dephosphorylation of threonine-14 and tyrosine-15. Proc Natl Acad Sci USA 90:3521–3524
Selenius M, Hedman M, Brodin D, Gandin V, Rigobello MP, Flygare J, Marzano C, Bindoli A, Brodin O, Bjornstedt M, Fernades AP (2012) Effects of redox modulation by inhibition of thioredoxin reductase on radiosensitivity and gene expression. J Cell Mol Med 16:1593–1605
Serrano M, Lin AW, McCurrach ME, Beach D, Lowe SW (1997) Oncogenic ras provokes premature cell senescence associated with accumulation of p53 and p16INK4a. Cell 88:593–602
Shackelford RE, Kaufmann WK, Paules RS (2000) Oxidative stress and cell cycle checkpoint function. Free Rad Biol Med 28:387–404
Sjöström J, Blomqvist C, Von Boguslawski K et al (2002) The predictive value of bcl-2, bax, bcl-xL, bag-1, fas, and fasL for chemotherapy response in advanced breast cancer. Clin Cancer Res 8:811–816
Starano S, Dell'Orso S, Di Agostino S, Fontemaggi G, Sacchi A, Blandino G (2007) Mutant P53: an oncogenic transcription factor. Oncogene 26:2212–2219
Starostina NG, Kipreos ET (2011) Multiple degradation pathways regulate versatile CIP/KIP CDK inhibitors. Trends Cell Biol 22:33–41
Suzuki K, Matsubara H (2011) Recent advances in p53 research and cancer treatment. J Biomed Biotechnol 1-7:e 978312
Suzuki YJ, Forman HJ, Sevanian A (1997) Oxidants as simulators of signal transduction. Free Rad Biol Med 22:269–285
Sykes JA, McCormack FX Jr, O'Brien TJ (1978) A preliminary study of the superoxide dismutase content of some human tumors. Cancer Res 38:2759–2762
Szatrowski TP, Nathan CF (1991) Production of large amounts of hydrogen peroxide by human tumor cells. Cancer Res 51:794–798

Taddei ML, Parri M, Mello T et al (2007) Integrin-mediated cell adhesion and spreading engage different sources of reactive oxygen species. Antioxid Redox Signal 9(4):469–481

Taylor WR, Stark GR (2001) Regulation of the G2/M transition by p53. Oncogene 20:1803–1815

Thiery JP (2002) Epithelial-mesenchymal transitions in tumor progression. Nat Rev Cancer 2:442–454

Tobar N, Villar V, Santibanez JF (2010) ROS-NFkappaB mediates TGF-beta1-induced expression of urokinase-type plasminogen activator, matrix metalloproteinase-9 and cell invasion. Mol Cell Biochem 340:195–202

Toledo F, Krummel KA, Lee CJ et al (2006) A mouse p53 mutant lacking the proline-rich domain rescues Mdm4 deficiency and provides insight into the Mdm2-Mdm4-p53 regulatory network. Cancer Cell 9:273–285

Trauth BC, Klas C, Peters AM et al (1989) Monoclonal antibody-mediated tumor regression by induction of apoptosis. Science 245:301–305

Turowski P, Franckhauser C, Morris MC, Vaglio P, Fernandez A, Lamb NJC (2003) Functional Cdc25C dual-specificity phosphatase is required for S-phase entry in human cells. Mol Biol Cell 14(2003):2984–2998

Ushio-Fukai M (2009) Compartmentalization of redox signaling through NaDPH oxidase-derived rOS. Antioxid Redox Signal 11(6):1289–1299

Valerie K, Yacoub A, Hagan MP, Curiel DT, Fisher PB, Grant S, Dent P (2007) Radiation-induced cell signaling inside-out and outside-in. Mol Cancer Ther 6:789–801

Van Remmen H, Ikeno Y, Hamilton M, Pahlavani M, Wolf N, Thorpe SR, Alderson NL, Baynes JW, Epstein CJ, Huang TT, Nelson J, Strong R, Richardson A (2003) Life-long reduction in MnSOD activity results in increased DNA damage and higher incidence of cancer but does not accelerate aging. Physiol Genomics 16:29–37

Verbon EH, Post JA, Boonstra J (2012) The influence of reactive oxygen species on cell cycle progression in mammalian cells. Gene 511:1–6

Vijayakumar S et al (2011) High-frequency canonical Wnt activation in multiple sarcoma subtypes drives proliferation through a TCF/β-catenin target gene, CDC25A. Cancer Cell 19:601–612

Vousden KH, Prives C (2009) Blinded by the light: the growing complexity of p53. Cell 137:413–431

Wang C, Li Z, Lu Y, Du R, Katiyar S, Yang J, Fu M, Leader JE, Quong A, Novikoff PM, Pestell RG (2006) Cyclin D1 repression of nuclear respiratory factor 1 integrates nuclear DNA synthesis and mitochondrial function. Proc Natl Acad Sci USA 103:11567–11572

Wang YT, Tzeng DW, Wang CY, Hong JY, Yang JL (2013) APE1/Ref1 prevents oxidative inactivation of ERK for G1-to-S progression following lead acetate exposure. Toxicology 305:120–129

Weydert C, Roling B, Liu J, Hinkhouse MM, Ritchie JM, Oberley LW, Cullen JJ (2003) Suppression of the malignant phenotype in human pancreatic cancer cells by the overexpression of manganese superoxide dismutase. Mol Cancer Ther 2:361–369

Wind S et al (2010) Comparative pharmacology of chemically distinct NADPH oxidase inhibitors. Br J Pharmacol 161(2010):885–898

Wu L et al (1994) Molecular cloning of the human CAK1 gene encoding a cyclin-dependent kinase-activating kinase. Oncogene 9(1994):2089–2096

Yamaoka-Tojo M, Tojo T, Kim HW, Hilenski L, Patrushev NA, Zhang L, Fukai T, Ushio-Fukai M (2006) IQGAP1 mediates VE-cadherin-based cell-cell contacts and VEGF signaling at adherence junctions linked to angiogenesis. Arterioscler Thromb Vasc Biol 26:1991–1997

Yamaura M et al (2009) NADPH oxidase 4 contributes to transformation phenotype of melanoma cells by regulating G2-M cell cycle progression. Cancer Res 69:2647–2654

Yarden Y (2001) The EGFR family and its ligands in human cancer. Signalling mechanisms and therapeutic opportunities. Eur J Cancer 37(Suppl 4):S3–S8

Yip KW, Reed JC (2008) Bcl-2 family proteins and cancer. Oncogene 27:6398–6406

Yonish-Rouach E, Resnitzky D, Lotem J, Sachs L, Kimchi A, Oren M (1991) Wild-type p53 induces apoptosis of myeloid leukaemic cells that is inhibited by interleukin-6. Nature 352:345–347

Yu J, Zhang L (2003) No PUMA, no death: implications for p53-dependent apoptosis. Cancer Cell 4:248–249

Yu J, Zhang L, Hwang PM, Kinzler KW, Vogelstein B (2001) PUMA induces the rapid apoptosis of colorectal cancer cells. Mol Cell 7:673–682

Yunbo L, Trush MA (1998) Dephenyleneiodonium, an NAD(P)H oxidase inhibitor, also potently inhibits mitochondrial reactive oxygen species production. Biochem Biophys Res Commun 253:296–299

Zhang Y, Zhao W, Zhang HJ, Domann FE, Oberley LW (2002) Overexpression of copper zinc superoxide dismutase suppresses human glioma cell growth. Cancer Res 62(2002):1205–1212

5 Cell Signaling and Gene Regulation by Oxidative Stress

Cells are characterized to perceive and correctly/quickly respond to their environment, which is the basis of development, tissue repair, and immunity as well as normal tissue homeostasis. Any plausible error in processing the cellular information can culminate in a disease. Thus, understanding the cell communication process called cell signaling may help in efficient treatment of the diseases. In this chapter general cell signaling elements, redox signaling, and oxidative stress-mediated modulation of the transcription factors will be presented.

Cell Signaling

Cell signaling, the most important aspect of the modern biochemistry and cell biology, is a part of the complex system of communication that governs the basic cellular activities and coordinates cell actions. Cells communicate with each other and respond to the extracellular stimuli through biological mechanisms called cell signaling or signal transduction. Signal transduction is a process enabling information to be transmitted from the outside of a cell to various functional elements inside the cell. It is triggered by the extracellular signals such as hormones, growth factors, cytokines, and neurotransmitters. Signals sent to the transcription machinery responsible for the expression of certain genes are normally transmitted to the cell nucleus by a class of proteins called transcription factors. By binding to the specific DNA sequences, these factors regulate the transcriptional activity. These signal transduction processes can induce various biological activities, such as muscle contraction, gene expression, cell growth, and nerve transmission.

All cells in a living body are exposed to a multiple form of the physiological environment in various tissues or organs. Cells sense and respond to their environment through a complex range of signaling pathways that are crucial for their survival. Understanding the biochemical basis for the transduction of extracellular signals into an intracellular event has long been the subject of enormous interest. Basically, a signal, which may be in different form, follows several events quickly in sequence: signal received by receptor proteins, transmission of signal by receptor into the cell, activation of the cell signaling cascade (signal transduction pathways), and finally response by the cell in alteration of activities in the cytoplasm and gene expression in the nucleus.

Signaling is classified based on the distance between the signaling and the target cells. If these two cells are touching, the signaling may simply be through pore in the membrane, such as gap junctions (juxtacrine signaling), or may be due to a membrane-bound ligand being identified by a receptor in the membrane of a neighboring cell (paracrine signaling). If cells are further apart, they may communicate via the release of

M. Bansal and N. Kaushal, *Oxidative Stress Mechanisms and their Modulation*,
DOI 10.1007/978-81-322-2032-9_5, © Springer India 2014

molecules which are then detected by the target cell (endocrine signaling) or via transmission of an electrical signal as in nerve cells.

The mechanisms of signal transduction from one cell to the next and subsequently into the interiors of the target cell often need to involve formation of the chains of signaling molecules, each passing on message to the next molecule in line (Dumont et al. 2001). An extracellular signal molecule or first messenger perceived by a cell often leads to the production of a small and transient signaling molecule inside the cell, often referred to as second messenger. Such intracellular messengers activate or alter the activity of the next component of transduction pathway, for example, a kinase (Ray and Sturgill 1987). Binding of ligand to the receptor switches on many molecules or enzymes on the plasma membrane. Such transduction may involve G proteins acting on enzymes such as adenyl cyclase, further producing many cAMP molecules causing amplification of the signal, activating many kinases which will phosphorylate a great many proteins, and causing yet further amplification of the signal.

Out of the most significant ways in which cells communicate with each other is by the release and detection of extracellular signal molecules such as hormones, cytokines, and growth factors (Arai et al. 1990). Such molecule can often be released at considerable distance from their point of action and have unspecific transport to the site of action. For instance, a hormone might be released, carried to the bloodstream, and supplied everywhere in the body. The specificity of the effect is determined by the presence of specific receptor molecule in detecting cells.

Detection of Extracellular Signal by the Receptors

Cells are usually surrounded with a large number of different signaling molecules even at very low concentration, and they respond through their specific selection system. Receptors on the cell surface or inside cell in the cytoplasm or nucleus detect such signals. A receptor provides specificity in detecting the signaling molecule(s) via (a) binding affinities and transmitting message of signaling molecules to the cell and (b) modulation of further components in the signaling cascade (Fig. 5.1). A wide number of receptors have evolved to fill the vital role of detection of the extracellular signals. Majority of the receptors fall into the following basic classes: G protein linked, ion channel linked, intrinsic enzyme activity, and intracellular type.

G-Protein-Linked Receptors

Largest family of receptors is the G-protein-coupled receptors, which depend on the GTP for their function. This receptor when activated by binding to its ligand results in the activation of G-protein which conveys message to the next component in signal pathway (Strader et al. 1994). Activity of these receptors is also regulated by their phosphorylation by the cAMP-dependent protein kinase (PKA) or by a class of kinase known as G-protein-coupled receptor kinases (GRKs). GRKs are known to phosphorylate these receptors on multiple sites, using threonine and serine residues as targets (Debburman et al. 1996). Phosphorylation deactivates the receptor. Many neurotransmitters, hormones, and small molecules bind to and activate the specific G-protein-coupled receptors (Lefkowitz 1993).

Ion Channel Linked Receptors

Binding of the ligand to receptor changes the ion permeability of plasma membrane as the receptor undergoes a conformation change which opens or closes an ion channel, allowing the efflux or influx of specific ions (Putney and McKay 1999). However, this is the only transient event with the receptor returning to its state very rapidly. These receptors contain several transmembrane polypeptide chains, often involved in the detection of neurotransmitter molecules, and are referred to as transmitter-gated ion channels.

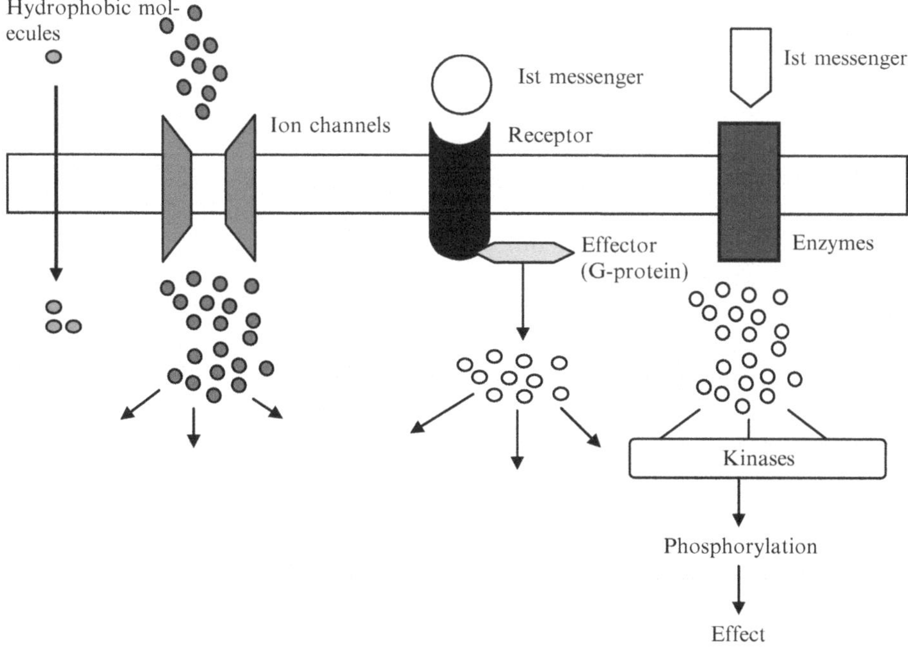

Fig. 5.1 Major pathways in general for cell signaling

Receptor with Intrinsic Enzymatic Activity

It is a heterogeneous class of receptors which is characterized by the presence of catalytic activity integral to the receptor polypeptide, and this activity is controlled by the ligand-binding event. The ligand-binding domain is found on the extracellular side of the membrane and the catalytic domain on the cytoplasmic side. The catalytic activity may be a guanylyl cyclase, a phosphatase, or a kinase. The receptors may contain a serine/threonine kinase (receptor serine/threonine kinases) or tyrosine kinase activity (receptor tyrosine kinases). Ligand binding to these receptors leads to the activation of the respective kinase activity of the receptor which causes phosphorylation of the receptor itself on the serine/threonine or the tyrosine residues. Several receptors do not themselves contain a tyrosine kinase domain, but on activation by the ligand binding, they stimulate the tyrosine kinases present normally in the cytoplasm (Fantl et al. 1993). Such kinases will recognize and bind to the changed conformation of the activated receptor on ligand binding. This class of receptors is commonly referred to as the cytokine receptor superfamily, as they are involved commonly in the recognition of the cytokines and growth factors.

Intracellular Receptors

These include receptors for the steroid hormones, thyroid hormones, retinoids, fatty acids, and prostaglandins. Some of these may be on the surface of plasma membrane or may be inside. Intracellular ligand binding for the extracellular signaling molecules allows them to get across to the intracellular responses. Receptor for these extracellular signals can be found in the cytoplasm or in the nucleus of the cell.

There are numerous nuclear receptors in the cells. Activated nuclear receptors attach to the DNA at receptor-specific hormone-responsive element (HRE) sequences, located in the promoter region of the genes activated by the hormone-receptor complex. Signal transduction via these receptors involves little proteins. Nucleic receptors have DNA-binding domains containing zinc fingers and a ligand-binding domain; the zinc fingers stabilize DNA binding by holding its phosphate backbone. DNA sequences that match the receptor are usually hexameric repeats of any kind; the sequences are similar but their orientation and distance differentiate them. The ligand-binding domain is additionally responsible for the dimerization of nuclei receptors prior to the binding and providing structures for transactivation used for communication with the translational apparatus.

Further, the activated receptor interacts with and/or activates several different types of proteins/enzymes which (due to conformation change) lead to a wide range of intracellular signaling cascades. The most common way of modifying the protein structure is via phosphorylation by kinases or via dephosphorylation by phosphatases at the primary amino acid sequence of a polypeptide (Barford 2001). Two main groups of kinases for phosphate addition are serine/threonine kinases that add phosphate to the serine and/or threonine and tyrosine kinases that add phosphate to the tyrosine. The serine/threonine kinase involves a large group of phosphorylating enzymes, including cAMP-dependent protein kinases (PKA), cGMP-dependent protein kinases (PKG), protein kinase C (PKC), Ca^{2+}/calmodulin-dependent protein kinase, phosphorylase kinase, and others. The mitogen-activated protein kinase (MAPK) cascades consist of four major MAPKs: the extracellular signal-related kinases (ERK1/2), the c-jun N-terminal kinases (JNK), and the p38 kinase (p38). These kinases play pivotal roles in the cellular responses to a wide variety of signals elicited by the growth factors, hormones, and cytokines, in addition to the genotoxic and oxidative sensors. MAPK pathways are composed of a three-rung kinase tier: MAPK kinase kinases (MAPKKK) phosphorylate and activate MAPK kinases (MAPKK), which in turn phosphorylate and activate MAPKs (Marshall 1994). MAP kinases transducer signals that are involved in a multitude of cellular pathways and functions in response to a variety of ligands and cell stimuli. The reversal of kinase action is performed by phosphatases, and again these are usually either serine/threonine-specific or tyrosine-specific, the latter being either cytosolic or receptor-linked. Several isoforms of each are known to exist.

Redox Cell Signaling: Cell to Cell Cross Talk

Due to a distant process of transcription and translation, eukaryotes unlike prokaryotes possess complex mechanisms for controlling the inducible gene expression and regulation. Free radicals (reactive oxygen species, ROS) though are frequently and predominantly implicated in causing the cell damage, but at the same time, they also play a major physiological role in several aspects of the intracellular signaling and regulation (Droge 2002). Cells are capable of generating ROS exogenously and constitutively, which are utilized in the induction and maintenance of the signal transduction pathways involved in the cell growth and differentiation. Most cell types have been shown to generate low concentrations of ROS when they are stimulated by cytokines, growth factors, and hormones (Thannickal and Fanburg 2000). ROS can thus play a very important physiological role as secondary messengers (Storz 2005).

Being initiators, transmitters, or modifiers of the cellular response, ROS occupy a significant place in the complex system of transmitting information along cell to target sensor and are thus regarded as essential participants in the cell signaling and gene regulation. They produce transient changes in the cellular redox state. ROS formed in the mitochondria and in the cytosol are important determinants of the redox state of the protein cysteine residues and thus constitute a regulatory mechanism in determination of the protein conformation and function. ROS-dependent redox cycling of the cysteine thiols is also critical for the establishment of the protein–protein and protein–DNA interactions that determines many aspects of the signal transduction pathway. Some ROS are capable of penetrating the plasma membrane and can thus directly modulate the activity of catalytic domain of the

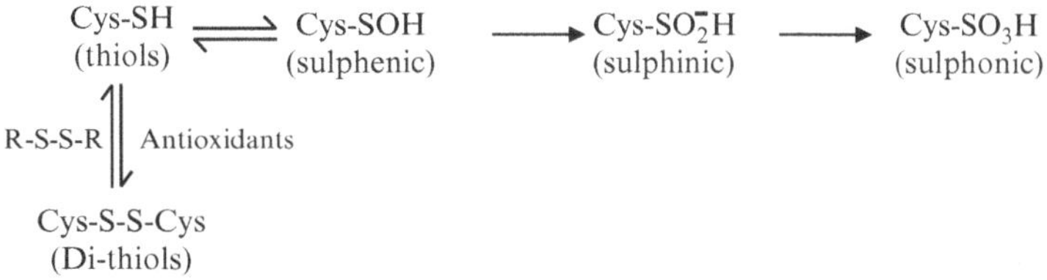

Fig. 5.2 Oxidative modification of a cysteine thiol moiety

Cys-SH (thiols) ⇌ Cys-SOH (sulphenic) → Cys-SO_2^-H (sulphinic) → Cys-SO_3H (sulphonic)

Cys-SH (thiols) ⇅ [R-S-S-R / Antioxidants] Cys-S-S-Cys (Di-thiols)

transmembrane receptors or the cytoplasmic signal-transducing enzymes, thus leading to activation of the key signal molecules such as transcription factors.

ROS concentration, which is controlled by the various enzymatic mechanisms in cells, influences the elements in signal transduction pathways involved in cell proliferation, differentiation, and apoptosis. ROS shows its influence as per their concentration. At the physiological low levels, ROS function as redox messengers (second messengers) in the intracellular signaling and stimulate the redox-sensitive signaling pathways to modify the cellular content of the cytoprotective regulatory proteins, whereas at the elevated levels, ROS may lead to more extensive and irreparable cell damage, resulting ultimately in cell death through necrosis or apoptosis. These pathological effects are usually mediated by the ion channel opening, lipid peroxidation, protein modifications, and DNA oxidation.

A great number of signaling pathways, especially those that can be modified by the free radicals and their oxidized products, have been delineated in recent years in eukaryotes. Basically ROS regulation of the signaling pathways is linked to the mechanisms of their interaction with the cellular components such as proteins through interaction with redox-reactive cysteine residues (also named redox switches). Signaling enzymes and proteins containing cysteine residues have been proposed as the potential targets for ROS (Fig. 5.2). Sulfur in the cysteine, present in such proteins, can be reversibly or irreversibly oxidized to a disulfide bond (-SSR), sulfenic acid (-SOH), sulfinic acid (-SO_2H), or sulfonic acid (-SO_3H) (Poli et al. 2004), and these initially were thought to be the markers of oxidative damage. Formation of the latter two compounds is irreversible and therefore is not involved in the signaling reaction, while disulfide bonds and protein sulfenic acids can easily be reduced by the reducing systems such as thioredoxin and peroxiredoxin and are often considered to be mediators of the redox signaling (Forman et al. 2002).

Also parallel to the ROS, RLS (reactive lipid species) play a significant role in redox signaling. Some lipid peroxidation product species formed through enzymatic and nonenzymatic pathways (Davies et al. 2004; Niki et al. 2005) are electrophilic in nature and therefore have affinity for the specific nucleophilic targets in proteins (called electrophilic-responsive proteome) and hence generate signaling response as is true with ROS. However, interesting differences have been explored in RLS signaling. Both reversible (Goetzi et al. 1995; Breyer et al. 2001) and irreversible interactions (Perez-Sala et al. 2002; Stamatakis and Perez-Sala 2006) of different electrophiles and nucleophiles, e.g., cysteine, histidine, and lysine, occur. Reversible signaling requires higher concentration of both and gives transient signal, whereas the irreversible signaling occurs with covalent interaction of electrophiles and nucleophiles (proteome) with time lag for saturation and sustains longer (also called covalent advantage) (Grunwald and Richards 2006; Schopler et al. 2011). Further, biological response depends on the hard/soft characteristics of electrophiles and nucleophiles (proteomes) (Carlson 1990). Soft electrophiles (RLS) react readily with the soft nucleophiles such as GSH and protein cysteinyl thiols which are biologically significant. Hard electrophiles include mutagenic compounds and often react with hard nucleophilic centers in purine/pyrimidine bases in DNA. Thus, site-specific modification of cysteine residue contributes to cell signaling through cysteine-rich proteins such as Keap1 protein (Levonen et al. 2004; Hong et al. 2005) as is true with ROS also. Further research developments and perspectives in RLS signaling have been well reviewed by Higdon et al. (2012).

Modulation of Cell Signaling by ROS

Earlier it has been discussed that normally various extracellular stimuli transduce signals through a variety of cellular signaling pathways. Protein phosphorylation and dephosphorylation plays a critical role in regulating many cellular responses by governing the multiple signal transduction pathways. Cellular target proteins are phosphorylated at specific cellular transduction sites (usually at serine/threonine or tyrosine residues), and this process is regulated by the protein kinases and phosphatases. Oxidants activate various signaling events which further lead to the modification of gene expression at the target level. This topic of research has been reviewed time to time by the various scientists (Kamata and Hirata 1999; Thannickal and Fanburg 2000; Pavlovic et al. 2002; Valko et al. 2007; Klaunig et al. 2010; Leonarduzzi et al. 2010; Ray et al. 2012; and more). A comprehensive detail of the modulation of major signaling elements with ROS is given below:

Protein Tyrosine Kinases (PTKs) and Protein Tyrosine Phosphatases (PTPs)

PTKs are a superfamily of enzymes consisting of both transmembrane-spanning receptors with intrinsic tyrosine kinase activity in their cytoplasmic domains and a wide range of subfamilies of the cytoplasmic tyrosine kinases, such as Src, Abl, or Janus kinase (JAK). Tyrosine kinase as a receptor is implicated in promoting the effects of growth factors, cytokines, and hormones (Arbabi and Maier 2002). These reactions lead to the activation of downstream signaling pathways such as the protein kinases of the MAP kinase cascade and PI3-kinase (Sun and Tonks 1994; Parson and Parson 1997). Redox-sensitive regulation occurs in the protein tyrosine kinase activity. Changes in the redox status of the cell govern this regulation which is determined by the content of thiol compounds in the cell, primarily by the content of the glutathione as the most widespread thiol compound in the cell.

Since H_2O_2 has been seen to induce the phosphorylation of tyrosine residues of numerous cell proteins, it is considered in the modulation of the activity of tyrosine kinase pathway in the cell signaling. Exposure of the cells to H_2O_2 stimulates numerous effects that are due to the activity of various extracellular ligands. For example, H_2O_2 simulates insulin effects through some of the signaling pathways, by the activation of tyrosine kinases responsible for the phosphorylation of receptor subunits and other intracellular proteins whose phosphorylation intensifies when affected by insulin, and by PIK3 and p38 MAPK activation (Koshio et al. 1988). Some of these pathways differ from those being activated only by the insulin. On similar lines, the exposure of lymphocytes to H_2O_2 also leads to the activation of a specific $p56^{lck}$ tyrosine kinase to initiate its auto phosphorylation at Tyr-394 position. Lck tyrosine kinases play a significant role in the activation of T lymphocytes.

On the other hand, protein tyrosine phosphatases (PTPs) are transiently inhibited by the oxidants, and their inhibition may directly or indirectly induce PTKs (Lee et al. 2007). Because all PTPs have reactive cysteine residue in their catalytic sites and these residues form a thiol–phosphate intermediate in catalysis, oxidation of these residues leads to their inactivation. Because the level of tyrosine phosphorylation of cellular proteins is determined by the balance of PTK and PTP activity, oxidant-induced inactivation of the PTPs results in an apparent enhancement of the tyrosine phosphorylation. H_2O_2 also have been reported to inactivate these enzymes by oxidizing the reactive cysteine residue in the active center and inhibiting their catalytic activity (Lu et al. 2007). Because the cysteine sulfenic acid is highly reactive, it can react with a thiol to form a catalytically inactive PTP disulfide (Peshenko and Shichi 2001). Also inactivation may occur by conversion of the cysteine residue into a mixed disulfide after reaction with the oxidized glutathione (GSSG).

Similarly, a reversible oxidation of PTPs during RTK stimulation by epidermal growth factor (EGF), platelet-derived growth factor (PDGF), or insulin has also been reported (Meng et al. 2002). Inhibition of PTPs also regulates the activation of MAPK pathways (Lee and Esselman 2002). H_2O_2 has also been found to inhibit dephosphorylation of the EGFR by inhibiting the tyrosine phosphatase, and consequently activating EGFR leads to the intracellular production of H_2O_2 which in turn induces inactivation of PTPs (Bae et al. 1997).

Serine/Threonine Kinases

ROS also induces the phosphorylation of various serine/threonine kinases.

Protein kinase C (PKC) is a family of structurally and functionally related proteins derived from the multiple genes and alternative splicing of the single RNA transcript, which regulate a variety of cell functions after the phosphorylation and translocation to the plasma membrane and are subject to cellular redox regulation. PKC is activated by the diacylglycerol or phorbol 12-myristate 13-acetate (PMA) and regulated in a variety of ways, such as by the phosphorylation, lipids, and Ca^{2+} (Newton 1997). Treatment of cells with H_2O_2 (Brawn et al. 1995) and redox-cycling quinines (Kass et al. 1989) lead to the stimulation of PKC activity. H_2O_2 was found to induce the tyrosine phosphorylation of PKC with enhanced activity (Konishi et al. 1997). PKCs contain cysteine-rich regions which can be modified by various oxidants (Gopalakrishna and Jaken 2000). Both the regulatory domain and catalytic site of the PKC in various cells are susceptible to the oxidative modulation induced by ROS, such as H_2O_2, with concomitant stimulation of their activity (Shukla et al. 2003). One possible mechanism of PKC activation might be tyrosine phosphorylation and conversion to the Ca^{2+}/phospholipid-independent form (Niwa et al. 2002). However, H_2O_2 might induce PKCδ activation independently, via PTP inhibition (Yamamoto et al. 2000). By regulating MAPKs, PKCδ may regulate cell apoptosis and survival in diverse cellular systems. It is shown that the activation of PKC is prerequisite for the NADPH oxidase-dependent ROS generation in numerous biological processes, and therefore PKC and ROS associate in amplifying the signals responsible for various pathological processes (Inoguchi et al. 2003). Experimentation in knockout diabetic mice shows that NADPH oxidase is activated via a PKC-dependent pathway and PKC-β is a major inducer of oxidative stress in diabetes (Ohshiro et al. 2006).

MAPKs

Mitogen-activated protein kinases (MAPKs) are strongly activated by ROS. MAPK family includes: the extracellular signal-related kinases (ERK1/2), c-jun N-terminal kinases (JNKs), p38 kinase (p38), and big MAP kinase 1 (BMK-1) which play pivotal roles in the cellular responses to a wide variety of signals elicited by the growth factors, hormones, and cytokines. MAPK pathways are considered one of the most important intracellular signaling systems to induce the optimal stress response. They can be regulated by oxidants, such as ROS, and are involved in the various cellular functions, including gene expression, proliferation, migration, differentiation, and apoptosis.

H_2O_2 stimulates members of the MAP kinase family such as the JNK, p38, and BMK-1 (Guyton et al. 1996). The expression of a protein phosphatase CL100, which is capable of dephosphorylating MAP kinase, is potently induced by the oxidative stress (Keyse and Emslie 1992), and therefore, the cell possesses a negative feedback loop for the ROS-induced activation of MAP kinase. MAPK activity is regulated by MAPK phosphatases through the cysteine residues (Zhang et al. 2002), and these phosphatases specifically dephosphorylate threonine or tyrosine residues in the MAPKs (Krautwald et al. 1995). ROS have been found to sustain JNK signaling for TNFα-induced cell death (Kamata et al. 2005) and cytotoxicity (Chen et al. 2007) via the oxidation and inactivation of JNK phosphatase. ROS have also been shown to induce apoptosis through the activation of p38 MAPK (Lee et al. 2008; Ranawat and Bansal 2009). Experimental studies on the upregulation of MAPKs by H_2O_2 treatment have shown that the activation of each signaling pathway is type- and stimulus-specific. For example, it has been reported that endogenous H_2O_2 production by the respiratory burst induces ERK but not p38 kinase activity (Iles and Forman 2002), however exogenous H_2O_2 activates p38 kinase, but not ERK, in rat alveolar macrophages. The ERK pathway has most commonly been associated with the regulation of cell proliferation. The balance between ERK and JNK activation is a key factor for the cell survival, since both a decrease in ERK and an increase in JNK are required for the induction of apoptosis.

ROS-mediated MAPK activation is also reported to involve an upstream event at the level of growth factor receptors, Src kinases (Aikawa et al. 1997), PKC (Lee and Esselman 2002), and/or $p21^{ras}$ (Muller et al. 1997). c-Src is directly activated by the H_2O_2 being responsible for JNK but not for ERK1/2 or p38 activation (Yoshizumi et al. 2000). On the contrary, ERK1/2 activation by H_2O_2 was related to Fyn through JAK2 and Ras activation. Moreover, ROS play a central role in sustained PKC/ERK activation, leading to cell migration of the hepatoma cells (Wu et al. 2006). ROS activate protein kinases of the MAP kinase cascade (Fialkow et al. 1994; Chen et al. 1995) also through a Ras-dependent mechanism (Guyton et al. 1996). Ras, a small G protein, can mediate activation of the NADH/NADPH oxidase induced by the oxidants and lead to the generation of intracellular ROS (Irani et al. 1997). Also Ras itself may be activated via the oxidative modification of cysteine residues by oxidative stress (Kuster et al. 2006).

ASK1

In addition, among the members of the MAPK cascades, apoptosis signal-regulated kinase 1 (ASK1) is an upstream MAPKKK (MAPK cascades) that regulates the JNK and p38 pathways leading to apoptosis. ASK1 is activated under various stress conditions such as the oxidative stress (Tobiume et al. 2001). ASK1 selectively activates the JNK and p38 MAPK pathways that regulate ROS-mediated cell death in several human diseases (Nagai et al. 2007). Deletion of ASK1 eliminates JNK activation in response to the H_2O_2 and renders cells resistant to apoptosis induced by the oxidants (Tobiume et al. 2001). A positive feedback mechanism may exist in the ASK1/p38/TNFα pathway, which enhances ROS-mediated apoptosis.

In related experiments, ASK1-associated proteins, specially the redox protein thioredoxin (Trx), were found interacting at *N*-terminal in the reduced form and inhibit its kinase activity (Saitoh et al. 1998). Upon treatment of cells with ROS such as H_2O_2, bound thioredoxin gets oxidized through a disulfide bridge between Cys-32 and Cys-35 in the active center and dissociates from ASK1. This allows the N-terminal hemophilic interaction and complete oligomerization of the ASK1, which is enhanced by the binding of TNF-α receptor-associated factors (TRAF) (Fujino et al. 2007). ASK1 oligomer subsequently undergoes autophosphorylation of a conserved threonine residue (Human: Thr-838, Mouse: Thr-845) located in the activation loop of ASK1, which is inactivated by protein phosphatase 5 (Morita et al. 2001; Tobiume et al. 2002). In addition to the homo-oligomerization of ASK1, it hetero-oligomerizes with ASK2, another ASK family serine/threonine MAPKKK. ASK2 binds to the C-terminal domain of ASK1, and this interaction stabilizes ASK2, resulting in the autophosphorylation of ASK2 at the conserved threonine (Human: Thr 806, Mouse Thr 807) in the activation loop. ASK1 is then phosphorylated at Thr-838 by ASK2, resulting in the activation of the hetero-oligomer. ASK1-deficient mouse embryonic fibroblast was shown to be less susceptible to TNF-α or H_2O_2-induced cytotoxicity along with the decreased JNK and p38 MAPK activation, suggesting that ASK1 plays pivotal role in promoting cell death under the oxidative stress. However, ROS-activated ASK1 mediates p38 signaling leading to the non-apoptotic outcomes also, such as differentiation and immune signaling (Choi et al. 2011).

PI3K

PI3K (phosphoinositide 3-kinase) plays a key role in cell proliferation and survival in response to the growth factor, hormones, and cytokine stimulation. PI3K tightly couples with the receptor tyrosine kinase (RTKs), and it activates tyrosine-phosphorylated RTK dimers through a disulfide domain in its regulatory subunit. PI3K catalyzes the synthesis of the second messenger PIP3 (phosphatidylinositol 3, 4, 5 triphosphate) from PIP2 (phosphatidylinositol 4, 5-biphosphate). Membrane-bound PIP3 then serves as a signaling molecule to recruit proteins containing the pleckstrin homology domain. These proteins, such as the phosphoinositide-dependent protein kinase (PDK) and protein kinase B (AKT) serine/threonine kinases, are thus activated and mediate further downstream signaling events (Cantrell 2001). The end result of AKT activation is stimulation of the growth pathways and inhibition of apoptotic pathways. For example, activation of the angiogenic growth factor, VEGF (vascular endothelial growth factor), by ROS occurs through the PI3K/Akt pathway (Gao et al. 2002).

Synthesis of PIP3 is regulated primarily by the phosphatase and tensin homology (PTEN) phosphatase, which dephosphorylates PIP3 back to PIP2 (Leslie and Downes 2002). Through PTEN, the PI3K pathway is reversibly redox regulated by the ROS generated by growth factor stimulation. H_2O_2 was shown to oxidize and inactivate the human PTEN through disulfide bond formation between the catalytic domain of Cys-124 and Cys-71 residues (Kwon et al. 2004). Also, the endogenously generated ROS following treatment with the peptide growth factors such as insulin, EGF, and PDGF causes oxidation of PTEN leading to the activation of the PI3K pathway (Seo et al. 2005). Thus, the PI3K pathway is regulated by ROS in a similar manner as the MAPK pathways; at the oxidative interface, protein phosphatases are directly oxidized by ROS resulting in the sustained activation of the signaling pathways. Various oxidants activate transcription of a battery of antioxidant genes through a PI3K-Nrf2-ARE mechanism, where PTEN knockdown enhances transcription of ARE-regulated antioxidant genes (Sakamoto et al. 2009). The PI3K/Akt pathway is stimulated by a variety of growth factors, cytokines, cell–matrix interactions, and lipid products generated by the PI3K activity.

Further several experimental findings also suggest that the PI3K pathway is critical for ROS generation (Kim et al. 2005). Activation of PI3K may be induced by tyrosine kinases, for example, by the oncogenic Bcr-Abl tyrosine kinase (Rodrigues et al. 2008), and its activation is not only important for cell transformation but also for ROS production. For example, PI3K activation is essential in the PDGFR-dependent production of H_2O_2 (Bae et al. 2000).

Transcription Factors (TFs)

ROS are capable of penetrating the plasma membrane and can thus directly modulate the activity of catalytic domain of transmembrane signal-transducing enzymes leading to the activation of key signaling molecules such as transcription factors, which further precisely control the "on" and "off" switch of the target gene expression. These factors bind to the consensus cis element in the promoter of target genes, and then transactivation domain facilitates the stimulation of transcription.

As the antioxidant enzymes play a major role in reducing the ROS levels, therefore, redox regulation of the transcription factors is significant in determining the gene expression profile and cellular response to oxidative stress (Fig. 5.3). Eukaryotes possess complex mechanisms for controlling the cell signaling in initiation of stress-inducible gene expression. Redox-sensitive transcription factors must mediate the cellular mechanisms that initiate changes in the gene expression. Several ARE/TRE sites (in GST, yGCS, and GSH-Px) and kB sites (in iNOS, IkBα, and MnSOD) are present in the promoter/enhancer of the antioxidant genes. ARE possesses structural and biological features that characterize its unique responsiveness to the oxidative stress. It is activated not only in response to the H_2O_2 but specifically by chemical compounds with the capacity to either undergo redox cycling or be metabolically transformed to a reactive or electrophilic intermediate. Moreover, the compounds that have propensity to react with the sulfhydryl groups are also potent inducers of

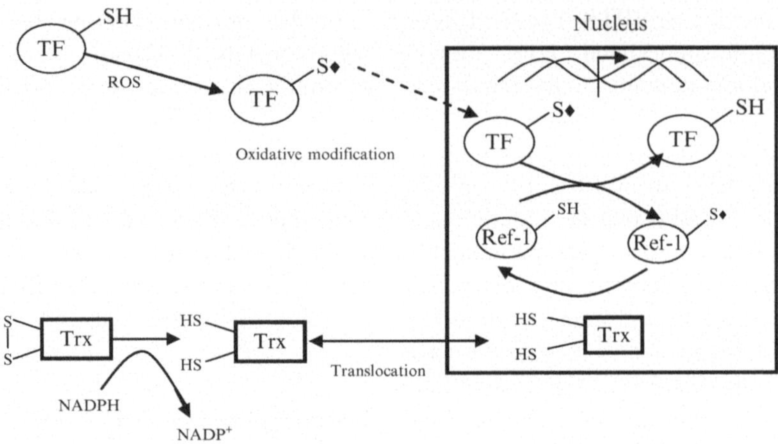

Fig. 5.3 Redox regulation of transcription factors

ARE activity. Thus, alteration of the cellular redox status due to elevated levels of ROS and electrophilic species and/or a reduced antioxidant capacity (e.g., glutathione) appears to be an important signal for triggering the transcriptional response (Fig. 5.4).

AP-1

AP-1 (activator protein 1) was first identified as a transcription factor that contributes both to the basal gene expression (Lee et al. 1987) and phorbol ester (TPA)-inducible gene expression (Angel et al. 1987). It is a collection of the dimeric bZip proteins that belong to the Jun (c-Jun, JunB, JunD), Fos (FosB, Fra-1, Fra-2), Maf (musculoaponeurotic fibrosarcoma), and ATF (activating transcription factor) subfamilies, all of which can bind to the TPA or cAMP response elements (CRE) (Chinenov and Kerppola 2001). AP-1 is a potent transcriptional regulator and is involved in the cell growth and differentiation. Mechanism of activation of AP-1 by the free radicals is one of the best explained mechanisms in eukaryotes. ROS production induced by the TNFα and basic fibroblast growth factors act as common signal to stimulate the cfos gene (Lo and Cruz 1995) and also oxidative stress caused by the ionizing radiation, and H_2O_2 is a potent inducer of c-jun expression (Collart et al. 1995). AP-1 activity is sensitive to the antioxidants, phenolic antioxidants substantially increases the expression of c-jun and cfos (Choi and Moore 1993), whereas thiol antioxidants ameliorate the induction caused by H_2O_2 and radiation.

Studies indicate that the oxidative stress can also result in the inactivation of AP-1, by oxidation of key SH-groups (Cys 272 and Cys 154) in c-jun and cfos, respectively, and causes the reversible inactivation (Morel and Barouki 1999). Also experimental selenium (an essential trace element) deficiency in mice was demonstrated to create oxidative stress in testis, and on further analysis, c-Jun/cFos expression was found decreased comparing adequate selenium status (Shalini and Bansal 2005). AP1-DNA-binding studies indicated that its binding affinity can be enhanced by the thioredoxin (Trx) as well as by the Ref-1 and inhibited by GSSG in several cell types suggesting that oxidation of sulfhydryl groups and disulfide bond formation affects the binding (Hirota et al. 1997). Such responses allow the cells to adopt to environmental changes and in maintenance of the normal physiology.

AP-1 activity can be induced by H_2O_2, cytokines, and other physical and chemical stressors. AP-1 activity is regulated by the redox state of Cysteine-64, which is located at the interface between the two c-jun subunits and therefore important in regulation of the redox status on gene transcription (Klatt et al. 1999). AP-1 activity is regulated both at the posttranscriptional and posttranslational levels (Hunter and Karin 1992;

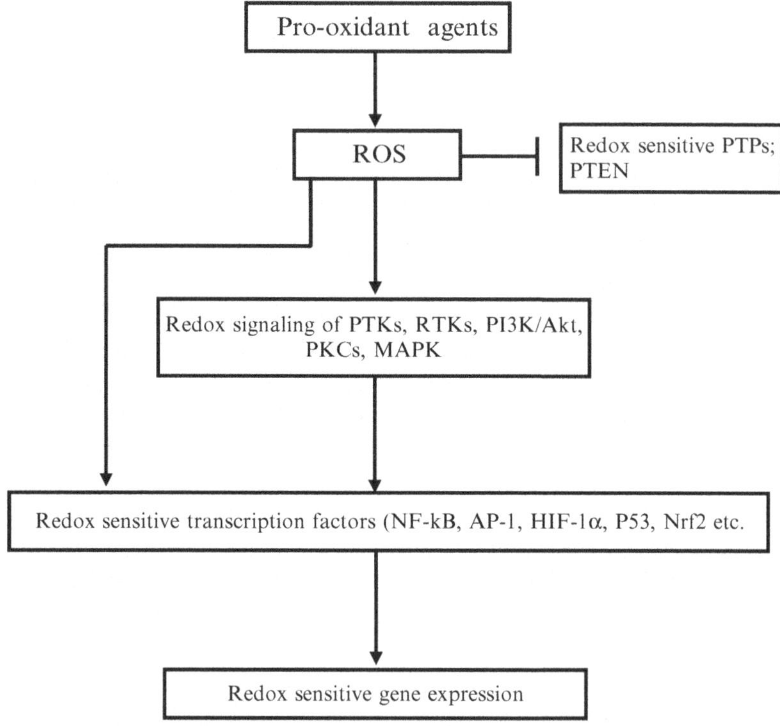

Fig. 5.4 ROS-dependent signaling pathways

Karin 1995). The exposure of HeLa cells to H_2O_2 or UV radiation leads to a significant increase in DNK-binding activity of AP-1, irrespective of the fos and jun protein synthesis. Here AP-1 is activated by the phosphorylation of specific residues of AP-1 subunits (Ser-63 and Ser-73 in jun subunit) by involving JNK (Karin 1995). Oxidative stress also influences at the posttranslational level of AP-1 activity regulation by activating signaling via JNK protein kinases (Go et al. 1999). Fos protein in AP-1 is also activated by the phosphorylation of the threonine residue (Thr-232) due to fos-regulatory kinase activated by $p21^{ras}$ protein (Deng and Karin 1994). Also the phosphorylation of jun protein (Thr-231, Ser-243, and Ser-149) by constitutive protein kinases, casein kinases II, and DNK-dependent protein kinase (Lin et al. 1992) results in inhibiting the binding of AP-1 to DNK. Dephosphorylation of the threonine and serine residues of jun protein increases the affinity of AP-1 for binding to DNK. This transcription factor is activated due to PKC that initiates the dephosphorylation of jun following activation of the phosphatases. Thus, this transcription factor works in multiple ways in different environmental conditions.

NF-kB

NF-kB belongs to the Rel-family of pluriprotein transcription factors and is a regulatory protein that controls the expression of numerous inducible and tissue-specific NF-kB responsible genes (Ghosh et al. 1998). Several reports show that ROS act as messengers for the activation of NF-kB, a peptide that is principally involved in the inflammatory response but also in the cell survival, differentiation, and growth (Pande and Ramos 2005). NF-kB is a redox-sensitive transcription factor, and in fact it was the first eukaryotic transcription factor shown to respond directly to the oxidative stress (Shalini and Bansal 2007). NF-kB is activated by a large number of conditions and agents such as inflammatory cytokines, mitogens, bacterial products, protein synthesis inhibitors, ROS, UV light, and phorbol esters (Schulze-Osthoff et al. 1995). However, it has been observed in most cases that these inducers of the NF-kB rely on the production of ROS. Evidence to this comes from the several previous studies. Firstly, H_2O_2 directly activates NF-kB in several cells (Flohe et al. 1997). Secondly, most of the inducers of the NF-kB activation such as LPS, TNFα, and IL-1β produce oxidative stress in cells (Iuvone et al. 1998). Thirdly, treatment with several antioxidants such as N-acetyl cysteine, α-lipoic acid, metallothionein, and pyrrolidine dithiocarbamate (PDTC) blocks NF-kB activation (Sakurai et al. 1999). Fourthly, overexpression of catalase, an enzyme that scavenges H_2O_2, as well as overexpression of the glutathione peroxidase (Kretz-Remy et al. 1996) that scavenges both H_2O_2 and organic peroxides inhibits NF-kB activation.

Activated form of the NF-kB is a heterodimer consisting of p65/RelA and p50 subunits (Ghosh et al. 1998). ROS act as a second messenger involved in the NF-kB activation and promote activation of a critical redox-sensitive kinase. NF-kB inducing kinase (NIK) or IKK causes phosphorylation of the critical serine residues in IkB resulting in liberation of the RelA/p50 heterodimers. Some other studies indicate that ROS interfere with the DNA-binding activity of NF-kB. A cysteine residue, Cys 62, located in the N-terminal region of p50 is redox sensitive, and oxidation of the SH group decreases NF-kB activity. The genes activated by NF-kB encode the synthesis of various cytokines, their receptors, and cell adhesion molecules. It has been found that inducible nitric oxide synthase (iNOS) gene promoter possesses an NF-kB binding site. Signaling molecules, such as TNF-α, endotoxin, interleukin 1β, mitogens, lipopolysaccharides, agents promoting oxidative stress, lectin, Ca^{2+}-ionophores, and UV radiation, activate NF-kB (Dalton et al. 1999). Free oxygen radicals (FORs), which are formed on the respiratory chain of mitochondria to act as second messengers, mediate NF-kB activation promoted by TNF-α and interleukin-1β.

Initially NF-kB is present in the cytoplasm as a complex with its inhibitor, IκB, then binding of the stimuli to the cell receptors is followed by phosphorylation of IκB at position Ser-32 and Ser-36 catalyzed by the IκB kinase (Traenckner et al. 1995), and further modification of inhibitors allows for dissociation of NF-kB from the complex, degradation of inhibitors, and rapid translocation of NF-kB into the nucleus where it binds to the target DNA elements and positively regulates the transcription of the various genes. NF-kB activation is stimulated by the pro-oxidative cell status, especially by an increased presence of H_2O_2. The exact signaling cascade seems to be due to the activation of MAP kinase pathway. Low concentration of the thiol compounds in the cell, primarily glutathione, plays a key role in positive regulation of the NF-kB activity (Traenckner et al. 1995). ROS regulate NF-kB activity and modify some of the links in a complex activating kinase cascade of NF-kB.

Since NF-kB has a ubiquitous role in controlling the cytokine activity and immunoregulatory genes, the inhibition of NF-kB activity by steroid hormones, antioxidants, nonsteroid anti-inflammatory drugs, and protease inhibitors represents adjuvant therapy in numerous diseases (Yamamoto and Gaynor 2001).

Nrf2

A major mechanism in the cellular defense against oxidative or electrophilic stress is the activation of the Nrf2-ARE signaling pathway, which controls the expression of genes whose protein products are involved in the detoxification and elimination of the reactive oxidants and electrophilic agents through conjugative reactions and by enhancing the cellular antioxidant capacity (Nguyen et al. 2009). Nrf2 is regarded as a master regulator of the redox homeostatic gene regulatory network. Under the oxidative and electrophilic stresses, Nrf2 signaling pathway is activated to enhance the expression of a multitude of antioxidants and phase II enzymes that restore the redox homeostasis. Nrf2 binds to the antioxidant response element (ARE) in promoter of the target antioxidant genes and tightly regulates its transcription (Osbern and Kensler 2008). It is a basic region-leucine zipper (bzip)-type primary transcription factor (Itoh et al. 1999), which further heterodimerizes with members of the small Maf family of transcription factors (Nguyen et al. 2000). During unstressed conditions, majority of the Nrf2 resides in the cytoplasm in association with a dimeric repressor protein, Kelch ECH-associated protein-1 (Keap1) (Itoh et al. 1999), and promotes its proteasomal degradation through another protein, Cul 3 (Furukawa and Xiong 2005; Villeneuve et al. 2010), and maintains a low basal level of Nrf2. In mouse keap1 cysteines (Cys-151, -273, and -288) (Dinkova-Kostova et al. 2002; Zhang and Hannink 2003) are redox sensors and upon oxidation by ROS become less effective at promoting Nrf2 degradation, resulting in the dissociation of Nrf2 from keap1/cul3 which allows Nrf2 translocation into the nucleus. Another mechanism of translocation was also proposed through the activation of protein kinases, such as PKC results in phosphorylation of the Nrf2, which enhances the stability and/or release of Nrf2 from Keep1 (Huang et al. 2002).

The nuclear export signal (NES), located in the transactivation domain of Nrf2 functions to shuffle Nrf2 out of the nucleus, is also redox sensitive. It contains a cysteine residue at position 183 that is modified under the oxidative stress, which weakens the NES activity, leading to increased retention of Nrf2 in the nucleus (Li et al. 2006). Accumulating Nrf2 in nucleus dimerizes with the small Maf proteins and binds ARE enhancer and activates ARE-dependent transcription of target genes which serve as antioxidants. The Nrf2-Keap-1 system has been observed in virtually all the vertebrates, suggesting that Nrf2 is a highly conserved cellular defense mechanism.

Nrf2 contains a conserved cysteine located in the DNA-binding domain in human (Cys-514) which is the site of Ref-1-mediated redox regulation. Also, GSH, a ubiquitous small molecular thiol antioxidant, biosynthesis is tightly controlled by its rate-limiting enzyme, y-glutamyl cysteine ligase (yGCL), which is transcriptionally regulated by Nrf2 (Chen et al. 2008). Deficiency of Nrf2 is implicated in the impairment of GSH production and thereby alters the intracellular redox state (Chan and Kwong 2000).

Ref-1

Redox factor-1 (Ref-1), a 37-kDa protein that stimulates fos-jun DNA-binding activity (Xanthoudakis and Curren 1992), was shown to be identical to an apurinic/apyrimidinic (AP)-endonuclease named APE (AP endonuclease) (Demple et al. 1991) or human AP endonuclease 1 (HAP1) (Robson and Hickson 1991). Thus, Ref-1 (also named as APE1/Ref-1 in literature) is a multifunctional protein that not only regulates transcription factor activity but also mediates base excision repair. The transcriptional regulatory function of Ref-1 is mediated through its redox activity on several transcription factors such as AP-1, p53, NF-kB, and hypoxia-inducible factor 1 (HIF-1) by regulating their redox states (Tell et al. 2009). The N-terminus region of Ref-1 is responsible for the redox activity while the AP-endonuclease activity domain is located at the C-terminal region. Cys-65 of human (Cys -64 of mouse) Ref-1 appears to be a major redox active site (along with Cys-93) that is required for the reduction and increased DNA binding of the targeted transcription factors. Ref-1 activates the AP-1, through redox regulation of cysteine residues (Cys-154 in fos and Cys-272 in jun) in the fos-jun DNA-binding domains (Abate et al. 1990; Xanthiooudakis et al. 1992). This cysteine is highly conserved in various human bzip transcription factors, and all may be regulated in a redox-dependent manner by Ref-1, resulting in the increased DNA binding and transcriptional activation of target genes.

Further, reduction of Ref-1 appears to be regulated by the thioredoxin. In response to phorbol myristate acetate or ionizing radiation, the thioredoxin (Trx) translocates into the nucleus and interacts with Ref-1, resulting in the activation of AP-1 transcriptional activity under reducing conditions (Hirota et al. 1997; Ueno et al. 1999; Wei et al. 2000). The interaction of Trx with Ref-1 and subsequent activation of Ref-1 target protein appears to be regulated by the redox active Cys-32 and Cys-35 residues of Trx which are responsible for its reducing activity. A related study (Ando et al. 2008) states that apart from that Ref-1 activates DNA-binding activity of many redox-sensitive transcription factors by directly reducing their cysteine residues, a novel activity of Ref-1 termed redox chaperone activity was reported, by which Ref-1 regulates DNA-binding activity of various transcription factors through promoting the reduction of their critical cysteine residues by other reducing molecules such as GSH and Trx (Droge et al. 1994; Mitomo et al. 1994; Nishi et al. 2002). Redox chaperone activity seems to be mediated by the direct interactions between Ref-1 and target transcription factors and does not require high concentration of Ref-1 for its redox activity (in vitro studies).

HIF-1

Hypoxia-inducible factor (HIF) is known as the master regulator of the cellular response to hypoxia and is of pivotal importance during development as well as in human disease, particularly cancer. Hypoxia-inducible factor 1(HIF-1) is a heterodimer of two proteins, HIF-1α and HIF-1β. HIF-1α is inducible and HIF-1β accumulate constitutively only in hypoxic cells (Semenza 2000). Activity of HIF-1α is affected by the oxygen concentration (Cash et al. 2007), and its stabilization is regulated by NADPH oxidase-derived ROS, mainly H_2O_2, produced during hypoxia (Haddad 2003). HIF-1α regulates the expression of many cancer-related genes including VEGF, aldolase, enolase, lactate dehydrogenase, and others. HIF-1 is induced by the expression of oncogenes such as Src and Ras and is overexpressed in many cancers. In the HIF-1 regulated protein, VEGF plays an important role in tumor progression and angiogenesis.

High levels of HIF-1 expression are in particular correlated with the tumorigenesis, because this factor regulates the expression of many cancer-related genes. It has been shown that overexpression of Rac1 increases HIF-1 and PAI-1 expression in response to hypoxia, through ROS-dependent mechanisms, thus suggesting that Rac1/NADPH oxidase/ROS pathways are important for the redox-dependent upregulation of HIF-1. Further, it has also been found that both HIF-1 and PAI-1 expression are regulated by the ROS production and c-Src activation in VSMCs playing a key role in angiogenesis and thrombosis in atherosclerotic vasculature (Sato et al. 2005). Increased expression of HIF-1α and VEGF, downstream of HIF-1, may also be induced by EGF-regulated ROS, through PI3K/Akt/p706K pathway, which is involved in tumorigenesis and angiogenesis (Liu et al. 2006). High ROS concentrations in primary leukemic cells upregulate VEGF as well as HIF-1, required in order to stimulate growth signals.

p53

Another intracellular mediator of ROS is the tumor suppressor p53 (Liu et al. 2008), a nuclear factor involved in apoptosis. p53 plays a key role in protecting a cell from tumorigenesis (Hofseth et al. 2004), and due to its ability to halt the cell cycle or initiate apoptosis if cell is damaged, it is often called a tumor suppressor. Mutations in p53 leading to its inactivation have been found in more than half of human cancers (Hofseth et al. 2004). Several cysteine residues in the central domain of the protein are critical for the p53 binding to the specific sequence.

ROS can modulate the redox status of a critical cysteine residue in the DNA-binding domain of p53 and influences its DNA-binding capacity (Meplan et al. 2000). Also p53 can be activated by ROS through cross-talk with other signaling pathways. For instance, both JNK and p38, activated by ROS, are capable of phosphorylating p53, and both have been implicated in regulating p53 activity by stabilizing the p53 protein under conditions of oxidative stress (Buschmann et al. 2001). The tumor suppressor p53 can also be upregulated in response to H_2O_2 in T cells in an NF-kB-dependent manner. In addition to the generation of ROS, p53 induces the expression of p85, which may function as a signaling molecule during ROS-mediated p53-dependent apoptosis. Moreover, p53 is recognized as a modifier of the angiogenic response. It interacts with HIF-1 but also has direct effects on the angiogenesis regulators and/or factors such as VEGF and fibroblast growth factor (Galy et al. 2001). Conversely, activated p53 results in generation of ROS, suggesting that an important consequence of the oxidant-induced activation of p53 is to further increase of the oxidative stress levels.

Others (Sp-1, Ets-1, Myb, NFAT)

Sp-1 (specificity protein 1) factor is also regulated by redox mechanism at the level of cysteine residues in their DNA-binding domain. ROS-generating enzyme, NOX1, has been found to mediate Ras-induced upregulation of VEGF and angiogenesis, by activating Sp-1 through Ras/ERK-dependent phosphorylation of Sp-1 (Komatsu et al. 2008). Ets-1 (E 26 transcription factor) is activated at low concentration of H_2O_2, which is involved in endothelial cell proliferation in vascular system, via ARE (Wilson et al. 2005). It is a critical regulator of AngII-mediated ROS generation and induction of the NADPH oxidase p47phox (Ni et al. 2007). It also regulates the expression of genes involved in extracellular matrix degradation, including MMPs (matrix metalloproteinase) and uPA (urokinase plasmin activator), and in the migration of cells (Zhan et al. 2005). For example, Ets-1 was found to upregulate MMP-9 triggered by TGFβ1 via MAPK signaling (Huang et al. 2005). These activities are related to metastasis in cancer.

Myb (myeloblastosis) is a photo-oncogene product that activates the transcription of several genes involved in cell-cycle progression. Myb possesses a conserved cysteine residue in a region of the helix-turn-helix domain of which the reduced state is essential for its DNA-binding and transformation activity (Myrset et al. 1993). The redox state of this residue is related to conformation of the DNA-binding domain; thus, it could function as a molecular sensor of the redox state.

NFAT (nuclear factor of activated T cells) family of nuclear transcription factors regulates muscle growth and differentiation, cytokine formation, and angiogenesis. Most NFAT proteins are calcium dependent (Rao et al. 1997) and activated by phosphatase calcineurin, which is in turn activated by high intracellular calcium levels. Various ROS/metals are known to increase intracellular calcium and activate NFAT.

Major Molecular/Metabolic Pathways Affected by ROS: Inflammatory Pathways, Stress Response Proteins, Gene Activation/Repressions, and Antioxidant Strategies

Multiple molecular mechanisms alter the cell metabolism to provide the need of the dividing cells especially for cancer cells, like ATP generation, biosynthesis of macromolecules, and maintenance of the cell redox status. In tumor cells, Warburg effect is a shift from ATP generation through oxidative phosphorylation to glycolysis. This effect is regulated by PI3K, HIF, P53, MYC, and AMPK-liver kinase B1 (LKB1) pathway (reviewed in Cairns et al. 2011). During altered cancer metabolism, NADPH is produced which functions as a reducing power in many enzymatic reactions that are crucial for the macromolecule biosynthesis.

Further, oxidative stress traditionally been viewed as a process of cell damage resulting from the aerobic metabolism and antioxidants has been viewed simply as free radical scavengers. However, now ROS are widely used as secondary messengers to propagate the pro-inflammatory or growth-stimulatory signals. New pharmacological strategies are aimed at supplementing the antioxidative defense system while antagonizing redox-sensitive signal transduction. This may allow improved clinical management of antioxidant therapy. Activation of redox-sensitive transcription factors, such as AP-1, p53, and NF-kB, regulates the expression of pro-inflammatory and other cytokines, cell differentiation, and apoptosis. Under normal conditions NF-kB is held inactive but under conditions of stress activates expression of pro-inflammatory and other cytokines. Understanding of the role of mitochondria and ROS they produce in inflammation is also growing. In a recent study, mitochondrial ROS enhances pro-inflammatory cytokine production through the regulation of the MAPK pathways. In addition, ROS modulate various other signaling pathways and block the dephosphorylation of MAPKs.

The role of ROS in inflammatory responses is also studied in TNF receptor-associated periodic syndrome (TRAPS), an autoinflammatory disease (Bulua et al. 2011). Enhanced inflammation in TRAPS is linked to increased activation of the MAPKs, p38, and JNK. Reducing the levels of mitochondrial ROS may be a potential therapeutic strategy for patients with TRAPS.

Another interesting indirect link is the use of nanoparticles (NP). Nanoparticles are used basically for the diagnosis and the therapeutic level in various systems. However, in many cases, these particles exhibit toxicity. Oxidative stress has largely been reported to be implicated in NP-induced toxicity, and it could activate a wide variety of cellular events such as cell-cycle arrest, apoptosis, inflammation, and induction of antioxidant enzymes at its localized site in the system. Specific site uptake of NPs can be controlled by using specific ligands and size. The responses occur after the activation of different cellular pathways, as MAP kinase cascades (ERK, p38, and JNK) as well as redox-sensitive transcription factors such as NF-kB and Nrf2. The induction of apoptosis is closely related to the modulation of signaling pathways induced by NPs. In one of the recent experimental study, various metal NPs were demonstrated to induce inflammatory signals in macrophages in cell culture (Nishanth et al. 2011). Recent activity is to study the interaction of NPs and other body fluids, cellular microenvironment, intracellular compounds, or secreted cellular proteins such as cytokines, growth factors, and enzymes and use of engineered NPs to target various signal transduction pathways in cancer therapy. These interactions could lead to a sustained modulation of specific signaling in the target cells. These studies are well reviewed in a recent article (Marano et al. 2011).

Further, ROS influence is observed through activation and repression of genes. There is full of literature on this account and also many events are sited in this chapter even. Few typical examples: Sublethal concentration of H_2O_2 elicits a decrease in IL-2 mRNA. Moreover, they repress the transcription of a reporter gene driven by the IL-2 gene promoter, whereas they activate the promoter of the c-jun gene (Beiging et al. 1996). It was then shown that this repression is mediated by inhibition of the activity of an NFAT transcription factor, through alteration of its binding to DNA. Another interesting experimentation showed that norepinephrine causes epigenetic repression of PKCε gene in rodent hearts by activating NOX1-dependent ROS production (Xiong et al. 2012). The repression was due to the PKCε promoter methylation at Egr-1 and Sp-1 transcription factor binding sites.

Normally, cells defend themselves against ROS damage with enzymes such as SOD, catalase, and GPx and a number of nonenzymatic antioxidants and small proteins including GSH, vitamins (A, E, and C), carotenoids, and polyphenols (flavonoids, curcumin, resveratrol, and others). Because of the role played by ROS as signaling molecules, antioxidant compounds may thus significantly interfere with cell signal transduction, not only by simply quenching ROS generation and propagation but also by intercepting reactive species at the level of critical signaling pathways. Antioxidant supplements are popularly consumed and certain dietary choices are made to modulate the potential oxidative damage caused by ROS. Recent advances made in understanding redox homeostasis maintained via, for example, the Keap1/Nrf2 signaling pathway may replace the concept of artificially supplying the body with antioxidants. Under normal or unstressed conditions, Nrf2 remain in cytoplasm with association with other proteins, but under stress conditions like oxidative stress, Nrf2 is able to translocate into the nucleus, bind to ARE, and express antioxidative genes.

Nrf2 induces many cytoprotective proteins, such as: NAD(P)H quinine oxidoreductase 1 (Nqo1 catalyzes the reduction and detoxification of highly reactive quinines that can cause redox cycling and oxidative stress) (Venugopal and Jaiswal 1996), glutamate-cysteine ligase (catalytic/modifier subunit from a heterodimer for glutathione synthesis which is a powerful endogenous antioxidant) (Solis et al. 2002), heme oxygenase-1 (HO-1 catalyzes breakdown of heme into the antioxidant biliverdin and the anti-inflammatory agent and protects from a variety of pathologies) (Jarnmi and Agarwal 2009; Wang and Dore 2007), glutathione S-transferase (GST, a cytosolic, mitochondrial, and microsomal enzyme, catalyzes conjugation of GSH with endogenous and xenobiotic electrophiles, a detoxification activity) (Hayes et al. 2000), UDP-glucuronosyl transferase (UGT catalyze conjugation of glucuronic acid moiety to a variety of endogenous and exogenous substances and readily excreted, major substrate bilirubin and acetaminophen) (Yueh and Tukey 2007), and multidrug resistance-associated proteins (Mrps, important membrane transporters that efflux various compounds from various organs and into bile or plasma, with subsequent excretion in feces or urine, respectively) (Maher et al. 2007).

Finally, in this chapter we have tried to explain initially the basics of the signaling tracts and then various pathways followed in different physiological conditions specially the oxidative stress conditions. Activation and inhibition of various transcription factors have been discussed as per literature, and in the end few specific examples of signaling in metabolic and pathological conditions related to oxidative stress are described. This compact information will provide the basis of understanding the pathology of various systems as explained separately in other chapters in this collection.

References

Abate C, Patel L, Rauscher FJ III, Curran T (1990) Redox regulation of fos and jun DNA-binding activity in vitro. Science 249:1157–1161

Aikawa R, Komuro I, Yamazaki T, Zou Y, Kudoh S, Tanaka M, Shiojima I, Hisoi Y, Yazaki Y (1997) Oxidative stress activates extracellular signal-regulated kinases through Src and Ras in cultured cardiac monocytes of neonatal rats. J Clin Invest 100:1813–1821

Ando K, Hirao S, Kabe Y, Ogura Y, Sato I, Yamaguchi Y, Wada T, Handa H (2008) A new APE1/Ref-1-dependent pathway leading to reduction of NF-kB and AP-1, and activation of their DNA-binding activity. Nucleic Acid Res 36:4327–4336

Angel P, Imagawa M, Chiu R, Stein B, Imbra RJ, Rahmadorf HJ, Jonat C, Herrlich P, Karin M (1987) Phorbol ester-inducible genes contain a common cis element recognized by a TPA-modulated trans-acting factor. Cell 49:729–739

Arai KI, Lee F, Miyajima A, Miyatake S, Arai N, Yokota T (1990) Cytokines: coordination of immune and inflammatory responses. Annu Rev Biochem 59:783–836

Arbabi S, Maier RV (2002) Mitogen-activated protein kinases C. Crit Care Med 30:S74–S79

Bae YS, Kang SW, Seo MS, Baines IC, Tekle E, Cook PB, Rhee SG (1997) Epidermal growth factor (EGF)-induced generation of hydrogen peroxide. Role in EGF receptor-mediated tyrosine phosphorylation. J Biol Chem 272:217–221

Bae YS, Sung JY, Kim OS, Hur KC, Kazlauskas, Rhee SG (2000) Platelet derived growth factor-induced H2O2 production requires the activation phosphatidylinositol 3-kinase. J Biol Chem 275:10527–10531

Barford D (2001) The mechanism of protein kinase regulation by protein phosphatases. Biochem Soc Trans 29:385–391

Beiging L, Chen M, Whisler RL (1996) Sublethal levels of oxidative stress stimulate transcriptional activation of c-Jun and suppress IL-2 promoter activation in Jurkat T cells. J Immunol 157:160–169

Brawn MK, Chiou WJ, Leach KL (1995) Oxidant-induced activation of protein kinase C in UCIIMG cells. Free Radic Res 22:23–37

Breyer RM, Bagdassarian CK, Mayers SA, Breyer MD (2001) Prostanoid receptors: subtypes and signaling. Annu Rev Pharmacol Toxicol 41:661–690

Bulua AC, Simon A, Maddipati R, Pelletier M, Park H, Kim KY, Sack MN, Kaster DL, Siegel RM (2011) Mitochondrial reactive oxygen species promote production of proinflammatory cytokines and are elevated in TNFR1-associated periodic syndrome (TRAPS). J Exp Med 208:519–533

Buschmann T, Potapova OA, Bar-Shira VN, Ivanov SY, Fuchs S, Henderson S, Frad VA, Minamoto T, Alarcon-Vargas D, Pincus MR, Gaarde WA, Holbrook NJ, Shiloh Y, Ronai Z (2001) Jun NH2-terminal kinase phosphorylation of p53 on Thr-81 is important for p53 stabilization and transcriptional activities in response to stress. Mol Cell Biol 21: 2743–2754

Cairns RA, Harris IS, Mak TW (2011) Regulation of cancer cell metabolism. Nat Rev Cancer 11:85–95

Cantrell DA (2001) Phosphoinositide 3-kinase signaling pathways. J Cell Sci 114:1439–1445

Carlson RM (1990) Assessment of the propensity for covalent binding of electrophiles to biological substrates. Environ. Health Perspect 87:227–232

Cash P, Pan Y, Simon MC (2007) Reactive oxygen species and cellular oxygen sensing. Free Radic Biol Med 43:1219–1225

Chan JY, Kwong M (2000) Impaired expression of glutathione synthetic enzyme genes in mice with targeted deletion of the Nrf2 basic-leucine zipper protein. Biochim Biophys Acta 1517:19–26

Chen Q, Olashaw N, Wu J (1995) Participation of ROS in the lysophosphatidic acid-stimulated mitogen-activated protein kinase kinase activation pathway. J Biol Chem 270:28499–28502

Chen CC, Young JL, Monzon RI, Chen N, Todorovic V, Lau LF (2007) Cytotoxicity of TNF alpha is regulated by integrin-mediated matrix signaling. EMBO J 26:1257–1267

Chen L, Dai AG, Hu RC (2008) Roles of NRF2 regulating gamma-glutamylcysteine synthetase in lung of rats with chronic obstructive pulmonary disease. Zhongguo Ying Yong sheng li xue za zhi =zhongguo shenglixue zahi. Clin J Appl Physiol 24:339–342

Chinenov Y, Kerppola TK (2001) Close encounters of many kinds: fos-jun interactions that mediate transcription regulatory specificity. Oncogene 20:2438–2452

Choi HS, Moore DD (1993) Induction of c-fos and c-jun gene expression by phenolic antioxidants. Mol Endocrinol 7:1596–1602

Choi TG, Lee J, Ha J, Kim SS (2011) Apoptosis signal-regulating kinase 1 is an intracellular inducer of P38 MAPK-mediated myogenic signaling in cardiac myoblasts. Biochim Biophys Acta 1813:1412–1421

Collart FR, Horio M, Huberman E (1995) Heterogeneity in c-jun gene expression in normal and malignant cells exposed to either ionizing radiation or hydrogen peroxide. Radiat Res 142:188–196

Dalton PT, Shertzer HG, Puga A (1999) Regulation of gene expression by reactive oxygen. Annu Rev Pharmacol Toxicol 39:67–101

Davies SS, Amarnath V, Roberts LJ II (2004) Isoketals: highly reactive γ-ketoaldehydes formed from the H2-isoprostane pathway. Chem Phys Lipids 128:85–99

Debburman SK, Ptasienski J, Benovie JL, Hosey MM (1996) G protein coupled receptor kinase GRK2 is a phospholipid dependent enzyme that can be conditionally activated by G protein beta/gamma subunits. J Biol Chem 271:22552–22562

Demple B, Herman T, Chen DS (1991) Cloning and expression of APE, the cDNA encoding the major human apurinic endonuclease: definition of a family of DNA repair enzymes. Proc Natl Acad Sci U S A 88:1145–11454

Deng T, Karin M (1994) c-Fos transcriptional activity stimulated by H-ras-activated protein kinase distinct from JNK and ERK. Nature 371:171–175

Dinkova-Kostova AT, Holtzclaw WD, Cole RN, Itoh K, Wakabayashi N, Katoh Y, Yamamoto M, Talalay P (2002) Direct evidence that sulfhydryl groups of keap1 are the sensors regulating modulation of phase 2 enzymes that protect against carcinogen and oxidants. Proc Natl Acad Sci U S A 99:11908–11913

Droge W (2002) Free radicals in the physiological control of cell function. Physiol Rev 82:47–95

Droge W, Schulze-Osthoff K, Mihm S, Galter D, Schenk H, Eck HP, Roth S, Gmunder H (1994) Function of glutathione and glutathione disulfide in immunology and immunopathology. FASEB J 8:1131–1138

Dumont JE, Pecasse F, Maenhant C (2001) Crosstalk and specificity in signalling. Are we crosstalking ourselves into general confusion? Cell Signal 13:457–463

Fantl WJ, Jonson DE, Williams LT (1993) Signalling by receptor tyrosine kinases. Annu Rev Biochem 62:453–481

Fialkow L, Chan CK, Rotin D, Grinstein S, Downey GP (1994) Activation of the mitogen-activated protein kinase signaling pathway in neutrophils. Role of oxidants. J Biol Chem 269:31234–31242

Flohe L, Brigelius-Floe R, Saliou C, Traber MG, Packer L (1997) Redox regulation of NFkB activation. Free Radic Biol Med 22:1115–1136

Forman HJ, Torres M, Fukuto J (2002) Redox signaling. Mol Cell Biochem 234–235:49–62

Fujino G, Noguchi T, Matsuzawa A, Yamauchi S, Saitoh M, Takeda K, Ichijo H (2007) Thioredoxin and TRAF family protein regulate ROS-dependent activation of ASK1 through reciprocal modulation of the N-terminal homophilic interaction of ASK1. Mol Cell Biol 27:8152–8163

Furukawa M, Xiong Y (2005) BTB protein keap1 targets antioxidant transcription factor Nrf2 for ubiquitination by the cullin 3-Roc1 ligase. Mol Cell Biol 25:162–171

Galy B, Créancier L, Zanibellato C, Prats AC, Prats H (2001) Tumour suppressor p53 inhibits human fibroblast growth factor 2 expression by a post-transcriptional mechanism. Oncogene 20:1669–1677

Gao N, Ding M, Zheng JZ, Zhang Z, Leonard SS, Liu KJ, Shi X, Jiang BH (2002) Vanadate induced expression of hypoxia –inducible factor 1 alpha and vascular endothelial growth factor through phosphotidylinositol 3-kinase/Akt pathway and reactive oxygen species. J Biol Chem 277:31963–31971

Ghosh S, May M, Koop E (1998) NFkB and Rel proteins: evolutionarily conserved mediators of immune responses. Annu Rev Immunol 16:225–260

Go YM, Patel RP, Maland MC, Park H, Beckman JS, Darley-Usmar VM, Jo H (1999) Evidence for peroxynitrite as a signaling molecule in flow-dependent activation of c-jun NH(2)-terminal kinase. Am J Physiol 277(4 Pt 2):H1647–1653

Goetzl EJ, An S, Smith WL (1995) Specificity of expression and effects of eicosanoid mediators in normal physiology and human diseases. FASEB J 9:1051–1058

Gopalakrishna R, Jaken S (2000) Protein kinase C signaling and oxidative stress. Free Radic Biol Med 28:1349–1361

Grunwald EW, Richards MP (2006) Mechanisms of heme protein-mediated lipid oxidation using hemoglobin and myoglobin variants in raw and heated washed muscle. J Agric Food Chem 54:8271–8280

Guyton KZ, Liu Y, Gonospe M, Xu Q, Holbnook NJ (1996) Activation of mitogen-activated protein kinase by H_2O_2. Role in cell survival following oxidant injury. J Biol Chem 271:4138–4142

Haddad JJ (2003) Science review redox and oxygen-sensitive transcription factors in the regulation of oxidant-mediated lung injury role for hypoxia-inducible factor-1 alpha. Crit Care 7:47–54

Hayes JD, Chanas SA, Henderson CJ, McMahon M, Sun C, Moffat GJ, Wolf CR, Yamamoto M (2000) The Nrf2 transcription factor contributes both to the basal expression of glutathione S-transferases in mouse liver and to their induction by the chemopreventive

synthetic antioxidants, butylated hydroxyanisole and ethoxyquin. Biochem Soc Trans 28:33–41
Higdon A, Diers AR, OH JY, Landar A, Darley-Usmar VM (2012) Cell signaling by reactive lipid species: new concepts and molecular mechanisms. Biochem J 442:453–464
Hirota K, Matsui M, Iwata S, Nishyama A, Mori K, Yodol J (1997) AP-1 transcriptional activity is regulated by a direct association between thioredoxin and Ref-1. Proc Natl Acad Sci U S A 94:3633–3638
Hofseth LJ, Hussain SP, Harris CC (2004) P53: 25 years after its discovery. Trends Pharmacol Sci 25:177–181
Hong F, Freeman ML, Liebler DC (2005) Identification of sensor cysteines in human keap 1 modified by the cancer chemopreventive agent sulforaphane. Chem Res Toxicol 18:1917–1926
Huang HC, Liu SY, Liang Y, Liu Y, Li JZ, Wang HY (2005) Transforming growth factor-beta 1 stimulates matrix metalloproteinase – a production through ERK activation pathway and upregulation of Ets-1 protein. Zhonghua YiXueZa Zhi 85:328–331
Hunter T, Karin M (1992) The regulation of transcription by phosphorylation. Cell 70:375–387
Iles KE, Forman HJ (2002) macrophage signaling and respiratory burst. Immunol Res 26:95–105
Inoguchi T, Sonta T, Tsubouchi H, Etoh T, Kakimoto M, Sonoda N, Sata N, Sekiguchi N, Kotayashi K, Sumimote H, Utsumi H, Nawata H (2003) Protein kinase-c dependent increase in reactive oxygen species (ROS) production in vascular tissues of diabetes; role of vascular NAD(P)H oxidase. J Am Sci Nepherol 14:S227–S232
Irani K, Xia Y, Zweier JL, Sollott SJ, Der CJ, Fearon ER, Sundaresan M, Finkel T, Goldschmidt-Clermont PJ (1997) Mutagenic signaling mediated by oxidants in Ras-transformed fibroblasts. Science 275:1649–1652
Itoh K, Wakabayashi N, Katoh Y, Ishii T, Igarashi K, Engel JD, Yamamoto M (1999) Keap1 represses nuclear activation of antioxidant responsive element to Nrf2 through binding to the amino-terminal Nrf2 domain. Genes Dev 13:76–86
Iuvone T, D'Acquisto F, VanOsselaer N, Rosa MD, Carnuccio R, Herman AG (1998) Evidence that inducible nitric oxide synthase is involved in LPS-induced plasma leakage in rat skin through the activation of nuclear factor induced leakage in rat skin through the activation of nuclear factor-induced plasma leakage in rat skin through the activation of nuclear factor kappaB. Br J Pharmacol 123:1325–1330
Jarmi T, Agarwal A (2009) Heme oxygenase and renal disease. Curr Hypertens Rep 11:56–62
Kamata H, Hirata H (1999) Redox regulation of cellular signaling. Cell Signal 11:1–14 (Review)
Kamata H, Honda S, Maeda S, Chang L, Hirata H, Karin M (2005) ROS promote TNF-α-induce death and sustained death and sustained JNK activation by inhibiting MAP kinase phosphatases. Cell 120:649–661
Karin M (1995) The regulation of AP-1 activity by mitogen-activated protein kinases. J Biol Chem 270:16483–16486
Kass GE, Duddy SK, Orrenius S (1989) Activation of hepatocyte protein kinase c by redox-cycling quinones. Biochem J 260:499–507
Keyse SM, Emslie EA (1992) Oxidative stress and heat shock induce a human gene encoding a potent tyrosine phosphatase. Nature 359:644–647
Kim JH, Chu SC, Gramlich JL, Pride YB, Babendeier E, Chaudhan SR, Podar K, Griffin JD, Sattler M (2005) Activation of the PI3K/mTOR pathway by BCR-ABL contribute to increased production of reactive oxygen species. Blood 105:1717–1723
Klatt P, Molina EP, DeLacoba MG, Padlilla CA, Martinez-Galesteo E, Barcena JA, Lamas S (1999) Redox regulation of c-jun DNA binding by reversible S-glutathiolation. FASEB J 13:1481–1490
Klaung JE, Kamendulis, Hocevar BA (2010) Oxidative stress and oxidative damage in carcinogenesis. Toxicol Pathol 38:96–109, Review
Komatsu D, Kato M, Nakayama J, Miyagawa S, Kamata T (2008) NADPH oxidase 1 plays a critical mediating role in oncogenic Ras-induced vascular endothelial growth factor expression. Oncogene 27:4724–4732
Konishi H, Tanaka M, Takemura Y, Martsuzaki H, Ono Y, Kikkawa U, Nishizuka Y (1997) Activation of protein kinase C by tyrosine phosphorylation in response to H_2O_2. Proc Natl Acad Sci U S A 94:11233–11237
Koshio O, Akanuma Y, Kasuga M (1988) Hydrogen peroxide stimulates tyrosine phosphorylation of the insulin receptor and its tyrosine kinase activity in intact cells. Biochem J 250:95–101
Krautwald S, Buscher D, Dent P, Ruthenberg K, Baccasini M (1995) Suppression of growth factor-mediated MAP kinase activation by v-raf in macrophages: a putative role for the MKP-1 phosphatase. Oncogene 10:1187–1192
Kretz-Remy C, Mehlen P, Mirault ME, Arrigo AP (1996) Inhibition of Ikappa B-alpha phosphorylation and degradation and subsequent NFkappaB activation by glutathione peroxidase overexpression. J Cell Biol 133:1083–1093
Kuster GM, Siwik DA, Pimentel DR, Colucci WS (2006) Role of reversible, thioredoxin –sensitive oxidative protein modifications in cardiac monocytes. Antioxid Redox Signal 8:2153–2159
Kwon J, Lee SR, Yang KS, Ahn Y, Kim YJ, Stadman ER, Rhee SG (2004) Reversible oxidation and inactivation of the tumor suppressor PTEN in cells stimulated with peptide growth factors. Proc Natl Acad Sci U S A 101:16419–16424
Lee K, Esselman WJ (2002) Inhibition of PTP by H_2O_2 regulates the activation of distinct MAPK pathways. Free Radic Biol Med 33:1121–1132
Lee W, Mitchell P, Tjian R (1987) Purified transcription factor AP-1 interacts with TPA-inducible enhancer elements. Cell 49:741–752
Lee JK, Edderkaoui M, Truong P, Ohno I, Jang KT, Berti A, Pandoi SJ, Gukovskaya AS (2007) NADPH oxidase promotes pancreatic cancer cell survival via inhibiting JAK2 dephosphorylation by tyrosine phosphatases. Gastroenterology 133:1637–1648

Lee KB, Lee JS, Park JW, Huh TL, Lee YM (2008) Low energy proton beam induces tumor cell apoptosis through reactive oxygen species and activation of caspases. Exp Mol Med 40:118–129

Lefkowitz RJ (1993) G-protein-coupled receptor kinases. Cell 74:409–412

Leonarduzzi G, Sottero B, Poli G (2010) Targeting tissue oxidative damage by means of cell signaling modulators: the antioxidant concept revisited. Pharmacol Ther 128:336–374, Review

Leslie NR, Downes CP (2002) PTEN: the down side of PI3-kinase signalling. Cell Signal 14:285–295

Levonen AL, Lander A, Ramachandran A, Ceaser EK, Dickinson DA, Zanoni G, Morrow JD, Darley-Usmar VM (2004) Cellular mechanisms of redox cell signaling: role of cysteine modification in controlling antioxidant defences in response to electrophilic lipid oxidation products. Biochem J 378:373–382

Li W, Yu SW, Kong AN (2006) Nrf2 possesses a redox – sensitive nuclear exporting signal in the Neh5 transactivation domain. J Biol Chem 281:27251–27263

Lin A, Frost J, Deng T, al-Alawi N, Smeal T, Kikkawa U, Hunter T, Brenner D, Karin M (1992) Casein kinase II is negative regulator of c-jun DNA binding and AP-1 activity. Cell 70:777–789

Liu LZ, Hu XW, Xia C, He J, Zhou Q, Shi X, Fang J, Jiang BH (2006) Reactive oxygen species regulate epidermal growth factor-induced vascular endothelial growth factor and hypoxia-inducible factor-1alpha expression through activation of AKT and P70S6K1 in human ovarian cancer cells. Free Radic Biol Med 41: 1521–1533

Liu B, Chen Y, St Clair DK (2008) ROS and p53: a versatile partnership. Free Radic Biol Med 44:1529–1535

Lo YYC, Cruz TF (1995) Involvement of reactive oxygen species in cytokine and growth factor induction of c-fos expression in chondrocytes. J Biol Chem 270:11727–11730

Lu D, Chen J, Hai T (2007) The regulation of ATF3 gene expression by mitogen-activated protein kinases. Biochem J 401:559–567

Maher JM, Dieter MZ, Aleksunes LM, Slitt AL, Guo G, Tanaks Y, Scheffer GL, Chan JY, Klaassen CD (2007) Oxidative and electrophilic stress induces multidrug resistance-associated protein transporters via the nuclear factor-E2-related factor-2 transcriptional pathway. Hematology 46:1597–1610

Marano F, Hussain S, Rodrigues-Lima F, Baeza-Squiban A, Boland S (2011) Nanoparticles: molecular targets and cell signalling. Arch Toxicol 85:733–741, Review

Marshall CJ (1994) MAP kinase kinase kinase, MAP kinase kinase and MAP kinase. Curr Opin Genet Dev 4:82–89

Meng TC, Fukada T, Tonks NK (2002) Reversible oxidation and inactivation of protein tyrosine phosphatases in vivo. Mol Cell 9:387–399

Méplan C, Richard MJ, Hainaut P (2000) Redox signalling and transition metals in the control of the p53 pathway. Biochem Pharmacol 59:25–33

Mitomo K, Nakayama K, Fujimoto K, Sun X, Seki S, Yamamoto K (1994) Two different cellular redox systems regulate the DNA-binding activity of the p50 subunit of NFkB in vitro. Gene 145:197–203

Morel Y, Barouki R (1999) Repression of gene expression by oxidative stress. Biochem J 342:481–496

Morita K, Saitoh M, Tobiume K, Matsuura H, Enomoto S, Nishitoh H, Ichijo H (2001) Negative feedback regulation of ASK1 by protein phosphatase 5 (PP5) in response to oxidative stress. EMBO J 20: 6028–6036

Muller JM, Cahill MA, Rupec RA, Baeuele PA, Nordheim A (1997) Antioxidants as well as oxidants activate c-fos via Ras-dependent activation of extra-cellular-signal –regulated kinase 2 and Eik-1. Eur J Biochem 244:45–52

Myrset AH, Bostad A, Jamin N, Lirsac PN, Toma F, Gabrielsen OS (1993) DNA and redox state induced conformational changes in the DNA-binding domain of the Myb oncoprotein. EMBO J 12:4625–4633

Nagai H, Noguchi T, Takeda K, Ichijo H (2007) Pathophysiological roles of ASK1-MAP kinase signaling pathways. J Biochem Mol Biol 40:1–6

Newton AC (1997) Regulation of protein kinase C. Curr Opin Cell Biol 9:161–167

Nguyen T, Huang HC, Pickett CB (2000) Transcription regulation of the antioxidant response element activation by Nrf2 and repression by Mafk. J Biol Chem 275:15466–15473

Nguyen T, Nioi P, Pickett CB (2009) The Nrf2-antioxidant response element signaling pathway and its activation by oxidative stress. J Biol Chem 284:13291–13295

Ni W, Zhan Y, He H, Maynard E, Balschi JA, Oettgen (2007) Ets-1 is a critical transcriptional regulator of reactive oxygen species and p47(phox) gene expression in response to angiotensin II. Circ Res 101:985–994

Niki E, Yoshida Y, Saito Y, Noguchi N (2005) Lipid peroxidation: mechanisms, inhibition, and biological effects. Biochem Biophys Res Commun 338:668–676

Nishanth RP, Jyotsna RG, Schlaqes JJ, Hussain SM, Reddanna P (2011) Inflammatory responses of Raw264.7 macrophages upon exposure to nanoparticles: role of ROS-NF-kB signaling pathways. Nano Toxicol 5:502–516

Nishi T, Shimizu N, Hiramoto M, Sato I, Yamaguchi Y, Haesegawa M, Aizawa S, Tanaka H, Kataoka K, Watanabe H, Hand H (2002) Spatial redox regulation of a critical cysteine residue of NFkB in vivo. J Biol Chem 277:44548–44556

Niwa K, Inami O, Yamamori T, Hamsu T, Karino T, Kuwabara M (2002) Roles of protein kinase c delta in the accumulation of p53 and the induction of apoptosis in H2O2-treated bovine endothelial cells. Free Radic Res 36:1147–1153

Ohshiro Y, Ma RC, Yasuda Y, Hiroka-Yamamoto J, Clermont AC, Isshiki K, Yagi K, Ankawa E, Kern TS, King GL (2006) Reduction of diabetes-induced oxidative stress, fibrotic cytokine expression and renal dysfunction in protein kinase cbeta-null mice. Diabetes 55:3112–3120

Osbern WO, Kensler TW (2008) Nrf2 signaling an adaptive response pathway for protection against environmental toxic insults. Mutat Res 659:31–39

Pande V, Ramos MJ (2005) NF-kB in human disease: current inhibitors and prospects for de novo structure base design of inhibiter. Curr Med Chem 12:357–374

Parson JT, Parson SJ (1997) Src family protein tyrosine kinases: co-operating with growth factor and adhesion signaling pathways. Curr Opin Cell Biol 9:187–192

Pavlovic D, Dordevic V, Kocic G (2002) A "cross-talk" between oxidative stress and redox cell signaling. Facta Univ Ser Med Biol 9:131–137 (Review)

Perez-Sala D, Cerunuda-Morollon E, Pineda-Molina E, Canada FJ (2002) Contribution of covalent protein modification to the anti-inflammatory effects of cyclopentenone prostaglandins. Ann N Y Acad Sci 973:533–536

Peshenko IV, Shichi H (2001) Oxidation of active center cysteine of bovine 1-cys peroxiredoxin to the cysteine sulfenic acid form by peroxide and peroxynitrite. Free Radic Biol Med 31:292–303

Poli G, Leonardduzzy G, Biasi F, Chiarpotto E (2004) Oxidative stress and cell signaling. Curr Med Chem 11:1163–1182

Putney JJW, McKay RR (1999) Capacitative calcium entry channels. Bioessays 21:38–46

Ranawat P, Bansal MP (2009) Apoptosis induced by modulation in selenium status involves p38 MAPK and ROS: implications in spermatogenesis. Mol Cell Biochem 330:83–95

Rao A, Luo C, Hogan PG (1997) Transcription factors of the NFAT family: regulation and function. Annu Rev Immunol 15:707–747

Ray LB, Sturgill TW (1987) Rapid stimulation by insulin of a serine/threonine kinase in 3T3-L1 adipocytes that phosphorylates microtubule-associated protein2 in vitro. Proc Natl Acad Sci U S A 84:1502–1506

Ray PD, Huang BW, Tsuji Y (2012) Reactive oxygen species (ROS) homeostasis and redox regulation in cellular signaling. Cell Signal 24:981–990

Robson CN, Hickson ID (1991) Isolation of cDNA clones encoding a human apurinic/apyrimidinic endonuclease that corrects DNA repair and mutagenesis defects in E. Coli Xth (exonuclease III) mutants. Nucleic Acid Res 19:5519–5523

Rodrigues MS, Reddy MM, Sattler M (2008) Cell cycle regulation by oncogenic tyrosine kinases in myeloid neoplasias from molecular redox mechanisms to health implications. Antioxid Redox Signal 10:1813–1848

Saitoh M, Nishitoh H, Fujii M, Takeda K, Toblume K, Sawada Y, Kawabata M, Miyazono K, Khijo (1998) Mammalian thyroxin is a direct inhibitor of apoptosis signal-regulating kinase (ASK). EMBO J 17: 2596–2606

Sakamoto K, Iwasaki K, Suglyama H, Tsuji Y (2009) Role of the tumor suppressor PTEN in antioxidant response element-mediated transcription and associated histone modifications. Mol Biol Cell 20:1606–1617

Sakurai A, Hara S, Okano N, Kondo Y, Inoue J, Imura N (1999) Regulatory role of metallothionein in NFkB activation. FEBS Lett 455:55–58

Sato H, Sato M, Kanai H, Uchiyama T, Iso T, Ohyama Y, Sakamoto H, Tamura J, Naqai R, Kurabayashi M (2005) Mitochondrial reactive oxygen species and c-Src play a critical role in hypoxic response in vascular smooth muscle cells. Cardiovasc Res 67:714–722

Schopler FJ, Cipollina C, Freeman BA (2011) Formation and signaling actions of electrophilic lipids. Chem Rev 111:5997–6021

Schulze-Osthoff K, Los M, Baeuerle PA (1995) Redox signaling by transcription factors NFkB and AP1 in lymphocytes. Biochem Pharmacol 50:735–741

Semenza GL (2000) HIF-1: mediator of physiological and pathological responses to hypoxia. J Appl Physiol 88:1474–1480

Seo JH, Ahn Y, Lee SR, Yeol Yeo C, Chung Hur K (2005) The major target of the endogenously generated ROS in response to insulin stimulation is phosphatase and tensin homolog and not phosphoinositide-3 kinase (PI3 kinase) in the PI-3 kinase/Akt pathway. Mol Biol Cell 16:348–357

Shalini S, Bansal MP (2005) Role of selenium in regulation of spermatogenesis: involvement of activator protein 1. Biofactors 23:151–162

Shalini S, Bansal MP (2007) Alterations in selenium influences reproductive potential of male mice by modulation of transcription factor NFkB. BioMetals 20:49–59

Shukla A, Stern M, Lounsbury KM, Flanders T, Mossaman BT (2003) Asbestos induced apoptosis is protein kinase C delta-dependent. Am J Respir Cell Mol Biol 29:198–205

Solis WA, Dalton TP, Dieter MZ, Freshwater S, Harrer JM, He L, Shertzer HG, Nebert DW (2002) Glutamate-cysteine ligase modifier subunit: mouse Gclm gene structure and regulation by agents that cause oxidative stress. Biochem Pharmacol 63:1739–1754

Stamatakis K, Perez-Sala D (2006) Prostanoids with cyclopentenone structure as tools for the characterization of electrophilic lipid-protein interactions. Ann N Y Acad Sci 1091:548–570

Storz P (2005) Reactive oxygen species in tumor suppression. Front Biosci 10:1881–1896

Strader CR, Fong TM, Tota MR, Underwood D (1994) structure and function of G-protein-coupled receptors. Annu Rev Biochem 63:101–132

Sun H, Tonks NK (1994) The coordinated action of protein tyrosine phosphatases and kinases in cell signaling. Trends Biochem Sci 19:480–485

Tell G, Quadrifoglio F, Tinbelli C, Kelley MR (2009) The many functions of APE-1/Ref-1: not only a DNA repair enzyme. Antioxid Redox Signal 11:601–620

Thannickal VJ, Fanburg BL (2000) Reactive oxygen species in cell signaling. Am J Physiol Lung Cell Mol Physiol 279:L 1005–L 1028, Review

Tobiume K, Matsuzawa A, Takahashi T, Nishitoh H, Morita K, Takeda K, Minowa O, Miyazono K, Noda T, Khijo H (2001) ASK1 is required for sustained activation of JNK/P38 MAP kinase and apoptosis. EMBO Rep 2:222–228

Tobiume K, Saitoh M, Ichijo H (2002) Activation of apoptosis signal-regulating kinase 1 by the stress-induced activating phosphorylation of pre-formed oligomer. J Cell Physiol 191:95–104

Traenckner EMB, Pahl HL, Schmidt KN, Wilk S, Baeuele PA (1995) Phosphorylation of human OkB on serine 32 and 26 controls IkB-α proteolysis and NFkB activation in response to diverse stimuli. EMBO J 14:2876–2883

Ueno M, Masutani H, Arai RJ, Yamauchi A, Hirota K, Sakai T, Inamoto T, Yamuoko Y, Yodol J, Nikaido T (1999) Thioredoxin-dependent redox regulation of p53-mediated p21 activation. J Biol Chem 274: 35809–35815

Valko M, Leibfritz D, Moncol J, Cronin MT, Mazur M, Telser J (2007) Free radicals and antioxidants in normal physiological functions and human disease. Int J Biochem Cell Biol 39:44–84, Review

Venugopal R, Jaiswal AK (1996) Nrf1 and Nrf2 positively and c-fos and Fra1 negatively regulate the human antioxidant response element-mediated expression of NSD(P)H: quinine oxidoreductase1 gene. Proc Natl Acad Sci U S A 93:14960–14965

Villeneuve NF, Lau A, Zhang DD (2010) Regulation of the Nrf2-Keap1 antioxidant response by the ubiquitin proteasome system: an insight into cullin-ring ubiquitin ligases. Antioxid Redox Signal 13:1699–1712

Wang J, Dore S (2007) Heme oxygenase-1 exacerbates early brain injury after intracerebral haemorrhage. Brain 130:1643–1652

Wei SJ, Botero A, Hirota K, Bradbury CM, Markovina S, Laszlo A, Spitz DR, Goswami PC, Yodoi J, Gius D (2000) Thioredoxin nuclear translocation and interaction with redox factor-1 activates the activator protein-1 transcription factor in response to ionizing radiation. Cancer Res 60:6688–6695

Wilson LA, Gemin A, Espiritu R, Singh G (2005) Ets-1 is transcriptionally up-regulated by H2O2 via an antioxidant response element. FASEB J 19:2085–2087

Wu WS, Trai RK, Chang CH, Wang S, Wu J, Chang YX (2006) Reactive oxygen species mediated sustained activation of protein kinase C alpha and extracellular signal-regulated kinase for migration of human hepatoma cell Hepg2. Mol Cancer Res 4:747–758

Xanthioudakis S, Miao G, Wang F, Pan YC, Curren T (1992) Redox activation of Fos-Jun DNA binding activity is mediated by a DNA repair enzyme. EMBO J 11:3323–3335

Xanthoudakis S, Curren T (1992) Identification and characterization of Ref-1, a nuclear protein that facilitates AP-1 DNA-binding activity. EMBO J 11:633–665

Xiong F, Xiao D, Zhang L (2012) Norepinephrine causes epigenetic repression of PKCε gene in rodent hearts by activating NOX1-dependent ROS production. FASEB J 26:2753–2763

Yamamoto Y, Gaynor RB (2001) Therapeutic potential of inhibition of NFkB pathway in the treatment of inflammation and cancer. J Clin Invest 107:135–142

Yamamoto T, Matsuzaki H, Konishi H, Ono Y, Kikkawa U (2000) H2O2 –induced tyrosine phosphorylation of protein kinase cdelta by a mechanism independent of inhibition of protein-tyrosine phosphatase in CHO and COS-7 cells. Biochem Biophys Res Commun 273: 960–966

Yoshizumi M, Abe J, Haendeler J, Huang Q, Berk BC (2000) Src and Cas mediate JNK activation but not ERK ½ and p38 kinases by reactive oxygen species. J Biol Chem 275:11706–11712

Yueh MF, Tukey RH (2007) Nrf2-Keap1 signaling pathway regulates human UGT1A1 expression in vitro and in transgenic UGT1 mice. J Biol Chem 282: 8749–8758

Zhan Y, Brown C, Maynard E, Anshelevich A, Ni W, Ho IC, Oettgen P (2005) Ets-1 is a critical regulator of Ang II-mediated vascular inflammation and remodeling. J Clin Invest 115:2508–2516

Zhang DD, Hannink M (2003) Distinct cysteine residues in keap 1 are required for keap 1-dependent and for stabilization of Nrf2 by chemopreventive agents and oxidative stress. Mol Cell Biol 23:8137–8151

Zhang ZY, Zhou B, Xie L (2002) Modulation of protein kinase signaling by protein phosphatases and inhibitors. Pharmacol Ther 93:307–317

Managing Oxidative Stress/ Targeting ROS

6

Free radicals are produced naturally and continuously within the cell. In order to prevent their accumulation and possible deleterious effects, the antioxidant systems act as ROS scavengers. The "steady state" concentrations of free radicals are determined by the balance between their rates of production and removal by the various antioxidants (Fig. 6.1). Thus, the redox state of a cell and its alterations determine the cellular functioning. Different ROS-mediated activities in fact protect the cells against ROS-induced damage and reestablish or maintain "redox balance" termed also "redox homeostasis." But ROS in higher concentration is responsible for the cellular damage. To protect the cells and organ systems of body against the free radicals, humans have evolved an extremely sophisticated and complex antioxidant protection system.

Antioxidants are classified in many ways: enzymatic (superoxide dismutase, SOD; catalase, CAT; glutathione peroxidase, Gpx; glutathione reductase, GR) and nonenzymatic (metabolic: glutathione, GSH; lipoic acid; L-arginine, coenzyme Q_{10}, melatonin, uric acid, bilirubin, metal-chelating proteins, transferrin; Nutrients: vitamin E, vitamin C, carotenoids; trace metals – selenium, manganese, zinc, flavonoids, omega-3 and omega-6 fatty acids), endogenous (bilirubin, glutathione, lipoic acid, N-acetyl cysteine, NADPH and NADH, ubiquinone coenzyme Q_{10}, uric acid, enzymes SOD/CAT/GSH-Px/GR) and exogenous, or dietary (vitamin C, vitamin E, β-carotene, other carotenoids/ oxycarotenoids; lycopene and lutein; polyphenols – flavonoids, flavones, flavonols, proanthocyanidins) and metal-binding proteins (albumin/copper, ceruloplasmin/copper, metallothionein/copper, ferritin/iron, myoglobin/iron, transferrin/iron).

Endogenous: Cellular Antioxidant Defense System

Several biologically important compounds have been reported to have the antioxidant functions, named endogenous antioxidants. They include the proteins (or enzymes), such as SOD, CAT, GPx, thioredoxin reductase, NOS, heme oxygenase-1, peroxidases, and metallothionein, and nonenzymatic antioxidants such as polyamines, melatonin (5-methoxy-N-acetyltryptamine), NADPH, adenosine, urate, coenzyme Q-10 (ubiquinol), GSH, cysteine, homocysteine, taurine, methionine, s-adenosyl-L-methionine, nitroxides, etc. (Mates et al. 2008; Andre et al. 2010).

In the normal condition, cell is capable to prevent the free radical-induced diseases by the generation of its own endogenous antioxidants. Several enzyme systems detoxify the ROS (Fig. 6.2). Antioxidant response elements (ARE) in the genome include gene sequence of enzymes for the glutathione synthesis and metabolism (γ-glutamylcysteine synthetase, glutathione peroxidase, glutathione reductase, sulforedoxin, and thioredoxin), stress proteins such as heme oxygenase-1, and drug detoxification enzymes

M. Bansal and N. Kaushal, *Oxidative Stress Mechanisms and their Modulation*,
DOI 10.1007/978-81-322-2032-9_6, © Springer India 2014

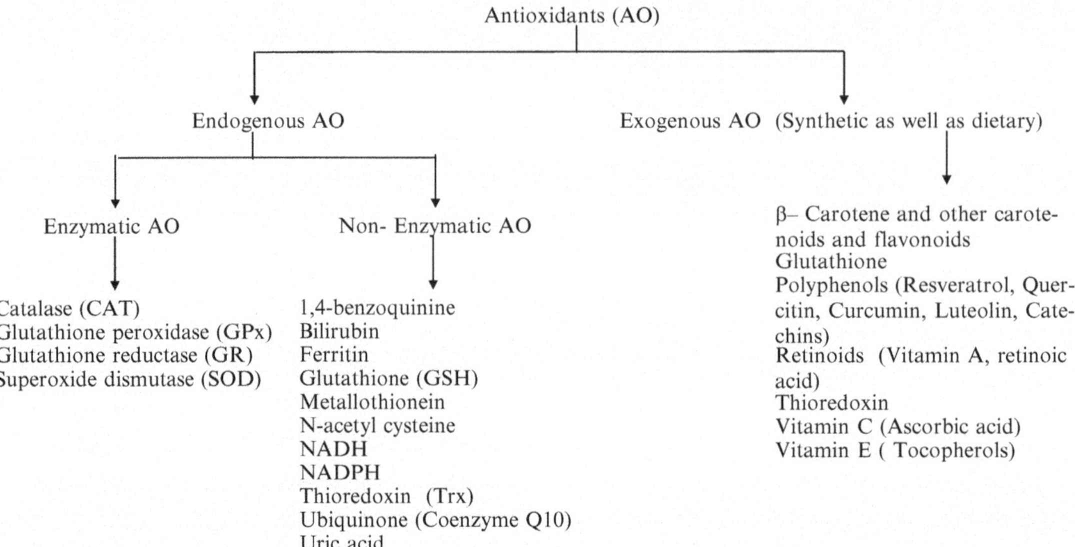

Fig. 6.1 Various groups of antioxidants

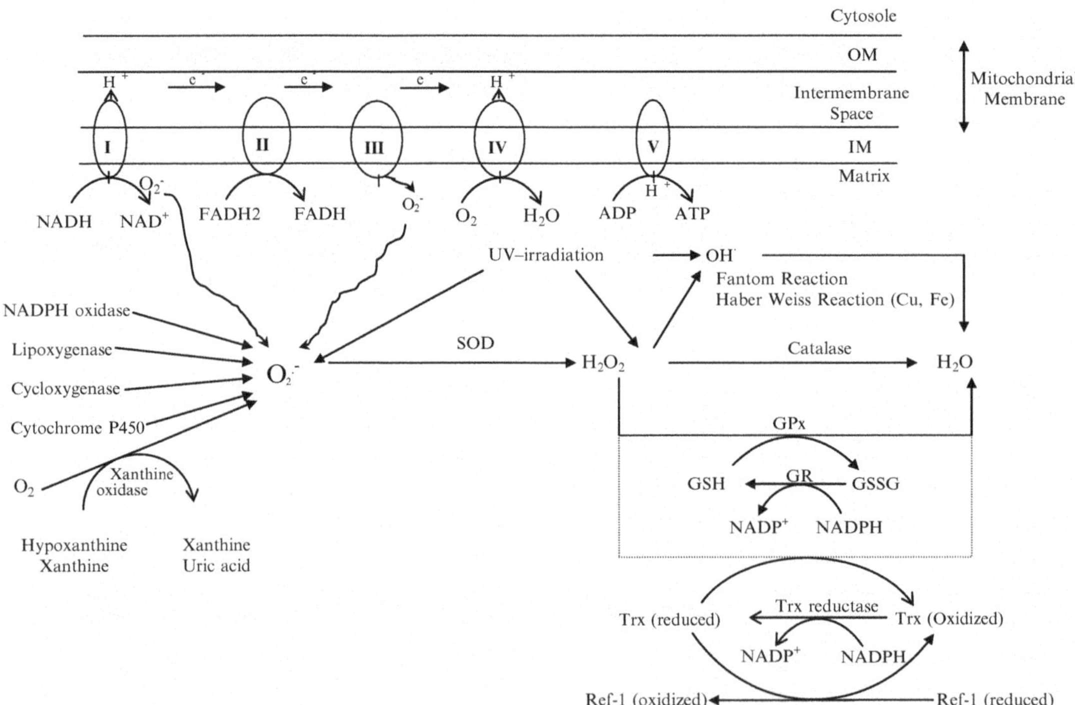

Fig. 6.2 Major ROS production and their scavenging pathways

such as the glutathione transferase and NADPH-quinone oxidoreductase, and cytokines such as interleukin-6. Catalase is a common and most efficient enzyme found in the cell. Each catalase molecule can decompose the millions of H_2O_2 molecules to water and oxygen in every second. Superoxide dismutase (SOD) is also an important endogenous antioxidant enzyme and can exist in several common forms: SOD1, in cytoplasm with Cu/Zn; SOD2, in mitochondria with Mn; and SOD3, outside cells with Cu/Zn. Glutathione peroxidase (GPx), in the cytoplasm of cells, protects against oxidative injury caused by the H_2O_2. GPx also protects from endogenous lipid peroxides, and that is why it is different from catalase and prevents the formation of hydroxyl radical from H_2O_2. It has four protein subunits; each of which contains one atom of the element selenium at its active site and catalyzes the reduction of peroxides (ROOH: including H_2O_2) into alcohol (ROH), using the reducing potential of glutathione. Glutathione reductase (GR) is a flavoprotein enzyme and important cellular antioxidant necessary for the conversion of GSH. Oxidized glutathione (GSSG), which is reduced to glutathione (GSH) by the enzyme glutathione reductase, uses NADPH as an electron donor. The ratio of GSH/GSSG known as "redox ratio" is an important measure of the oxidative stress of an organism, and very high concentration of GSSG may damage many enzymes oxidatively (Fang et al. 2002).

Thiol-containing moieties (such as the cysteine residue in glutathione) have a reducing power, abolishing oxidative power of the ROS by supplying them with electrons, and play a key role in maintaining the intracellular redox equilibrium. Glutathione is a powerful antioxidant and is the major soluble, nonenzymatic antioxidant in various cell compartments. It is the most abundant low-molecular-weight intracellular nonprotein thiol compound (NPSH) synthesized intracellularly from cysteine, glycine, and glutamate. Under the oxidizing conditions, depletion of GSH may occur, and also 1–5 % of the total GSH pool is in the oxidized disulfide form, GSSG. GSH is important in maintaining the –SH groups in other molecules including proteins, regulating thiol–disulfide status of the cell, and detoxifying foreign compounds and free radicals. Sulfur-containing amino acids like methionine and cysteine are the precursors of GSH but also provide –SH groups to react with the H_2O_2 and $OH^\cdot$ radicals and may prevent tissue damage. GSH is capable of scavenging the hydroxyl radical ($OH^\cdot$) and singlet oxygen ($^1O_2$) directly or detoxifying H_2O_2 and lipid peroxides by the catalytic action of the glutathione peroxidase. GSH is also involved in the amino acid transport through the plasma membrane and regeneration of some important antioxidants; for example, GSH can reduce the tocopherol radical of vitamin E directly or indirectly, via reduction of the semidehydro-ascorbate to ascorbate.

GSH may also modulate the cell signaling and a variety of cellular events. Through the thiol–disulfide exchange reactions, GSH is responsible for the protein modifications that regulate the cellular function and survival. Under conditions of oxidative stress, the degree of GSH depletion, affected GSH pool in cytosol/endoplasmic reticulum/mitochondria/nucleus, and the types of stimuli determine the final redox status (Biswas and Rehman 2009). Another, a group of small ubiquitous proteins containing two redox-active cysteine residues in the catalytic site is the thioredoxin family of proteins which are functionally associated with GSH redox system. This activity not only enables them to reversibly reduce the disulfide bonds in oxidized protein but also to reversibly bind to the signaling molecules, whose activity is modulated in a redox-sensitive manner (Circu and Aw 2008).

Thioredoxin reductase (TrxR) in conjunction with thioredoxin (Trx) is a ubiquitous oxidoreductase system with the antioxidant and redox regulatory role. Mammalian TrxR containing selenocysteine residue has a highly reactive site, making it highly reductive. In addition to reducing Trx, it reduces several other substrates, e.g., lipoic acid, lipid hydroperoxides, ebselen, etc., and hence has use in antioxidant defense (Nordberg and Arner 2001). TrxR catalyzed regeneration of several antioxidant compounds including the ascorbic acid, selenium-containing substances, lipoic acid, and ubiquinone (Q10). Lipoic acid, another endogenous antioxidant, characterized as a "thiol" or "biothiol," is a sulfur-containing molecule. Lipoic acid and its reduced form, dihydrolipoic acid, has been called a "universal antioxidant," as this is capable of quenching the free radicals in both lipid and aqueous domains. Lipoic acid may also produce its antioxidant effect by chelating with prooxidant metals (Demir et al. 2003).

The cysteine-rich metallothionein protein also displays the antioxidant property. Other enzymes, including the quinine reductase and heme oxygenase, can prevent the formation of oxygen-derived radicals. These enzymes are induced as part of a concerted response to the oxidative stress. Cells also protect themselves with the antioxidant systems involving a cascade of functional redox molecules, such as thioredoxin (Trx) and redox factor 1 (Ref-1), or the radical-scavenging vitamin C (cytosolic) and E (membrane-bound). Endogenous compounds (glutathione, ubiquinol, uric acid, and bilirubin) make major contributions to the detoxification of ROS. Uric acid, which is a product of purine metabolism, contributes 60–70 % of plasma antioxidant capacity (Duplancic et al. 2011). It has been shown to act as an intracellular free radical scavenger and is active in reducing the oxidative stress by reacting with the ROS including nitric oxide (NO), peroxyl radicals (ROO·), and hydroxyl radicals (OH).

Coenzyme Q10 (CoQ10) is a naturally occurring antioxidant and a prominent component of the mitochondrial electron transport chain. It is recognized as an obligatory cofactor for the function of uncoupling proteins and a modulator of transition pores. CoQ10 have been found to affect the expression of genes involved in the human cell signaling, metabolism, and transport (Ernster and Forsmark 1993), and some of its effects of exogenous administration may be due to this property. CoQ10 also influences on the endogenous antioxidant defense by increasing the superoxide dismutase and glutathione peroxidase (Kim et al. 2007).

Exogenous: Essential Trace Elements, Vitamins, Dietary Supplements, and Their Modes of Action

Main source of the exogenous antioxidants is through food. Bioactive food components with antioxidant activity include vitamin C, vitamin E, N-acetyl cysteine (NAC), carotenoids, CoQ_{10}, alpha-lipoic acid, carotenoids, lycopene, selenium (Se), flavonoids, etc. (Bouayed and Bohn 2010; Andre et al. 2010). Endogenous and exogenous antioxidants act synergistically to maintain or reestablish the redox homeostasis, such as during regeneration of vitamin E by glutathione or vitamin C to prevent the lipid peroxidation process (Bouayed et al. 2009). Dietary or natural antioxidants play an important role in helping the endogenous antioxidants in scavenging the excess of free radicals. However, the dietary antioxidants can only have beneficial effects in the radical scavenging or effects on the redox potential if they are present in tissues or body fluids at sufficient concentrations. For many dietary components, absorption is limited or metabolism into derivatives that can be easily assimilated reduces the antioxidant capacity. It is important to know that the various antioxidants in fruits and vegetables influence their defense against the free radicals individually. Also some specific antioxidants have limited function because of their inability to penetrate the blood–brain barrier, poor absorption and conversion to the prooxidants (e.g., ascorbates and carotenoids) under certain physiological conditions (Poljšak et al. 2005). The antioxidants that are reducing agents can also act as prooxidants, since they are capable of reacting with the molecular oxygen (e.g., ascorbic acid), and generate superoxide radicals under aerobic conditions. These will dismutase to the H_2O_2 that can enter cells and react with superoxide or reduced metal ions to form the highly damaging hydroxyl radicals (OH) (Anderson et al. 2003). The most common antioxidant mechanisms in vitro involve the hydrogen-atom transfer, electron donation, or metal chelation (Leopoldini et al. 2011), although the carotenoids are also effective singlet oxygen ($^1O_2$) quenchers, which can be important in tissues, e.g., skin, where activation of oxygen may occur (Omoni and Aluko 2005). Plasma antioxidant capacity or related measures of the antioxidant effects of the dietary phytochemicals have often been considered but are limited to saturation or excretion, e.g., in fruits, vegetables, fish oil, and green tea extract (Young et al. 2002).

In addition to the concentration of the above-mentioned antioxidants, the presence of metal ions has been reported to play an important role in the detoxification mechanisms. Dietary antioxidants such as phenolics can display prooxidant activities in the presence of metal ions owing to their reducing capacity and form chelates such as with the transition metals, iron, and copper (Galati and O'Brien 2004). Mechanism of the antioxidative action of the natural compounds as explained is of two types: (a) direct action of antioxidants on the free radicals by a scavenging process characterized by the donation of hydrogen atoms or electrons and (b) as deactivators of singlet oxygen or by converting hydroperoxides to nonradical species when the antioxidants absorb UV radiation or intervene in anti-oxidation process as chelators of the transition metal ion catalysts (Maisuthisakul et al. 2007). Also, the strong reducing power of antioxidants may also affect metal ions especially Fe^{3+} and Cu^{2+}, increasing their ability to form the highly reactive hydroxyl radical concentrations, potentially harmful radicals, originating from the peroxides via the Fenton reaction, discussed in Chap. 1 (Valko et al. 2007). Further, phenolics, when scavenging the free radicals, can form less reactive phenoxyl radicals, which are stabilized by the delocalization of the unpaired electrons around the aromatic ring (Rice-Evans et al. 1996). However, even though these radicals are relatively stable, they can also display prooxidant activities inducing cellular damage (Galati and O'Brien 2004). It is well established that one of the chemopreventive mechanisms of the polyphenols (or fruits and vegetables rich in antioxidants) against cancer development is the inhibition of the initiation, the first step of carcinogenesis occurring following oxidative DNA damage leading to the mutagenesis (Lee et al. 2004). Further, as the polyphenols have been evidenced as antioxidants or prooxidant in vivo in the human (Halliwell 2008), some of their beneficial effects have been explained due to their prooxidant effects through inducing the endogenous protective enzymes by exerting mild oxidative stress. In the polyphenol, curcumin is the most active component of turmeric, a spice derived from dried rhizome of the plant *Curcuma longa*. Curcumin, recognized as an antioxidant compound, acts as metal chelator (Baum and Ng 2004) and free radical scavenger, by lowering lipid peroxidation and preserving cellular antioxidant enzymes and GSH (Rahman et al. 2006).

Vitamins C and E are the important nonenzymatic antioxidants, which react with free radicals to form the radicals themselves which are less reactive than the original radicals. They break radical chain reactions within the cell membrane by trapping peroxyl and other reactive radicals. Vitamin E, first classic antioxidant, is a very strong lipophilic chain-breaking antioxidant. It is a chiral compound with eight stereoisomers in nature, but only α-tocopherol is the most bioactive form in humans. Vitamin E can transfer its phenolic hydrogen to a peroxyl free radical of a peroxidized PUFA; thus, it breaks the radical chain reaction and prevents peroxidation of PUFA in cellular and subcellular membrane phospholipids. Continuous recycling of the cellular α-tocopherol and its reduction with L-ascorbic acid makes this a key player in preventing the generation and propagation of oxidative stress in the biomembranes. Vitamin C as a reducing agent reacts with vitamin E radical to yield a vitamin C radical while regenerating vitamin E. Like vitamin E radical, the vitamin C radical is not a reactive species as its unpaired electron is energetically stable.

Under the physiological conditions, vitamin C predominantly exists in its reduced form as ascorbic acid and also exists in trace quantities in the oxidized form as dehydroascorbic acid (DHA). In all cells, vitamin C is transported as DHA via facilitative glucose transporters, and once inside the cell, DHA is rapidly reduced and accumulates as ascorbic acid (Vera et al. 1993). In the plasma and cells, ascorbic acid is a powerful antioxidant, quenching ROS and RNS (Halliwell and Gutteridge 1999). Intracellular vitamin C can prevent cell death and inhibit mutations induced by the oxidative stress (Lutsenko et al. 2002). A large population has low plasma concentrations of vitamin C, and its supplementation may be beneficial for them. Health benefits of vitamin C include antiatherogenic, anticarcinogenic, and immunomodulator. It is beneficial in reducing the incidence of stomach cancer, preventing lung and colorectal cancer (Fang et al. 2002; Demir et al. 2003).

Carotenoids are considered to be beneficial in the prevention of a variety of major human diseases. They favor apoptotic death of different types of cancer cells, modulate the immune system, and influence cell function through changes in the membrane fluidity and cell–cell communication (Elliott 2005). It is considered as a strong antioxidant and the best quencher of singlet oxygen ($^1O_2$). It provides antioxidant protection to the lipid-rich tissues. β-carotene, a fat soluble carotenoid, considered as pro-vitamin can be converted into active vitamin A (retinol), which is essential for the vision. Retinoids include retinol (absorbed from foods), retinal (reversibly oxidized form), and retinoic acid (irreversibly oxidized metabolite, which only retains some of the vitamin's properties). Lycopene is another carotenoid with antioxidant activity. It is found beneficial against prostate cancer. Lycopene is present in many fruits and vegetables, with tomatoes and processed tomato products being among the richest sources. Several recent studies suggest that dietary lycopene is able to reduce the risk of chronic diseases such as cancer (Giovannocci 1999) and cardiovascular diseases (Rao 2002). Because of its high numbers of conjugated dienes, lycopene is one of the most potent antioxidants, with a single-oxygen-quenching ability twice as high as that of β-carotene and ten times higher than that of α-tocopherol (DiMascin et al. 1989). Lycopene as carotenoid may chemically interact with ROS and undergo oxidation and may prevent ROS-induced cell damage (Palozza et al. 2010).

Leafy vegetables are important dietary sources of the minerals, trace elements, and phytochemicals (absorbed from soil) with health-protective and immune-strengthening properties. Molecular evidence suggests that the trace elements and antioxidant molecules in the green leafy vegetables lower the risks of cancer and cardiovascular diseases (CVD) through the mechanisms that modulate free radical attack on the nucleic acids, proteins, and PUFAs (Borek 2003). It is demonstrated that the leafy vegetables have a higher antioxidative capacity than either fruits or root crops (Lako et al. 2007). For protection, plants manufacture the organic detoxification molecules in which iron, zinc, and selenium feature as essential structural components. Dark-green leafy vegetables, therefore, are the primary sources of the minerals, trace elements, and antioxidant molecules, such as the polyphenols and carotenoids, all of which function in the enzymatic and/or nonenzymatic-mediated plant defenses against radiation-induced oxidative stress (VanDuyn and Pivonka 2000). Selenium (Se) is an essential trace mineral required to form the active site of several antioxidant enzymes including glutathione peroxidase. At low dose, it is important for its antioxidant, anticarcinogenic, and immunomodulatory activity. Some compounds contribute to an antioxidant defense by chelating transition metals and preventing them from catalyzing the production of free radicals in the cell. Metal-chelating antioxidants such as transferrin, albumin, and ceruloplasmin avoid radical production by inhibiting the Fenton reaction catalyzed by the copper and iron. Particularly important is the ability to sequester iron, which is the function of iron-binding proteins such as transferrin and ferritin (Imlay 2003). Cells do not have a well-characterized pool of low-molecular-weight iron (Voogd et al. 1992). If these come into contact with ascorbate, the prooxidant effects may occur. Ascorbic acid reduces Fe(III) to Fe(II) which reduces the oxygen to hydroxyl radical (Halliwell and Gutteridge 2005).

Further, the principal health benefits of fruits and vegetables lie in the phytonutrients such as the polyphenols, flavonoids, terpenoids, and quinines. Many phytochemicals occur at low concentrations in the plasma and tissues and have many physiological effects. Flavonoids are polyphenolic compounds found in most plants with the antioxidant activity. Different flavanoids are reported to prevent or delay a number of chronic and degenerative ailments such as cancer, cardiovascular diseases, arthritis, aging, cataract, memory loss, stroke, Alzheimer's disease, inflammation, and infection. Flavonoids are derivatized extensively by the glucuronidation, methylation, and sulfation in the intestinal mucosa and liver. Antioxidant activity of the derivatives is commonly less than that of the parent flavonoid, as shown for the methylation or sulfation of the quercetin (Lotito et al. 2011). The flavonoids (subclasses: flavonols, flavones, isoflavones, flavanones, flavonols, and anthocyanidins) possess in general the strong antioxidant activity, because of their free radical scavenging and the metal-chelating capabilities (Leopoldini et al. 2011), as well as their ability to interact with the enzymatic and nonenzymatic mechanisms of redox balance regulation within the cells and tissues (Aron and Kennedy 2008). Under certain conditions, the flavonoids may also exert a marked prooxidant activity, thus being potential cytotoxic compounds. They undergo transition metal- or peroxidase-catalyzed reactions, leading to the formation of ROS and highly reactive phenoxyl radicals, which can damage biological molecules such as proteins and DNA (Galati and O'Brien 2004). Among the polyphenolic compounds, resveratrol (3,4′,5-trihydroxystilbene) has shown anticarcinogenic, anti-inflammatory, and cardioprotective properties. Due to its hydroxylated structure and its potential for the electron delocalization across the conjugated structure, the resveratrol is a well-recognized and potent antioxidant. It can act both as free radical scavenger and as the metal chelator. It also modulates diverse enzymes involved in the redox status regulation, including catalase, SOD, GR, NADPH oxidase, xanthine oxidase, myeloperoxidase, and LOX (Delmas et al. 2005; Pervaiz and Holme 2009).

Several scavengers and neutralizers of the ROS or RNS are present in the natural or dietary products, but their therapeutic potential as neutralizing agent is low. However, for many dietary phytochemicals, direct antioxidant effects may be less important for the health than other effects including effects on the cell signaling and gene expression in vivo. These effects can be demonstrated at low concentrations.

Oxidative Stress-Scavenging Strategies/Targeting: Endogenous and Exogenous

In the presence of endogenous and exogenous antioxidants, human body has developed several strategies to protect itself from the oxidative stress and ROS damage. With these activities, the normal oxidative metabolism and ROS-mediated cellular responses continue.

Antioxidants are defined as the substances which at low concentration significantly inhibit or delay the oxidative process and often get oxidized themselves. Endogenous and exogenous antioxidants are used to neutralize the free radicals and to protect the body from free radicals by maintaining the redox balance. Although the antioxidant supplements are ineffective in various pathological conditions, sometimes they can pose adverse effects. However, the reports in this direction are conflicting and contradictory. Considering these situations, Poljsak (2011) has discussed that the homeostatic mechanisms in cells govern the amount of allowable antioxidative activity. Intake of an exogenous antioxidant could influence the complex endogenous antioxidative defense of the cells, while the overall antioxidative capacity remains unaffected. In other words, dosing cells with the exogenous antioxidants might decrease the rate of synthesis or the uptake of endogenous antioxidants, so that the total "cell antioxidant potential" remains unaltered. Thus, this model named as "oxidative stress compensation model" explains why the dietary supplements of antioxidants have minimum effect on the longevity (Cutler 2003). Further, it is explained that most humans maintain their required normal level of the oxidative stress, in spite of consuming the excess antioxidant supplements in their diet.

Antioxidant effects of the plants occur because of their acute, but transient, prooxidant (electrophilic and/or ROS-generating) effects. Electrophiles abstract electrons (oxidized) from the other molecules; this results in the antioxidant defenses and innate immunity activation. Potent phytonutrients such as the curcumin in turmeric, quercetin in onions, sulforaphane in broccoli, and epigallocatechin gallate in green tea produce their antioxidant effects by acting acutely as prooxidants in the cell and inducing metabolic hormesis. Especially the electrophilic chemical nature of these compounds upregulates the Keap/Nrf2-mediated transcription of various antioxidant defense genes that contain antioxidant

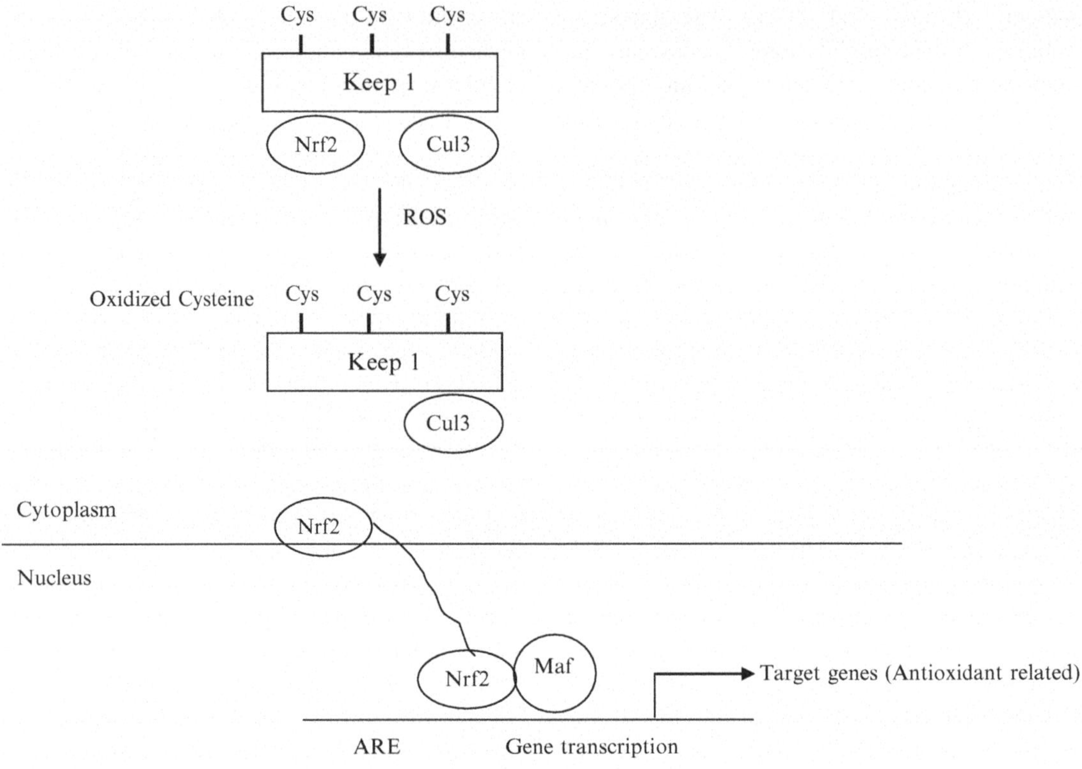

Fig. 6.3 Redox regulation of Nrf2-ARE pathways

response elements (electrophile response elements) (Fig. 6.3). Phytonutrients have evolved to activate the mammalian cellular targets such as sirtuins, cytoskeleton, tyrosine protein kinases, phosphoinositide 3 kinase, MAP kinases, and transcription factors such as keep1 and NRf2 (Murakami and Ohnish 2012). Presence of the redox-cycling metal ions with the antioxidants might result in a synergistic effect, resulting in an increased free radical formation. So whether an antioxidant functions as an antioxidant or prooxidant is at least determined by the redox potential of the cellular environment, the presence/absence of transition metals, and local concentrations of that antioxidant.

The phytochemicals have potentials for regulating and modulating the human health, as shown by both the experimental and epidemiological approaches. Several plant antioxidants (e.g., sulforaphane, resveratrol, curcumin, flavonoids, green tea catechins, and diallyl sulfides) exhibit hermetic properties by acting as "low-dose stressors" that may prepare cells to resist the more severe stress (Poljsak and Milisav 2012). Low doses of these phytochemicals are usually present in the plants, activate the various cell signaling pathways, and provide resistance but are cytotoxic at high doses. These so-called adaptive responses and their molecular mechanisms induced by the most known plant hormone antioxidants are well reviewed (Spenciale et al. 2011). Several protein kinases were identified as binding proteins of flavonoids, including myricetin, quercetin, and kaempferol; isothiocyanates, sulfur-containing phytochemicals present in cruciferous plants, are well known to target the Keap1 for activating the Nrf2 for inducing self-defensive and antioxidant gene expression. Recently CD36 as a cell surface receptor for the ursolic acid, a triterpenoid ubiquitously occurring in plants (Murakami and Ohnish 2012). Importantly, these target proteins are indispensable for the phytochemicals to exhibit, at least in part, their bioactivities.

In addition to the therapeutic use of natural or synthetic molecules as antioxidants, small molecules that mimic antioxidant enzymes are of interest for the efficient treatment of many diseases requiring antioxidants. These developments are well reviewed in Mates et al. (2012). Some examples include specific spin traps like α-phenyl-*N*-tert-butyl nitrone (an EC-SOD mimetic); porphyrins (Mn(III)tetrakis(4-benzoic acid) porphyrin chloride, so-called MnT-BAP); the SOD mimetic M40419, also named tempol (4-hydroxy-2,2,6,6-tetramethylpiperidine-1-oxyl) (Samai et al. 2007); Mn(II)-pentaazamacrocycles (Maroz et al. 2008); copper complexes of nicotinic–aromatic carboxylic acids (Suksrichavalit et al. 2008); iNOS and myeloperoxidase inhibitors; and lipid peroxidation inhibitors/blockers edaravone and lazaroids/tirilazad (Rahman 2012). Synthetic catalytic scavengers of ROS with SOD, CAT, and peroxidase mimetic activity of the Eukarion (EUK) structure have been shown to have anticancer activity in lung models (Vorotnikova et al. 2010). Further, modulation of the intracellular ROS level by overexpression of Mn-SOD or Cu–Zn-SOD inhibited ras-induced transformation (Yang et al. 2002). On the other hand, it has been used as a recombinant adenoviral vector expressing the radical-scavenging enzyme Mn-SOD (Ad-Mn-SOD), showing antitumor effect by itself, but this effect is more pronounced in the presence of the commonly used anticancer drug, 1,3-bis(2-chloroethyl)-1-nitrosurea (BCNU) (Yang et al. 1999). Additionally, a study has been carried on human cancer protection with intravesical injection of Mn-SOD plasmid liposomes (Tarhini et al. 2011).

The targeted delivery of the modified antioxidants is also another strategy to combat oxidative stress. Examples include: providing the protection against mitochondrial oxidative damage, by the triphenylphosphonium-conjugated antioxidants, e.g., plastoquinone. It is a very effective electron carrier and the antioxidant of chloroplasts. It was conjugated with the decyltriphenylphosphonium to obtain the action which easily penetrates through the membranes. Very low (nano- and subnanomolar) concentrations of 10-(6′-plastoquinonyl) decyltriphenylphosphonium (SKQ1) were found to prolong the life span of a fungus (*Podospora anserina*), a crustacean (*Ceriodaphnia affinis*), an insect (*Drosophila melanogaster*), and a mammal (mouse) (Skulachev et al. 2011). These peptides scavenge H_2O_2 and peroxynitrite and inhibit lipid peroxidation. By reducing mitochondrial ROS, they inhibit the mitochondrial permeability transition and cytochrome c release, thus preventing oxidant-induced cell death (Rocha et al. 2010). Another example of targeted activity is a transduced fusion protein for the SOD with the specific nuclear localization signal and membrane translocation sequence signal (Kim et al. 2008). These purified SOD fusion proteins were efficiently translocated into mammalian cells with the enzymatic activities. Viability of the cells treated with paraquat was markedly increased by the transduced fusion proteins, and hence these peptides are useful for targeting the specific localization of the therapeutic proteins in various human diseases. Also a similar targeting strategy of SOD (also involved in many forms of vascular oxidative stress, including ischemia/reperfusion, hypertension, and inflammation against superoxide attack) and anti-PECAM (platelet-endothelial cell adhesion molecule)/SOD conjugates were created which specifically bind to the endothelial cells, but not PECAM-negative cells, and hence used for the management of vascular oxidative stress (Shuvaev et al. 2007). In one of the studies, two types of the cationized CAT deliveries, HMD- and ED-conjugated CAT, were developed to achieve hepatic delivery of catalase (CAT) for the prevention of CCl4-induced failure in mice. These compounds showed increased binding to the HepG2 cells and were rapidly taken by the liver (Ma et al. 2006). H_2O_2-induced cytotoxicity in the HepG2 cells was significantly prevented by the preincubation of the cells with cationized CAT derivatives.

Further, animal studies support a causative role for the oxidative stress in the pathogenesis of the hypertension, but there is no solid evidence in humans. However, biomarkers of the excess ROS are increased in patients with hypertension, and oxidative damage is important in the molecular mechanism associated with cardiovascular and renal injury in hypertension. Though the antioxidant clinical trials in hypertension are ineffective, the strategies that combat the oxidative stress by targeting NOXs (major source of ROS in cardiovascular system) may have therapeutic potential (Montezano and Touyz 2012).

Molecular Network and Modes of Actions of Antioxidants in Transcriptional Regulation of ROS and Oxidative Stress

Because of the clearly established role played by the ROS as signaling molecules, regulation of the gene expression by the antioxidants was an important area of study in modulation of the ROS-dependent disorders in biological systems. Many studies (mostly in the cell culture systems) have been reported on the topic which is of pharmaceutical interest currently. Some of these reports for various antioxidants are presented here:

Vitamin E

Vitamin E has been shown to significantly modulate the several signaling pathways and gene expression within the cells and thus might contribute to maintain the cellular behavior and function within the physiological range (Azzi et al. 2004). It might protect against the atypical cell proliferation and downregulate the inflammatory processes. Vitamin E modulates the activity of several enzymes involved in the signal transduction, such as protein kinase C, protein kinase B, protein tyrosine kinases, lipoxygenases, cyclooxygenase-2, phospholipaseA2, and diacylglycerol kinase. Activation of some of these enzymes after the stimulation of cell surface receptors with growth factors or cytokines can be normalized by vitamin E (Zingg 2007). Tocopherols have shown a constitutive activation of the PI3K signaling pathway and phosphorylation of the downstream effectors, namely, phosphate-dependent kinases (PDK) 1/2 and Akt in the human mastocytoma cell lines, leading to the inhibition of cell proliferation. A number of experimental studies show that vitamin E is able to inhibit cortical neurons against H_2O_2-induced apoptosis, likely through activation of PI3K, Akt, and ERK1/2, with the subsequent overexpression of antiapoptotic protein Bcl-2. Many in vitro studies have characterized the effects of tocopherols on the NF-κB activation in many different cell types. Vitamin E derivatives inhibit the TNF-α induced NF-kB activation in the human Jurkat T cells in concentration-dependent manner (Suzuki and Packer 1993). Vitamin E acts by quenching ROS-dependent signaling at various molecular levels, e.g., at the NADPH oxidase and/or PKC activation steps. In both these cases, α-tocopherol and/or its derivatives have been demonstrated to directly downregulate the translocation of enzymes or enzyme's subunits to cell plasma membrane (Zingg 2007).

Incubation of the human keratinocytes with α-tocopherol prevented the transactivation of AP-1 provoked by the UV irradiation. The molecular mechanism by which α-tocopherol inhibits the activity of AP-1 and also of NF-κB was investigated in the human peripheral blood T cells. It appears to interfere with the binding of two transcription factors to the promoter region of at least one of the dependent genes, namely, IL-4 (Li-Weber et al. 2002). α-tocopherol alone induces AP-1 DNA binding in a protein kinase C-independent manner in vascular smooth muscle cells (Stauble et al. 1994). Such an effect was not observed with β-tocopherol, suggesting that the effect may not be dependent on the antioxidant property of α-tocopherol.

Carotenoids

The carotenoids change cellular redox status through redox-sensitive cell signaling pathways and also by other non-antioxidant properties in vivo. In human macrophage cell line (HL-60) stimulated with phorbol 12-myristate 13-acetate (PMA), high concentrations of β-carotene (20 μM) significantly enhanced the release of two pro-inflammatory mediators (IL-8 and TNFα) while a lower concentration (2 μM) strongly prevented the inflammatory events stimulated by PMA (Yeh et al. 2009). This biphasic effect of the β-carotene is probably dependent on the cellular production of ROS at different situations, such as different effects on the PKC-dependent activation of plasma membrane NADPH oxidase activity. NF-kB DNA-binding activity increased in both the human leukemia and colon adenocarcinoma cells following treatment with β-carotene. Since both the α-tocopherol and N-acetylcysteine (NAC) were able to diminish the β-carotene-induced NF-kB DNA-binding activity, ROS production strongly suggests that the oxidative stress may contribute to regulation of this activity. Carotenoids regulate the MAPK cell signaling pathway also. In lung of ferrets chronically exposed to cigarette smoking, increase of the phosphorylated p38, JNK, and c-Jun were observed, and these changes were prevented if the animals were supplemented with low doses of the β-carotene, but not high doses, throughout the period of smoke exposure (Liu et al. 2004).

β-carotene has been reported to induce the oxidative stress in oral tumor cells, resulting in expression of the stress proteins (as hsp70/hsp90), which are nuclear binding proteins in apoptosis (Toba et al. 1997). Further, β-carotene downregulates the expression of COX-2 in colon cancer cells and follows apoptosis induction. Peroxisome proliferator-activated receptor (PPAR), a regulator of the COX-2, has been reported to be modulated by the carotenoids (Sharoni et al. 2002). There is a strong evidence for an involvement of the carotenoids in regulation of apoptosis through the modulatory effects on activation of the caspase cascade and on expression of Bcl-2 family proteins and transcription factors (Paola et al. 2006). Inhibitory or stimulatory actions at these pathways are likely to affect the cellular functions by altering the phosphorylation state of the target molecules and by modulating gene expression. The carotenoid has been reported to directly modulate the several redox-sensitive signaling pathways altered in cancer (Palozza et al. 2011), via their cell regulatory functions. The redox molecules regulated by lycopene involve antioxidant response elements (ARE), ROS-producing enzymes, small GTPases, MAPK, NF-kB, AP-1, and redox-sensitive proteins involved in cell growth, such as p53 and the Bcl-2 family of proteins. Lycopene and β-carotene have repeatedly been shown to reduce the expression and transcriptional activity of AP-1, but at high doses, they can upregulate the expression of AP-1 proteins (Sharoni et al. 2004). Carotenoids, particularly lycopene, have been demonstrated to induce the detoxification phase II enzymes by stimulating Nrf2 nuclear translocation and its interaction with the ARE transcription system. The ability of carotenoids to induce the phase II enzymes may be dependent on certain products of its oxidative metabolism. In human bronchial epithelial cells, the antioxidant and anticarcinogenic functions appear to be mediated by the apo-10′-lycopenoids, which can activate the Nrf2 pathway (Lian and Wang 2008). Lycopene also inhibited the MAPK activation, as well as NF-κB nuclear translocation, in murine dendritic cells challenged with the LPS. Lycopene also blocked the growth factor stimulation of in vitro migration of human retinal pigment epithelial cells, by quenching through the PI3K/Akt pathway and on the ERK and p38 phosphorylation (Chan et al. 2009).

Retinoids

A series of signaling events are initiated by the binding of retinoids to their specific receptors in nucleus of the target cells. The resulting complexes bind to the retinoic acid responsive elements (RAREs) in the promoters of RA-inducible genes to initiate the gene expression (Kastner et al. 1995). Retinoic acid has been shown to inhibit the upregulation of JAK1 and STAT3 activity induced by the IL-6 in a lymphoblastoid B-cell line (Zancai et al. 2004). Upregulation of the MKP-1 and MKP-2 by RA was demonstrated to occur in the rat neonatal cardiomyocytes in the primary cultures. RA-dependent upregulation of the MKP-1 has also been reported to play a role in counteracting the activation of the Src-MAPK signaling pathway induced by the HIV accessory protein Nef, detectable in the nephropathy associated with the HIV infection, and leading to the proliferation and dedifferentiation of the podocytes (Lu et al. 2008). In the SH-SY5Y cells, RA caused the PI3K phosphorylation by binding to its specific nuclear receptor, which in turn reacts directly with a subunit of the kinase (Masia et al. 2007).

Under the specific conditions, RA has been shown to stimulate the AP-1 rather than inhibit it: 24 h RA treatment of the cultivated Sertoli cells induced apoptotic death by upregulating the JNK/AP-1 pathway (Zanotto-Filho et al. 2008). At very low concentrations (1 nM–1 μM), RA has been found to inhibit the Nrf2-dependent gene transactivation in the human mammary MCF7 cell line (Wang et al. 2007). On the contrary, higher amounts of RA (1–50 μM) significantly stimulated the binding of this transcription peptide to ARE in the promoter of cytoprotective genes, in particular the glutamate-cysteine ligase gene (Tan et al. 2008).

Vitamin C

Ascorbic acid, an ROS scavenger, has been recognized as the cell signaling modulator by acting at the level of redox-sensitive molecular pathways and transcription factors, although it plays a less important part than the lipophilic antioxidants. It strongly counteracts the expression of NADPH oxidase's subunit p47phox and its activation in mouse microvascular ECs challenged in vitro with LPS or with H_2O_2 (Wu et al. 2007). Vitamin C-induced activation of the lipid signaling enzyme, phospholipase D, in the vascular endothelial cells was regulated by the upstream activation of another signaling phospholipase, phospholipase A2, cyclooxygenase (COX), and lipoxygenase (LOX) through the formation of arachidonic acid metabolites involving oxidative stress, calcium, and iron (Steinhour et al. 2008). A dual molecular action of the vitamin C in signal transduction provides a direct linkage between the redox state of the vitamin C and NF-kB signaling events. Ascorbic acid quenches the ROS intermediates involved in the activation of NF-kB and is oxidized to the dehydroascorbic acid which directly inhibits the IKKβ and IKKα enzymatic activity (Caramo et al. 2004). The increased phosphorylation of p38 and JNK but not the ERK was prevented by the cell pretreatment with ascorbic acid or NAC (Kyaw et al. 2002).

Several reports point to the intracellular vitamin C influencing the inflammatory, neoplastic, and apoptotic processes via inhibition of the redox-sensitive transcription factor, NF-kB, involving its upstream pathway activation, i.e., the phosphorylation of IκB, its inhibitory component (Carcamo et al. 2002). However, a non-antioxidant action also appears to be involved. This was demonstrated in human dendritic cells generated from the blood monocytes, by treatment with both the granulocyte monocyte-colony stimulating factor and IL-4, and finally being incubated in the presence of the inflammatory cytokines IL-1β, TNFα, and IL-6. Treatment with the ascorbate and α-tocopherol inhibited the intracellular ROS and subsequently lead to the upregulation of the PKC, p38 MAPK, and IκB phosphorylation, due to the pro-inflammatory stimuli (Tan et al. 2008). Effect of the ascorbate on the AP-1 has been reported to be partially or entirely quenching its upregulation by the oxidant stimuli. In vitro vitamin C supplementation inhibits the endothelin-1 mediated cell signaling and proliferation in the rat aortic SMCs. Pretreatment of the SMCs with ascorbate significantly attenuated the induction of hemoxygenase 1 (HO-1) by the oxidized LDL (oxLDL) (Anwar et al. 2005). Induction of the Nrf2 translocation from the cytoplasm to the cell nucleus is promoted by the oxidative stress-inducing agents whereas reducing compounds such as the ascorbic acid may actually quench Nrf2-driven gene expression and consequent cellular responses.

Curcumin

Curcumin has also emerged as a cell signaling regulator that targets diverse receptors, kinases, enzymes, cytokines, and transcriptional factors potentially involved in the pathological cellular events. In an experimental colitis rat model, the curcumin ameliorated the inflammatory status, decreasing the NF-κB and its upstream inducers toll-like receptor-4 and myeloid differentiation proteins (Lubbad et al. 2009). The ability of the curcumin to modulate MAPK signaling pathways might contribute to its inhibition of the inflammation and cancer cell growth. Curcumin also attenuated experimental colitis by reducing the activity of p38 MAPK. Further, in human astroglioma cells, PMA triggered the MMP-9 expression and also ERK, JNK, p38 kinases, NF-κB, and AP-1. All of these proteins were strongly downregulated by the curcumin, likely through the inhibition of the upstream PKC (Woo et al. 2005). It has been suggested that the curcumin affects PKC activity by competing with the Ca^{2+}-binding domains of the protein and by exerting either inhibitory or stimulatory influence at the lower or higher Ca^{2+} levels, respectively (Mahmmoud 2007). All these findings strongly confirm curcumin to be more than a simple antioxidant molecule, at least in regard to the kinase modulation.

It has been demonstrated that some curcumin analogues are able to inhibit the TNFα-dependent NF-κB activation, although they lack the antioxidant properties (Weber et al. 2006). It is suggested that the NF-κB downregulation by curcumin might occur through the inhibition of the p65 subunit nuclear translocation, in association with the sequential suppression of IKK phosphorylation, IKK activity, IκBα degradation, p65 phosphorylation, and/or acetylation (Hussain et al. 2008). Moreover, NF-κB binding to its consensus DNA sequences also appears to be reduced by the curcumin. Further, the NF-κB targeted effects of the curcumin might be due to its inhibition of the proteasome activity (Milacie et al. 2008). AP-1 inhibition is mainly due to the curcumin's interference with the AP-1 binding to its DNA-binding motif through downregulation of the JunD expression (Tomita et al. 2006). Curcumin also stimulated the activity of caspase-8, which initiates Fas signaling pathway of apoptosis. Curcumin-induced apoptosis appears to be dependent on the p53, a redox-sensitive pro-apoptotic factor. Both expression and translocation of the p53 to mitochondria were enhanced by the curcumin in human prostate cancer cells, where changes in the proapoptotic gene expression, mitochondrial functions, and activation of the caspase-3 were also observed (Shankar and Srivastava 2007). The upregulation and nuclear translocation of p53 was observed in human neuroblastoma cell lines with subsequent induction of the p21 and Bax expression. In one study, curcumin was shown to be potent inhibitor of the cell proliferation and an inducer of the apoptosis in the head and neck squamous cell carcinoma through the suppression of IKK-mediated NF-kB activation and of NF-kB-regulated gene expression (Aggarwal et al. 2004). Curcumin has been shown to upregulate the Nrf2 and enhance its nuclear translocation and binding to ARE, thus leading to the increased level and activity of detoxifying enzymes (Garg et al. 2008). In renal epithelial cells, curcumin-stimulated Nrf2 expression and binding to ARE may be due to the p38 involvement, which led to an increased synthesis of the cytoprotective protein HO-1 (Balogun et al. 2003). In curcumin-supplemented human bronchial epithelial cells, an elevation in the intracellular GSH content was also observed on the increased nuclear content and/or activation of the transcription factors, such as AP-1 and Nrf2.

Resveratrol

Resveratrol interfere with the ROS-dependent cell signal cascade, by binding to and/or interacting with a variety of the cell signaling molecules. It has been shown to inhibit the LPS-induced expression of NOX1 and the consequent ROS production in macrophages (Park et al. 2009). In the human monocytes, resveratrol affects the NADPH oxidase activity through inhibition of the PI3K activity and Akt phosphorylation. On the contrary, upregulation of the NOX1 and NOX4 has been reported to be induced in the human ECs by chronic resveratrol supplementation (Schilder et al. 2009). At low doses resveratrol also activates the NADPH oxidase in human leukemia cells. Resveratrol has shown the chemopreventive activity against the cancer by inhibiting cell proliferation. It induced cell-cycle arrest at the G2 phase through the inactivation of CDK7 and consequent low phosphorylation of the p34 CDC2 protein kinase (Liang et al. 2003). Resveratrol caused DNA damage and S phase arrest in the ovarian cancer cells, inducing phosphorylation of the CDC2 by regulating CDC25C tyrosine phosphatase (Tyagi et al. 2005). Moreover, in the normal prostatic epithelial cells, resveratrol efficiently counteracted the TNFα and IL-1β-stimulated expression of pro-inflammatory genes through the upregulation of the MKP-5 (Nonn et al. 2007). In the human prostate cancer cells, resveratrol suppressed the EGFR-dependent ERK1/2 activation by selectively inhibiting the PMA-induced PKCα membrane translocation and decreasing its autophosphorylation (Stewart and O'Brian 2004). However, in the human cervical cancer cells, PMA-induced MMP-9 expression was abrogated through the inhibition of PKCδ activation and repression of the downstream JNK/NF-κB and AP-1 signaling pathways (Woo et al. 2004).

AP-1 activity appears to be downregulated by the resveratrol, as has been demonstrated in the myeloid, lymphoid, and epithelial cells activated by the various stimuli and in the skin of PMA-treated mice. Also the level of AP-1 components appeared to be under the resveratrol control, since it inhibited c-fos and c-jun expression in PMA-treated mouse skin. Sp-1, another redox-sensitive transcription factor, might be influenced by the resveratrol. It has been found that the resveratrol-induced apoptosis is associated with the activation of p53, most likely through the p53 acetylation with subsequent activation of the pro-apoptotic genes (Kai et al. 2010). Resveratrol exhibited a variety of molecular events in the HT-29 colon cancer cells including the AMPK activation, inhibition of cell growth, induction of apoptosis, and ROS generation. ROS was found as upstream regulator of the AMP-activated protein kinase, AMPK (Hwang et al. 2007).

Flavonoids

Flavonoids exert its effects on the cell signaling and gene expression. An extract from the cocoa, rich in flavonoids, has been shown to suppress the TNF-α induced VEGF expression by inhibiting the PI3K and MEK1 activities (Kim et al. 2010). Several flavonoids are strong inhibitors of the superoxide ($O_2.^-$)-generating enzyme NADPH oxidase. Inhibition of the NADPH oxidase can be elicited in platelets by the quercetin plus catechin incubation. Flavonoids interfere with the MAPK family signaling pathway, usually by inhibiting one or more of its members, and this partly explains the reported modulation of NF-κB and AP-1 by the flavonoids (Khan and Mukhtar 2008). ERK activation in the PMA-stimulated carcinoma cells is reversed by the quercetin (Lin et al. 2008) which prevents the PKCδ and PKCα membrane translocation, respectively. Since both the PKC isoforms are ATP-dependent proteins and since flavonoids have been demonstrated to competitively bind to the ATP-binding site of several proteins, a direct conformational change of the two kinases by this class of compounds may occur (Spencer 2008). Flavonoids have been implicated in the neurodegeneration suppression through the MPPK signaling pathway (Schroeter et al. 2002).

Another target of the flavonoids is PI3K/Akt signaling pathway, which plays a key role in supporting the cell survival. Flavonoids and their metabolites are reported to act at the PI3-kinase, Akt/protein kinase B, tyrosine kinases, and PKC and MAPK signaling cascades (Williams et al. 2004). Inhibitory or stimulatory actions at these pathways are likely to affect the cellular function profoundly by altering the phosphorylation state of the target molecules and by modulating gene expression. Some flavonoids also induce stabilization of the p53 and upregulation of its transcriptional activity, further leading to the activation of downstream targets such as p21 or Bax, and the induction of apoptosis (Shin et al. 2008). On the contrary, some flavonoids appear to inhibit the nuclear transactivation of p53 (Choi et al. 2005) or to downregulate p53 expression in H_2O_2-exposed cells and to regulate the expression of apoptotic downstream genes, preventing apoptosis and promoting cell survival. Perturbations in the pathway of the Nrf2 and ARE/EpREs might be another mechanism through which flavonoids exert effects.

Thiol Antioxidants/GSH and Thioredoxin

Effects of GSH on the MAPK cascade appear to be organ-specific. Experimental induction of the GSH depletion in the mouse led to the p38 and c-Jun phosphorylation and NF-κB nuclear translocation in the liver, but only to the p38 and c-Jun phosphorylation in the kidney, while in the brain, activation of ERK2 and Nrf2 nuclear translocation were observed (Limón-Pacheco et al. 2007). Activation of the p38 MAPK pathway appears to be responsible for the GSSG-induced apoptosis in promonocytic U937 cells. Since the hydrophilic GSSG form is not able to enter cells, it was suggested that its action could be mediated by the thiol/disulfide exchange reactions with the membrane protein thiols, such as ASK1 upstream cysteine-rich receptors of the TNF receptor superfamily (Filomeni et al. 2003).

Thioredoxin, but not the GSH, has been found to downregulate the activity of ASK1. In unstimulated cells, the reduced thioredoxin bound to the ASK1 at its N-terminal region through the formation of the interchain disulfide bonds, and the resulting complex called ASK1 signalosome blocked the kinase activity but in a reversible way. Activity was restored with the oxidized thioredoxin which dissociates from the ASK1 (Fujino et al. 2007). Further, H_2O_2-dependent activation of ASK1 has been found to be associated to the intermolecular disulfide-bond-mediated oligomerization of the kinase, an event that was reversed by the thioredoxin through the reducing reactions (Nadeau et al. 2007). Alternatively, thioredoxin might inhibit ASK1-mediated apoptosis in a redox-independent way, mainly by associating with the ASK1 through a single cysteine and thus favoring ASK1 ubiquitination and degradation. The NF-κB system contains several thiols sensitive to the redox changes and effective modulation of its pathway by the GSH through the reversible formation of protein mixed disulfides. In the LPS-stimulated alveolar epithelial cells, GSH depletion, induced by the pharmacological inhibitors of GSH-related enzymes, was found to be associated to the reduction of IκBα phosphorylation and NF-κB nuclear translocation (Haddad et al. 2002).

Because of the biphasic influence, it appears that an intermediate optimal level of the intracellular GSSG is required for the effective NF-kB activation (Droge et al. 1994). It was shown that the GSH deficiency of T cells is associated with a suppression of the NF-kB function, and this effect is related to the very low levels of GSSG in GSH-deficient cells. Another physiological relevant thiol that plays a crucial role in the regulation of NF-kB function is reduced thioredoxin, an important cellular oxidoreductase with the antioxidant function. Transient expression and exogenous addition of the thioredoxin cause a dose-dependent inhibition of the phorbol ester-induced NF-kB activation in the human cervical carcinoma HeLa cells (Schenk et al. 1994). Further, it has been consistently found that the AP-1 transactivation is influenced by the intracellular thioredoxin and glutathione status (Schenk et al. 1994). However, some results available from the different experimental systems indicate that the mode of AP-1 regulation of the reduced thiols is opposite from the regulation of NF-kB. Overexpression of the thioredoxin in cells increased TPA-induced AP-1 transcription activity in a dose-dependent manner. The effect was shown to be specific to the antioxidant property of thioredoxin (Schenk et al. 1994).

Thioredoxin may also modulate the NF-κB pathway. In a study on UVB-irradiated human keratinocytes, thioredoxin associated directly with the p50 by intermolecular disulfide bond formation and rapidly translocated to the nucleus of the cells, where it potentiated NF-κB transcriptional activity by enhancing its ability to bind the DNA; however, in the cytoplasm, overexpression of the thioredoxin inhibited the degradation of IκB (Hirota et al. 1999). In human lung epithelial cells, the thioredoxin has been shown to favor the IκB degradation by upregulating the JNK pathway and eventually stimulating NF-κB nuclear translocation (Das 2001).

As regards the AP-1 interaction with GSH, a decrease in the GSH/GSSG ratio would allow the oxidation of c-Jun sulfhydryls by mechanisms including both protein disulfide formation and reversible S-glutathionylation. The latter mechanism specifically targets the cysteine residue located in the DNA-binding site of the protein and can thus sterically block DNA binding of the transcription factor AP-1 (Klatt et al. 1999). The DNA-binding activity of AP-1 is facilitated by a DNA repair enzyme, namely, redox factor-1 (Ref-1), which in turn is activated by the direct association with thioredoxin in the nucleus of PMA-treated HeLa cells. In addition, thioredoxin might negatively interfere with the AP-1 transcriptional activity, through a direct interaction with the C-terminal of Jun activation domain-binding protein 1 (Jab1), a co-activator of AP-1 whose expression is associated to the tumor progression (Hwang et al. 2004). Under conditions of the oxidant stress reproduced in human cancer cell lines, p53 appeared to be the target of reversible S-glutathionylation on cysteines of the proximal DNA-binding domain; this modification markedly weakened the p53–DNA association and interfered with the protein dimerization (Velu et al. 2007). ROS and RNS can oxidize the cellular glutathione or induce its extracellular export leading to the loss of intracellular redox homeostasis and activation of the apoptotic signaling cascade.

Finally, in this chapter we have classified the endogenous and exogenous antioxidants and their target in various systems. To increase the efficiency of the antioxidant function, modifications in the present antioxidant have been listed as per literature which is of interest in pharmacological industry. In the end, various antioxidants influencing the signaling pathways have been explained as per literature, mostly in vitro studies which could be a rich source for the pharmacology industry in designing efficient antioxidant products.

References

Aggarwal S, Takada Y, Singh S, Myers JN, Aggarwal BB (2004) Inhibition of growth and survival of human head and neck squamous cell carcinoma cells by curcumin via modulation of nuclear factor-kB signaling. Int J Cancer 111:679–692

Anderson RM, Bitterman KJ, Wood JG, Medvedik O, Sinclair DA (2003) Nicotinamide and PNC1 govern lifespan extension by calorie restriction in saccharomyces cerevisiae. Nature 423:181–185

Andre CM, Larondelle Y, Eners D (2010) Dietary antioxidants and oxidative stress from a human and plant perspective, a review. Curr Nutr Food Sci 6:2–12

Anwar AA, Li FY, Leake DS, Ishi T, Mann GE, Siow RC (2005) Induction of heme oxygenase 1 by moderately oxidized low-density lipoproteins in human vascular smooth muscle cells: role of mitogen-activated protein kinases and Nrf2. Free Radic Biol Med 39: 227–236

Aron PM, Kennedy JA (2008) Flavan-3-ols: nature, occurrence and biological activity. Mol Nutr Food Res 52:79–104

Azzi A, Gysin R, Kempona P, Munteanu A, Villacorta L, Visarius T, Zingg JM (2004) Vitamin E mediates cell signaling and regulation of gene expression. Ann N Y Acad Sci 1031:86–95

Balogun E, Hoque M, Gong P, Killeen E, Green CJ, Foresti R, Alam J, Motterlini R (2003) Curcumin activates the haem oxygenase-1 gene via regulation of Nrf2 and the antioxidant-responsive element. Biochem J 371:887–895

Baum L, Ng A (2004) Curcumin interaction with copper and iron suggests one possible mechanism of action in Alzheimer's disease animal models. J Alzheimers Dis 6:367–377

Biswas SK, Rehman I (2009) Environmental toxicity, redox signalling and lung inflammation: the role of glutathione. Mol Asp Med 30:60–76

Borek C (2003) Cancer prevention by natural dietary antioxidants in developing countries. In: Bahorun T, Gurib-Fakim A (eds) Molecular and therapeutic aspects of redox biochemistry. OICA International (UK) Limited, London, pp 259–269

Bouayed J, Bohn T (2010) Exogenous antioxidants-double edged swords in cellular redox state: health beneficial effects at physiological doses versus deleterious effects at high doses. Oxidative Med Cell Longev 3:228–237

Bouayed J, Rammal H, Soulimani R (2009) Oxidative stress and anxiety: relationship and cellular pathways. Oxidative Med Cell Longev 2:63–67

Caramo JM, Pedraza A, Borquez-Ojeda O, Zhang B, Sanchez R, Golde DW (2004) Vitamin C is a kinase inhibitor: dehydroascorbic acid inhibits IkBα kinase β. Mol Cell Biol 24:6645–6652

Cárcamo JM, Pedraza A, Bórquez-Ojeda O, Golde DW (2002) Vitamin C suppresses TNF alpha-induced NF kappa B activation by inhibiting I kappa B alpha phosphorylation. Biochemistry 41:12995–13002

Chan CM, Fang JY, Lin IL, Yang CY, Hung CF (2009) Lycopene inhibits PDGF-BB-induced retinal pigment epithelial cell migration by suppression of PI3K/Akt and MAPK pathways. Biochem Biophys Res Commun 388:172–176

Choi YJ, Jeong YJ, Lee YJ, Kwon HM, Kang YH (2005) Epigallocatechin gallate and quercetin enhance survival signaling in response to oxidant-induced human endothelial apoptosis. J Nutr 35:707–713

Circu ML, Aw TY (2008) Glutathione and apoptosis. Free Radic Res 42:689–706

Cutler RG (2003) Genetic stability, dysdifferentiation, and longevity determinant genes. In: Cutler RG, Rodriguez H (eds) Critical reviews of oxidative stress and damage. World Scientific, River Edge

Das KC (2001) C-Jun NH2-terminal kinase-mediated redox-dependent degradation of IκB. Role of thioredoxin in NF-κB activation. J Biol Chem 276: 4662–4670

Delmas D, Jannin B, Latruffe N (2005) Resveratrol: preventing properties against vascular alterations and ageing. Mol Nutr Food Res 49:377–395

Demir S, Yilmaz M, Koseoglu M, Akalin N, Aslan D, Aydin A, Turk J (2003) Role of free radicals in peptic ulcer and gastritis. Gastroenterology 14:39–43

DiMascin P, Kaiser S, Sies H (1989) Lycopene as the most efficient biological carotenoid singlet oxygen quencher. Arch Biochem Biophys 274:532–538

Droge W, Schulze-Osthoff K, Mihm S, Galter D, Schenek H, Eck HP, Roth S, Gmumder H (1994) Function of glutathione and glutathione disulfide in immunology and immunopathology. FASEB J 8:1131–1138

Duplancic D, Kukoc-Modum L, Modum D, Radic N (2011) Simple and rapid method for the determination of uric acid-independent antioxidant capacity. Molecules 16:7058–7067

Elliott R (2005) Mechanisms of genomic and non-genomic actions of carotenoids. Biochim Biophys Acta 1740:147–154

Ernster L, Forsmark-Andree P (1993) Ubiquinol: an endogenous antioxidant in aerobic organisms. Clin Invest 71:S60–S65

Fang YZ, Yang S, Wu G (2002) Free radicals, antioxidants and nutrition. Nutrition 18:872–879, Review

Filomeni G, Rotilio G, Circolo MR (2003) Glutathione disulfide induces apoptosis in U937 cells by a redox-mediated p38 MAP kinase pathway. FASEB J 17:64–66

Fujino G, Noguchi T, Matsuzawa A, Yamauchi S, Saitoh M, Takeda K, Ichijo H (2007) Thioredoxin and TRAF family proteins regulate reactive oxygen species-dependent activation of the N-terminal homophilic interaction of ASK1. Mol Cell Biol 27:8152–8163

Galati G, O'Brien PJ (2004) Potential toxicity of flavonoids and other dietary phenolics. Significance for their chemopreventive and anticancer properties. Free Radic Biol Med 37:287–303

Garg R, Gupta S, Maru GB (2008) Dietary curcumin modulates transcriptional regulators of phase I and phase II enzymes in benzo[a]pyrene-treated mice: mechanism of its anti-initiating action. Carcinogenesis 29:1022–1032

Giovannocci E (1999) Tomatoes, tomato-based products, lycopene, and cancer: review of the epidemiologic literature. J Natl Cancer Inst 91:317–331

Haddad JJ, Safieh-Garabedian B, Saadé NE, Lauterbach R (2002) Inhibition of glutathione-related enzymes augments LPS-mediated cytokine biosynthesis: involvement of an IkappaB/NF-kappaB-sensitive pathway in the alveolar epithelium. Int Immunopharmacol 2: 1567–1583

Halliwell B (2008) Are polyphenols antioxidants or pro-oxidants? What do we learn from cell culture and in vivo studies? Arch Biochem Biophys 476:107–112

Halliwell B, Gutteridge JMC (1999) Free radicals in biology and medicine, 3rd edn. Oxford University Press, Oxford

Halliwell B, Gutteridge JMC (2005) Free radicals in biology and medicine. Oxford University Press, Oxford

Hirota K, Murata M, Sachi Y, Nakamura H, Takeuchi J, Mori K, Yodoi J (1999) Distinct roles of thioredoxin in the cytoplasm and in the nucleus. A two-step mechanism of redox regulation of transcription factor NF-kappaB. J Biol Chem 274:27891–27897

Hussain AR, Ahmed M, Al-Jomah NA, Khan AS, Manogaran P, Sultana M, Abubaker J, Platanias LC, AJ-Kuraya KS, Uddin S (2008) Curcumin suppresses constitutive activation of nuclear factor-kappa B and requires functional Bax to induce apoptosis in Burkitt's lymphoma cell lines. Mol Cancer Ther 7:3318–3329

Hwang CY, Ryu YS, Chung MS, Kim KD, Park SS, Chae SK, Chae HZ, Kwon K (2004) Thioredoxin modulates activator protein 1 (AP-1) activity and p27Kip1 degradation through direct interaction with Jab1. Oncogene 23:8868–8875

Hwang JT, Kwak DW, Lin SK, Kim YM, Park OJ (2007) Resveratrol induces apoptosis in chemoresistant cancer cells via modulation of AMPK signaling pathway. Ann N Y Acad Sci 1095:441–448

Imlay JA (2003) Pathways of oxidative damage. Annu Rev Microbiol 57:395–418

Kai L, Samuel SK, Levenson AS (2010) Resveratrol enhances p53 acetylation and apoptosis in prostate cancer by inhibiting MTA1/NuRD complex. Int J Cancer 126:1538–1548

Kastner P, Mark M, Chambon P (1995) Nonsteroid nuclear receptors: what are genetic studies telling us about their role in real life? Cell 83:859–869

Khan N, Mukhtar H (2008) Multitargeted therapy of cancer by green tea polyphenols. Cancer Lett 269:269–280

Kim DW, Hwang IK, Kim DW, Yoo KY, Won CK, Moon WK, Won MH (2007) Coenzyme Q10 effects on manganese superoxide dismutase and glutathione peroxidase in the hairless mouse skin induced by ultraviolet B irradiation. Biofactors 30:139–147

Kim DW, Kim SY, Lee SH, Lee YP, Lee MJ, Jeong MS, Jang SH, Park J, Lee KS, Kang TC, Won MH, Cho SW, Kwon OS, Eum WS, Choi SY (2008) Protein transduction of an antioxidant enzyme: subcellular localization of superoxide dismutase fusion protein in cells. BMB Rep 29:170–175

Kim JE, Son JE, Jung SK, Kang NJ, Lee CY, Lee KW, Lee HJ (2010) Cocoa polyphenols suppress TNF-α-induced vascular endothelial growth factor expression by inhibiting phosphoinositide 3-kinase (PI3K) and mitogen-activated protein kinase kinase-1 (MEK1) activates epidermal cells. Br J Nutr 104:957–964

Klatt P, Molina EP, DeLacoba MG, Padilla CA, Martinez-Galesteo E, Barcena JA, Lamas S (1999) Redox regulation of c-jun DNA binding by reversible S-glutathiolation. FASEB J 13:1481–1490

Kyaw M, Yoshizumi M, Tsuchya K, Suzaki Y, Abe S, Hasegawa T, Tamaki T (2002) Antioxidants inhibit endothelin-1(1-31)-induced proliferation of vascular smooth muscle cells via the inhibition of mitogen-activated protein (MAP) kinase and activator protein-1 (AP-1). Biochem Pharmacol 64:1521–1531

Lako J, Trenerry VC, Wahlqvist M, Wattanapenpaiboon N, Sotheeswaran S, Premier R (2007) Phytochemical flavonols, carotenoids and the antioxidant properties of a wide selection of Fijian fruit, vegetables and other readily available foods. Food Chem 101:1727–1741

Lee KW, Lee HJ, Lee CY (2004) Vitamin phytochemicals, diets and their implementation in cancer chemoprevention. Crit Rev Food Sci Nutr 44:437–452

Leopoldini M, Russo N, Toscano M (2011) The molecular basis of working metabolism of natural polyphenolic antioxidants. Food Chem 125:288–306

Lian F, Wang XD (2008) Enzymatic metabolites of lycopene induce Nrf2-mediated expression of phase II detoxifying/antioxidant enzymes in human bronchial epithelial cells. Int J Cancer 123:262–1268

Liang YC, Tsai SH, Chen L, Lin-shiau SY, Lin JK (2003) Resveratrol-induced G2 arrest through the inhibition of CDK7 and p34CDC2 kinases in colon carcinoma HI29 cells. Biochem Pharmacol 65:1053–1060

Limón-Pacheco JH, Hernández NA, Fanjul-Moles ML, Gonsebatt ME (2007) Glutathione depletion activates mitogen-activated protein kinase (MAPK) pathways that display organ-specific responses and brain protection in mice. Free Radic Biol Med 43:1335–1347

Lin CW, Hou WC, Shen SC, Juan SH, Ko CH, Wang LM, Chen YC (2008) Quercetin inhibition of tumor invasion via suppressing PKCδ/ERK/AP-1-dependent matrix metalloproteinase-9 activation in breast carcinoma cells. Carcinogenesis 29:1807–1815

Liu C, Russell RM, Wang XD (2004) Low dose beta-carotene supplementation of ferrets attenuates smoke-induced lung phosphorylation of JNK, p38 MAPK, and p53 proteins. J Nutr 134:2705–2710

Li-Weber M, Giaisi M, Treiber MK, Krammer PH (2002) Vitamin E inhibits IL-4 gene expression in peripheral blood T cells. Eur J Immunol 32:2401–2408

Lotito SB, Zhang WJ, Young CS, Crozier A, Frei B (2011) Metabolic conversion of dietary flavonoids alters their anti-inflammatory and antioxidant properties. Free Radic Biol Med 51:454–463

Lu TC, Wang Z, Feng X, Chuang P, Fang W, Chen Y, Neves S, Maayan A, Xiong H, Liu Y, Iyngar R, Klotman PE, He JC (2008) Retinoic acid utilizes CREB and USFI in a transcriptional feed-forward loop in order to stimulate MKPI expression in human immunodeficiency virus-infected podocytes. Mol Cell Biol 28:5785–5794

Lubbad A, Oriowo MA, Khan I (2009) Curcumin attenuates inflammation through inhibition of TLR-4 receptor in experimental colitis. Mol Cell Biochem 322:127–135

Lutsenko EA, Carcamo JM, Golde DW (2002) Vitamin C prevents DNA, mutation induced by oxidative stress. J Biol Chem 277:16895–16899

Ma SF, Nishikawa M, Katsumi H, Yamashita AAF, Hashida M (2006) Liver targeting of catalase by cationization for prevention of acute liver failure in mice. J Control Release 110:273–282

Mahmmoud YA (2007) Modulation of protein kinase C by curcumin; inhibition and activation switched by calcium ions. Br J Pharmacol 150:200–208

Maisuthisakul P, Suttajit M, Pongswaimanit R (2007) Assessment of phenolic content and free radical-scavenging capacity of some Thai indigenous plants. Food Chem 100:1409–1418

Maroz A, Gf K, Smith RA, Ware DC, Anderson RF (2008) Pulse radiolysis investigation on the mechanism of the catalytic action of Mn(II)-pentaazamacrocycle compounds as superoxide dismutase mimetics. J Phys Chem A 112:4929–4935

Masiá S, Alvarez S, de Lera AR, Barettino D (2007) Rapid, nongenomic actions of retinoic acid on phosphatidylinositol-3-kinase signaling pathway mediated by the retinoic acid receptor. Mol Endocrinol 21:2391–2402

Mates JM, Segura JA, Alonso FJ, Marquez J (2008) Intracellular status and oxidative stress: implications for cell proliferation, apoptosis and carcinogenesis. Arch Toxicol 82:273–299

Mates JM, Segura JA, Alonso FJ, Marquez J (2012) Oxidative stress in apoptosis and cancer: an update. Arch Toxicol 86:1649–1665

Milacie V, Banerjee S, Landis-Piwowar KR, Sarkar FH, Majumdar AP, Dou QP (2008) Curcumin inhibits the proteasome activity in human colon cancer cells in vitro and in vivo. Cancer Res 68:7283–7292

Montezano AC, Touyz RM (2012) Molecular mechanisms of hypertension-reactive oxygen species and antioxidants: a basic science update for the clinician. Can J Cardiol 28:288–295

Murakami A, Ohnish K (2012) Target molecules of food phytochemicals: food science bound for the next dimension. Food Funct 3:462–476

Nadeau PJ, Charette SJ, Toledano MB, Landry J (2007) Disulfide mediated multimerization of Ask1 and its reduction by thioredoxin-1 regulate H_2O_2-induced c-Jun NH_2-terminal kinase activation and apoptosis. Mol Biol Cell 18:3903–3913

Nonn L, Duong D, Peehl DM (2007) Chemopreventive anti-inflammatory activities of curcumin and other phytochemicals mediated by MAP kinase phosphatase-5 in prostate cells. Carcinogenesis 28:1188–1196

Nordberg J, Arner SJ (2001) Reactive oxygen species, antioxidants and the mammalian thioredoxin system. Free Radic Biol Med 31:1287–1312

Omoni AO, Aluko RE (2005) The anti-carcinogenic and anti-atherogenic effects of lycopene: a review. Trends Food Sci Technol 16:344–350

Palozza P, Parrone N, Catalano A, Simone R (2010) Tomato lycopene and inflammatory cascade: basic interactions and clinical implications. Curr Med Chem 17:2547–2563

Palozza P, Parrone N, Smone R, Catalano A (2011) Role of lycopene in the control of ROS-mediated cell growth implications in cancer prevention. Curr Med Chem 18:1846–1860

Paola P, Simona S, Gabriella (2006) Carotenoids as modulators of intracellular signaling pathways. Curr Signal Transduct Ther 1:325–335

Park DW, Baek K, Kim JR, Lee JJ, Ryu SH, Chin BR, Baek SH (2009) Resveratrol inhibits foam cell formation via NADPH oxidase 1-mediated reactive oxygen species and monocyte chemotactic protein-1. Exp Mol Med 41:171–179

Pervaiz S, Holme AL (2009) Resveratrol: its biological targets and functional activity. Antioxid Redox Signal 11:2851–2897

Poljsak B (2011) Strategies for reducing or preventing the generation of oxidative stress. Oxid Med Cell Longev. 2011, Art ID 194586, 15 pages. doi:10.1155/2011/194586

Poljsak B, Milisav I (2012) The neglected significance of antioxidative stress. Oxid Med Cell Longev. 2012, Art ID 480895, 12 pages. doi:10.1155/2012/480985

Poljšak B, Gazdag Z, Jenko-Brinovec S, Fuis S, Pesty M, Belgui J, Plesnicar S, Raspor P (2005) Pro-oxidative vs antioxidative properties of ascorbic acid in chromium(VI)-induced damage: an in vivo and in vitro approach. J Appl Toxicol 25:535–548

Rahman I (2012) Pharmacological antioxidant strategies as therapeutic interventions for COPD. Biochim Biophys Acta 1822:714–728

Rahman I, Biswas SK, Kirkham PA (2006) Regulation of inflammation and redox signaling by dietary polyphenols. Biochem Pharmacol 72:1439–1452

Rao AV (2002) Lycopene, tomatoes and the prevention of coronary heart disease. Exp Biol Med (Maywood) 227:908–913

Rice-Evans CA, Miller NJ, Paganga G (1996) Structure-antioxidant activity relationships of flavonoids and phenolic acids. Free Radic Biol Med 20:933–956

Rocha M, Hernandez-Mijares A, Garcia-Malpartida K, Bañuls C, Bellod L, Victor VM (2010) Mitochondria-targeted antioxidant peptides. Curr Pharm Des 28: 3124–3131

Samai M, Shape MA, Gard PR, Chtterjee PK (2007) Comparison of the effects of the superoxide dismutase mimetics EUK-134 and tempol on paraquat-induced nephrotoxicity. Free Radic Biol Med 43:528–534

Schenk H, Klein M, Erdbrugger W, Droge W, Schulze-Osthoff K (1994) Distinct effects of thioredoxin and antioxidants on the activation of transcription factors NFkB and AP-1. Proc Natl Acad Sci U S A 91:1672–1676

Schilder YD, Heiss EH, Schachner D, Ziegler J, Reznicek G, Sorescu D, Dissch VM (2009) NADPH oxidases 1 and 4 mediate cellular senescence induced by resveratrol in human endothelial cells. Free Radic Biol Med 46:1598–1606

Schroeter H, Boyd C, Spencer JPE, Williams RJ, Cadenas E, Rice-Evans C (2002) MAPK signaling in neurodegeneration: influences of flavonoids and of nitric oxide. Neurobiol Aging 23:861–880

Shankar S, Srivastava RK (2007) Involvement of Bcl-2 family members, phosphatidylinositol 3′-kinase/AKT and mitochondrial p53 in curcumin (diferuloylmethane)-induced apoptosis in prostate cancer. Int J Oncol 30:905–918

Sharoni Y, Danilenko M, Walfisch S, Amir H, Nahum A, Ben-Dor A, Hirsch K, Khanin M, Steiner M, Agemy L, Zango G, Levy J (2002) Role of gene regulation in the anticancer activity of caretenoids. Pure Appl Chem 74:1469–1477

Sharoni Y, Danilenko M, Dubi N, Ben-Dor A, Levy J (2004) Carotenoids and transcription. Arch Biochem Biophys 430:89–96

Shin JI, Shim JH, Kim KH, Choi HS, Kim JW, Lee HG, Kim BY, Park SN, Park OJ, Yoon DY (2008) Sensitization of the apoptotic effect of gamma-irradiation in genistein-pretreated CaSki cervical cancer cells. J Microbiol Biotechnol 18:523–531

Shuvaev W, Tliba S, Nakada M, Albeida SM, Muzykantov VR (2007) Platelet-endothelial cell adhesion molecule-1-directed endothelial targeting of superoxide dismutase alleviates oxidative stress caused by either extracellular or intracellular superoxide. J Pharmacol Exp Ther 323:450–457

Skulachev MV, Antonenko YN, Anisimov VN, Chernyak BV, Cherepanov DA, Skuachev PV (2011) Mitochondrial-targeted plastoquinone derivatives. Effect on senescence and acute age-related pathologies. Curr Drug Targets 12:800–826

Spencer JP (2008) Flavonoids: modulators of brain function? Br J Nutr 99(E Suppl 1):ES60–ES77

Spenciale A, Chirafisi J, Saija A, Cimino F (2011) Nutritional antioxidants and adaptive cell responses: an update. Curr Mol Med 11:770–789

Stauble B, Boscoboinik D, Tasinato A, Azzi A (1994) Modulation of activator protein-1 (AP-1) transcription factor and protein kinase C by hydrogen peroxide and D-α-tocopherol in vascular smooth muscle cells. Eur J Biochem 226:393–402

Steinhour E, Sherwani SI, Mazerik JN, Ciapala V, O'connor Butler E, Cruff JP, Magalang U, Parthasarathy S, Sen CK, Marsh CB, Kuppusamy P, Parinandi NL (2008) Redox-active antioxidant modulation of lipid signaling in vascular endothelial cells: vitamin C induces activation of phospholipase D through phospholipase A2, lipoxygenase and cyclooxygenase. Mol Cell Biochem 315:97–112

Stewart JR, O'Brian CA (2004) Resveratrol antagonizes EGFR-dependent Erk1/2 activation in human androgen-independent prostate cancer cells with associated isozyme-selective PKCα inhibition. Invest New Drugs 22:107–117

Suksrichavalit T, Pracchayasittikul S, Piacham T, Isarankura-Na-Ayudhya C, Nantasenamat C, Prachayasittikul V (2008) Copper complexes of nicotinic-aromatic carboxylic acids as superoxide dismutase mimetics. Molecules 13:3040–3056

Suzuki YJ, Packer L (1993) Inhibition of NFkB activation by vitamin E derivatives. Biochem Biophys Res Commun 193:277–283

Tan KP, Kosuge K, Yang M, Ito S (2008) NRF2 as a determinant of cellular resistance in retinoic acid cytotoxicity. Free Radic Biol Med 45:1663–1673

Tarhini AA, Belani CP, Luketich JD, Argiris A, Ramalingan SS, Gooding W, Pennathur A, Petro D, Kane K, Liggitt D, Championsmith T, Zhang X, Epperly MW, Greenbager JS (2011) A phase I study of concurrent chemotherapy (paclitaxel and carboplatin) and thoracic radiotherapy with swallowed manganese superoxide dismutase plasmid liposome protection in patients with locally advanced stage III non-small-cell lung cancer. Hum Gene Ther 22:336–342

Toba T, Shidoji Y, Fujii J, Moriwaki H, Muto Y, Suzuki T, Ohishi K, Yagi K (1997) Growth suppression and induction of heat-shock protein-70 by 9-cis beta-carotene in cervical dysplasia-derived cells. Life Sci 61:839–845

Tomita M, Kawakami H, Uchihara JN, Okudaira T, Masuda M, Takasu N, Matsuda T, Ohta T, Tanaka Y, Mori N (2006) Curcumin suppresses constitutive activation of AP-1 by downregulation of JunD protein in HTLV-1-infected T-cell lines. Leuk Res 30:313–321

Tyagi A, Singh RP, Agarwal C, Siriwardana S, Sclafani RA, Agarwal R (2005) Resveratrol causes Cdc2-tyr15 phosphorylation via ATM/ATR-Chk1/2-Cdc25C pathway as a central mechanism for S phase arrest in human ovarian carcinoma ovcar-3 cells. Carcinogenesis 26:1978–1987

Valko M, Leibfritz D, Moncol J, Cronin MT, Mazur M, Telser J (2007) Free radicals and antioxidants in normal physiological functions and human disease. Int J Biochem Cell Biol 39:44–84

VanDuyn MAS, Pivonka E (2000) Overview of the health benefits of fruit and vegetables consumption for the dietaries professional: selected literature. J Am Diet Assoc 100:1511–1521

Velu CS, Niture SK, Doneanu CE, Pattabiraman N, Srivenugopal KS (2007) Human p53 is inhibited by glutathionylation of cysteines present in the proximal DNA-binding domain during oxidative stress. Biochemistry 46:7765–7780

Vera JC, Rivas CL, Fischbarg J, Golde DW (1993) Mammalian facilitative hexose transporters mediate the transport of dehydroascorbic acid. Nature 364:79–82

Voogd A, Sluiter W, Van Eijk HG, Koster JF (1992) Low molecular weight iron and the oxygen paradox in isolated rat hearts. J Clin Invest 90:2050–2055

Vorotnikova E, Rosenthal RA, Tries M, Doctrow SR, Braunhut SJ (2010) Novel synthetic SOD/catalase mimetics can mitigate capillary endothelial cell apoptosis caused by ionizing radiation. Radiat Res 173:748–759

Wang XJ, Hayes JD, Henderson CJ, Wolf R (2007) Identification of retinoic acid as an inhibitor of transcription factor Nrf2 through activation of retinoic acid receptor alpha. Proc Natl Acad Sci U S A 104:19589–19594

Weber WM, Hunsaker LA, Roybal CN, Bobrovnikova-Marjon EV, Abcouwer SF, Royer RE, Deck LM, Vander Jagat DL (2006) Activation of NFkappaB is inhibited by curcumin and related enones. Bioorg Med Chem 14:2450–2461

Williams RJ, Spencer JPE, Rice-evans C (2004) Flavonoids:antioxidants or signaling molecules. Free Radic Biol Med 36:838–849

Woo JH, Lim JH, Kim YH, Soh SI, Min DS, Chang JS, Lee YH, Park JW, Kwon TK (2004) Resveratrol inhibits phorbol myristate acetate-induced matrix metalloproteinase-9-expression by inhibiting JNK and PKC δ signal transduction. Oncogene 23:1845–1853

Woo MS, Jung SH, Kim SY, Hyun JW, Ko KH, Kim WK, Kim HS (2005) Curcumin suppress phorbol ester-induced matrix metalloproteinase-9 expression by inhibiting the PKC to MAPK signalling pathways in human astroglioma cells. Biochem Biophys Res Commun 335:1017–1025

Wu F, Schuster DP, Tyml K, Wilson JX (2007) Ascorbate inhibits NADPH oxidase subunit p47phox expression in microvascular endothelial cells. Free Radic Biol Med 42:124–131

Yang JQ, Li S, Domann FE, Buettner GR, Oberley LW (1999) Superoxide generation in v-Ha-ras-transduced human keratinocyte HaCaT cells. Mol Carcinog 26:180–188

Yang JQ, Buettner GR, Domann FE, Li Q, Engelhardt JF, Weydert CD, Oberley LW (2002) v-Ha-ras mitogenic signaling through superoxide and derived reactive oxygen species. Mol Carcinog 33:206–218

Yeh SL, Wang HM, Chen PY, Wu TC (2009) Interactions of β-carotene and flavonoids on the secretion of pro-inflammatory mediators in an in vitro system. Chem Biol Interact 179:386–393

Young JF, Dragsted LO, Haraldsdottir J, Daneshvar B, Kall MA, Loft S, Nilsson L, Nielsen SE, Mayer B, Skibsted LH, Huynh-Ba T, Hermetter A, Sanastrom B (2002) Green tea extract only affects markers of oxidative status postprandially: lasting antioxidant effect of flavonoid-free diet. Br J Nutr 87:343–355

Zancai P, Cariati R, Quaia M, Guldoboni M, Rizzo S, Boiocchi M, Dolcetti R (2004) Retinoic acid inhibits IL-6 dependent but not constitutive STAT3 activation in Epstein-Barr virus immortalized B lymphocytes. Int J Oncol 25:345–355

Zanotto-Filho A, Cammarota M, Gelain DP, Oliveira RB, Delgado-Cañedo A, Dalmolin RJ, Pasquall MA, Moreira JC (2008) Retinoic acid induces apoptosis by a non-classical mechanism of ERK1/2 activation. Toxicol In Vitro 22:1205–1212

Zingg JM (2007) Modulation of signal transduction by vitamin E. Mol Asp Med 28:481–506

7 Selenium: A Potent Natural Antioxidant

Selenium element description is included in this collection, as this essential element acts as a potent antioxidant through the selenoproteins and also its deficiency is linked in various pathological conditions. However, at high level, it acts as toxic but is not of interest in the present write-up. Here we have described the characteristics of various selenoproteins known till today and discussed there possible functions. In the end, epidemiological consideration of the selenium is discussed as per literature.

Selenium: An Essential Trace Element

Selenium (Se) was discovered in 1817 by Swedish chemist Jons Jacob Berzelius who named it Selene after the Greek goddess of the moon. Essentiality of selenium linked to its severe deficiency causing Keshan disease, a potentially fatal form of cardiomyopathy that was first found in northeast China and was found to be cured with selenium supplementation. Interest in selenium increased in 1957 with the discovery that demonstrated that selenium was an essential trace element for many life forms including man (Schwarz and Foltz 1957). Sources of selenium to living systems are from soil through the food chain (Fig. 7.1), and its availability in adequate quantity contributes in many biochemical and physiological processes including the biosynthesis of coenzyme Q (a component of the mitochondrial electron transport systems), regulation of ion fluxes across the membranes, maintenance of the integrity of keratins, stimulation of the antibody synthesis, and activation of the glutathione peroxidase (an enzyme involved in preventing oxidative damage to the cells) (Hammond and Beliles 1980).

Selenium compounds have been recognized as the most promising among the inhibitors of cancer induction (Combs and Liu 2001). They alter the carcinogen metabolism and provide the protection to DNA against carcinogen-induced damage. It is also involved in the detoxication of the metals and certain xenobiotics (Schrauzer 1992). Adequate supply of the selenium is claimed to inhibit the viral infection, slow down the aging process, modulate the immune system function (McKenzie et al. 2002), delay the progression of AIDS in the HIV-infected persons (Baum et al. 2001), and prevent the heart disease and other muscle disorders (Coppinger and Diamond 2001).

Metabolism and Deposition of Selenium

Selenium occurs in several oxidation states: −2 (hydrogen selenide, sodium selenide, dimethyl selenium, trimethyl selenium, and selenoamino acids such as selenomethionine), 0 (elemental selenium), +4 (selenium dioxide, selenious acid, and sodium selenite), and +6 (selenic acid and

M. Bansal and N. Kaushal, *Oxidative Stress Mechanisms and their Modulation*,
DOI 10.1007/978-81-322-2032-9_7, © Springer India 2014

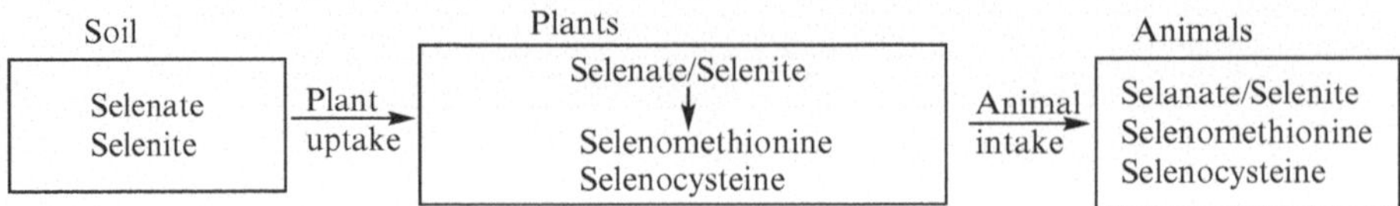

Fig. 7.1 Sources of selenium in biological system

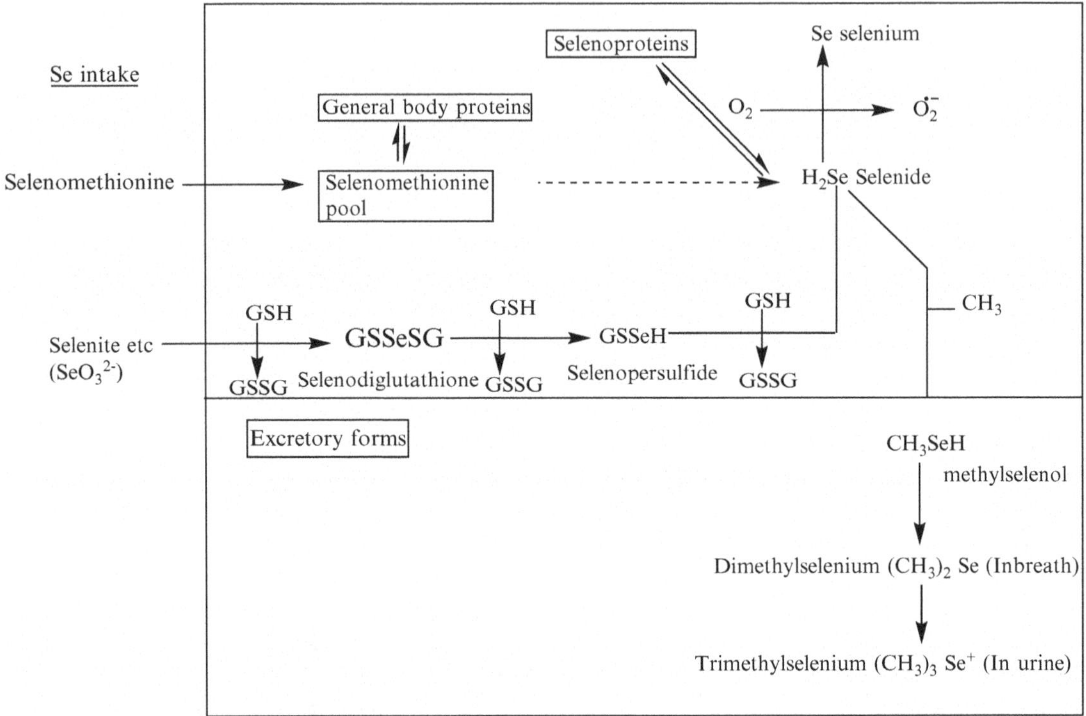

Fig. 7.2 Selenium metabolic pathways

sodium selenate). It is inactive in its elemental form but is highly reactive in the different oxidation states (−2 to +6).

Selenium is an essential component of the several major metabolic pathways, including antioxidant defense systems, thyroid hormone metabolism, and the immune function. Two major metabolic products of the selenite have been identified: dimethyl selenide and trimethylselenonium ion (Nakamuro et al. 1977; Jiang et al. 1983). In the formation of dimethyl selenide, selenite (H_2SeO_3) is first reduced nonenzymatically to the stable selenotrisulfide (GS-Se-SG) by the four glutathione (GSH) molecules. An NADPH-dependent reduction involving the glutathione reductase converts GS-Se-SG to the very unstable selenopersulfide (GSSeH). Further reduction by the NADPH and glutathione reductase converts the GSSeH to hydrogen selenide (H_2Se). Methyl groups donated by the S-adenosylmethionine are transferred by the methyltransferase to the hydrogen selenide to form the dimethyl selenide (Ganther 1979; Bopp et al. 1982).

Selenomethionine can be converted into the selenocysteine, which in turn is converted into the hydrogen selenide (H_2Se) which is a central metabolite in the utilization and excretion of the selenium and serves as a substrate for the biosynthesis of selenoproteins (Fig. 7.2). It can also be converted to the selenophosphate which is required for the selenoprotein biosynthesis.

Excess selenium is generally converted into the mono-, di- and trimethyl selenides and is excreted either through the breath or in urine (Birringer et al. 2002).

The role of selenium, as an antioxidant, is primarily mediated by the expression of selenoproteins. There are almost 25 selenoproteins known till date with a variety of pathophysiological functions. However, several other important roles of the selenium such as an anticarcinogenic activity cannot be solely ascribed to these selenoproteins. Also it has been shown (Shen et al. 2000) that during their metabolism, selenols (selenides) can enter a redox cycle, react with the glutathione, and generate oxidative stress (as shown below). Selenopersulfide anion, GSSeH, is formed directly in the reaction of selenite with GSH and further produces superoxides ($O_2^{-\cdot}$), which builds up the oxidative stress:

$$\underset{\text{(Selenite)}}{SeO_3^{2-}} \xrightarrow{4GSH \downarrow GSSG} \underset{\text{(Selenodiglutathione)}}{GSSeSG} \xrightarrow{GSH \downarrow GSSG} \underset{\text{(selenopersulfide)}}{GSSeH} \xrightarrow{GSH \downarrow GSSG} \underset{\text{(Selenide)}}{H_2Se} \xrightarrow{O_2 \downarrow O_2^{\cdot\cdot}} \underset{\text{(Selenium)}}{Se^o}$$

Metabolism of the various selenoamino acids L-selenomethionine and L-methylselenocysteine to the methylselenol also produces selenides. High cellular selenium concentration may therefore be pro-oxidative and result in greater amount of the free radical generation and oxidative stress. Selenocysteine (Sec) and the other selenium compounds like selenite react with the oxygen and mammalian thioredoxin and thioredoxin reductase, resulting in the rapid NADPH oxidation and ROS formation.

Selenium is found in all the tissues at concentrations that vary with amount ingested in the diet and type of the tissue. After the initial administration, selenium is taken up by the erythrocytes, metabolized to the selenides, and shifted to plasma proteins (Jenkins and Hidiroglou 1972). From here selenium in different tissues shows a hierarchy, with the preferential accumulation in the thyroid, brain, gonads, and pituitary and adrenal glands over liver and erythrocytes, heart, and muscles (Behne et al. 1988). Selenium is also concentrated more in the erythrocytes relative to plasma (Butler et al. 1990). As a result of the occupational exposures, high concentrations may be found in the peribronchial nodes, lung, hair, and nails (Diskin et al. 1979). The reproductive organs clearly appear to be a priority tissue since a major part of the administered dose of selenium accumulates in the reproductive organs and secretions (Behne et al. 1988).

Gastrointestinal absorption in humans for various selenium compounds ranges from about 44 % to 95 % of the ingested dose (Bopp et al. 1982). In studies on rats, mice, and dogs, the gastrointestinal absorption rates of 87 % or more have been reported for [^{75}Se]-selenite (selenious acid) (Bopp et al. 1982). Respiratory tract absorption rates of 97 and 94 % for aerosols of [^{75}Se]-selenite (selenious acid) have been reported, respectively, for the dogs and rats (Weissman et al. 1983). Absorption of the selenium in the mammalian system greatly depends upon the chemical form and mode of administration of the element. Selenomethionine is better absorbed and has more bioavailability than the sodium selenite when administered orally similar to the absorption of methionine (Combs and Combs 1984) whereas the selenite is passively but rapidly absorbed.

Selenoproteins and Their Functions

The actual research impetus into the effects of selenium came with discoveries of selenium containing enzymes and proteins called "selenoproteins." Several selenoproteins have been identified and characterized in almost all the living forms on earth such as bacteria, plant, and eukaryotes (Castellano et al. 2001; Fu et al. 2002). Studies in the mammals regarding these selenoproteins

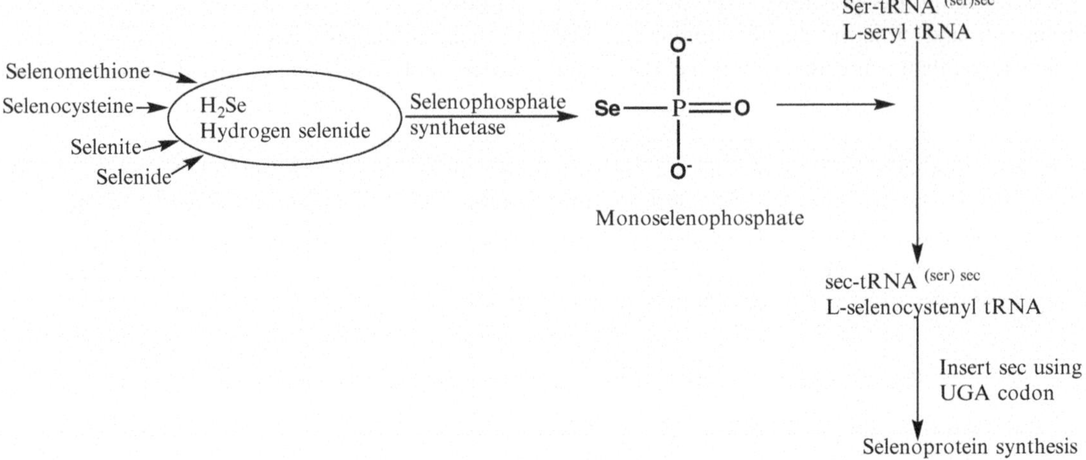

Fig. 7.3 Selenoprotein synthesis pathways (general)

have been shown to play important roles in the diverse biological functions.

The main form of selenium in mammalian proteins is selenocysteine (Sec) which differs from the cysteine by a single atom of selenium in place of the sulfur and has similar chemical properties, but the lower pK_a value and stronger nucleophilicity of Sec make it much more reactive. There is no free pool of Sec in the cells, and during protein catabolism, Sec is broken down to the elemental selenium. Incorporation of selenium as Sec into the selenoproteins requires a specific mechanism to decode the UGA codon in mRNA, which normally operates in the translation termination. Sec is co-translationally incorporated within the growing polypeptide chain by an unusually complex process first described in Escherichia coli and also characterized in mammalian cells. Sec biosynthesis occurs on its tRNA named as Sec $tRNA^{[Ser]Sec}$ (Fig. 7.3) (Bock 2001; Carlson et al. 2001), and this tRNA governs the expression of the entire class of selenoproteins. Further, Sec $tRNA^{[Ser]Sec}$ is initially aminoacylated with serine in both prokaryotes (Bock 2001) and eukaryotes (Carlson et al. 2001), and serine serves as the backbone for Sec synthesis (Bock 2001; Carlson et al. 2001). In E. coli, a pyridoxal phosphate-dependent Sec synthase catalyzes the removal of hydroxyl group from serine to form an aminoacrylyl intermediate which serves as the acceptor for activated selenium, resulting in the formation of selenocysteyl-$tRNA^{[Ser]Sec}$ (Bock 2001). In mammals, a minor seryl tRNA which decodes UGA (Hatfield and Portugal 1970) and formed phosphoseryl tRNA was also identified as Sec $tRNA^{[Ser]Sec}$ (Lee et al. 1989). Formation of phosphoserine is consistent with a Sec synthase-catalyzed reaction, as phosphorylated serine would have a better leaving group than serine in the Sec biosynthetic pathway. The active form of selenium that is donated to the intermediate in Sec biosynthesis was identified in the prokaryotes as monoselenophosphate, which is synthesized from the selenide and ATP by selenophosphate synthetase (Glass et al. 1993). Two selenophosphate synthetase genes in mammals, Sps1 and Sps2, were identified (Guimaracs et al. 1996). SPS2 is a selenoprotein involved in the autoregulation of its own biosynthesis (Guimaracs et al. 1996). Once the activated form of selenium is donated to the intermediate, biosynthesis of Sec on $tRNA^{[Ser]Sec}$ is completed. Besides Sec $tRNA^{[Ser]Sec}$ and the in-frame UGA codon in selenoprotein mRNA, there are several other factors that are required for the donation of Sec to protein. These include Sec insertion sequence (SECIS) element (Low and Berry 1996), SECIS-binding protein 2 (SBP2) (Copeland et al. 2001), and Sec-specific elongation factor (EFsec) (Fagegaltier et al. 2000).

Characteristics of Selenoproteins

There are extensive reviews on the existence of many different selenoproteins with their known function in the mammalian, eukaryotic, and human systems (Brown and Arthur 2001; Gromer et al. 2005; Gladyshev 2006). Following selenoproteins with their main functional characteristics are summarized here as per literature in the mammalian and human systems.

Glutathione Peroxidases (GPx)

Glutathione peroxidase (GPx) was the first known animal selenoprotein, first functional marker of the selenium status (Rotruck et al. 1973), and then a highly efficient antioxidant enzyme that catalyzes glutathione-dependent hydroperoxide reduction. Mammals contain eight glutathione peroxidase homologs, of which five are selenoproteins, including GPx1 (or cGPx), GPx2 (or GI-GPx), GPx3 (or pGPx), GPx4 (or PHGPx), and GPx6. In humans also, seven isoenzymes exist; GPx5/GPx7 are not selenoenzymes. The glutathione peroxidases reduce and thereby detoxify the different types of peroxides to their respective alcohols at the expense of the glutathione (R-OOH + 2GSH → R-OH + H_2O + GSSG) and play an important antioxidative role in the body.

Glutathione peroxidase 1 (GPx1), a selenoenzyme, is a ubiquitous homotetrameric cytosolic enzyme (cGPx) and is abundant in the liver and erythrocytes depending on the nutritional selenium status. Studies on the GPx1 knockout mice showed that GPx1 plays an active role as antioxidant only during the oxidative stress conditions (Fu et al. 1999).

Glutathione peroxidase 2 (GPx2) is found in the liver and gastrointestinal system often abbreviated as GI-GPx. It is a homotetrameric cytoplasmic enzyme that uses organic hydroperoxides as substrate such as t-butyl-, linolic acid- and cumene-hydroperoxides. GPx2 is considered the first line of defense against the ingested organic hydroperoxides (Winger et al. 1999) and found conserved even under the selenium-deficient conditions. GPx2 knockout mice do not have a unique phenotype; however, inflammatory bowel disease is typically observed in the GPx1-GPx2 double knockout (Chu et al. 2004).

Glutathione peroxidase 3 (GPx3) is located in the plasma (pGPx) and intestine (Tham et al. 1998). The physiological function of this homotetrameric glycoprotein may be regulatory. Hypoxia induces this protein, and its deficiency seems to correlate with the cardiovascular events and cancer (Sarto et al. 1999). Plasma GPx3 is primarily expressed in the renal proximal tubules and is used as a marker to monitor the tubular integrity (Whitin et al. 1998).

Glutathione peroxidase 4 (GPx4) is a monomeric enzyme. By using alternative initiation sites (Met^1 or Met^{28}), GPx4 synthesis can generate mitochondrial and cytoplasmic isoforms. GPx4 exhibits the broadest substrate specificity of all the glutathione peroxidases. It protects the membranes by reducing membrane-integrated hydroperoxides. It is also involved in the redox signaling and regulatory processes, such as inhibiting lipoxygenases and apoptosis (Brigelius-Floh 1999). In testis, GPx4 is the major selenium containing protein and forms a structural component of the sperm's midpiece (Foresta et al. 2002; Floh et al. 2002) which is required only for the sperm fertilization characteristics (Urisini et al. 1999).

Glutathione peroxidase 6 (GPx6) was discovered by using in silico approach, and its expression (as judged by mRNA) is shown in the olfactory epithelium and embryonic tissues (Kryukov et al. 2003).

Deiodinases

Mammals and humans have three deiodinases (DI1, DI2, and DI3) which activate or inactivate thyroid hormones by the reductive deiodination. Deiodinases cleave the specific iodine carbon bonds in the thyroid hormones, thereby changing their metabolic functions required for the normal growth and development. Three important thyroid hormones are thyroxine T_4 (3,3′,5,5′-tetraiodo-L-thyronine, t1/2 = 7 days), T_3 (3,3′,5-triiodo-L-thyronine, t1/2 = 1 day), and reverse T_3 (rT_3, 3,3′,5′-triiodo-L-thyronine). The normal thyroid function depends on the two trace elements: iodine and selenium. The thyroid gland has the highest per gram selenium content of all the organs (present not only in deiodinases but also in glutathione peroxidases), which are presumably required for the peroxide-dependent formation of T4. Three types of the deiodinases not only differ in sequence and structure but also catalyze the different reactions.

Deiodinase 1 (DIO1), a selenoenzyme, is a homodimeric plasma membrane protein and deiodinates the 5′-position of the phenolic ring in L-thyroxin, but under certain circumstances, it also deiodinate the 5-position. The -Se-H group of DIO1 (similarly in DIO2) gets converted into a –Se-I group, and further reduction releases iodine and regenerates the enzyme's selenol group. 5′-deiodination activity converts L-thyroxin (T4, the major form secreted by the thyroid) to T3 (the major thyroid hormone in peripheral circulation). DIO1 expression is high in the liver, kidney, thyroid, and pituitary gland whereas trace levels are found in most tissues. More than 80 % of the T4 is converted to T3 outside the thyroid, primarily in the kidney and liver (Kelly 2000). Reduced DIO1 levels are found in the low-T3 syndrome (accompanied by the elevated levels of rT3), a clinical condition occasionally seen in the clinically ill patients. DIO1 expression is induced by the elevated T4 and T3 levels.

Deiodinase 2 (DIO2), a selenoenzyme, has the functional SECIS (selenocysteine insertion sequence) element at an unusual far distance (5.4 kb) from the UGA codon in the human enzyme (Buettner et al. 1998). It is an ER-membrane protein that deiodinates the 5′-position with a preference for T4 over rT3. DIO2 is present in the CNS, pituitary/thyroid glands, skeletal/heart muscles, and in placental and brown adipose tissues. T3 production within the brain was possible because of the DIO2 presence in the brain, whereas blood–brain barrier (BBB) does not allow supply from the blood (Escobar-Morreale et al. 1999). Located inside the cell, primary function of DIO2 is the conversion of T4 into T3 in specific target tissues. Unlike DIO1, DIO2 is downregulated with the increasing T4 (as well as rT3) levels and rapidly degraded via ubiquitin-dependent pathways (T1/2 = minutes to 1 h). DIO2 knockout mice show the little gross phenotype abnormalities, although mild growth retardation and hearing loss are observed (Ng et al. 2004).

Deiodinase 3 (DIO3) deiodinates the 5-position of tyrosyl ring of T_4/T_3 and the resulting products cannot bind to the nuclear T3 receptor, and therefore, the primary physiological function of DIO3 is the inactivation of the T3 and T4. The brain, placenta, and pregnant uterus express considerably high amounts of the DIO3. However, persistently high levels of the DIO3 and low levels of T3 may have deleterious effects upon the CNS development and brain function (Salvatore et al. 1995). DIO3 is induced with increasing the T4 levels.

Thioredoxin Reductases (TR)

In mammals there are three thioredoxin reductase selenoproteins with Sec at the penultimate C-terminus as the additional active site that is itself a substrate for the *N*-terminal thiol–disulfide active site (Sandalova et al. 2001). TR1 (TrxR1, TxnRd1), a cytosolic protein, mainly control the reduced state of the thioredoxin and exhibit broad substrate specificity (Amer and Holmgren 2000). Thioredoxin/glutathione reductase (TGR, also known as TR2 and TrxR3) is a protein that compared to the other animal thioredoxin reductase has an additional N-terminal glutaredoxin (Grx) domain. TGR can catalyze many reactions specific for the thioredoxin and glutathione systems. This protein was implicated in the disulfide bonds formation during sperm maturation (Su et al. 2005). TR3 (TrxR2), a mitochondrial protein, reduces the mitochondrial thioredoxin and glutaredoxin 2. TR1 and TR3 are essential proteins in the mammals (Jakupoglu et al. 2005).

In humans, the thioredoxin reductases act as the thioredoxin system using NADPH and redox-active protein thioredoxin, Trx (TrxR; $TrxS_2 + NADPH + H^+ \rightarrow Trx\ (SH)_2 + NADP^+$). Three distinct human thioredoxin reductases are known (TrxR1 = TR1 = TRα; TrxR2 = TR3 = TRβ; TGR = TR2 = TrxR3).

Thioredoxin reductase 1 (TrxR1), a cytoplasmic enzyme, involves in the cellular redox regulation (Sun and Gladyshev 2002). It is capable of inducing apoptosis if the enzyme does not contain selenocysteine or if this residue is blocked, e.g., by a chemotherapeutic agent (Anestal and Arner 2003). TrxR1 is also secreted in the plasma (Soderberg et al. 2000).

Thioredoxin reductase 2 (Trx2) is located in the mitochondria (Miranda-Vizuete et al. 2000) with highest levels in the prostate, testis, liver, uterus, and small intestine and intermediate levels in the brain, skeletal muscle, heart, and spleen. TrxR2 knockout studies led to the early embryonic death with the sign of severe anemia, apoptosis in the liver, and heart abnormalities. A heart-specific knockout causes dilatative cardiomyopathy and early death, similar to the Keshan disease (Conrad et al. 2004).

Thioredoxin glutathione reductase (TGR) is a testis-specific enzyme, located in the ER (Sun et al. 2001). Unlike TrxR1 and TrxR2, it can reduce the glutathione disulfide.

Methionine-R-Sulfoxide Reductase 1 (MsrB1)

In mammals MsrB1 was initially identified using bioinformatics as the selenoprotein R/selenoprotein X and later was shown to catalyze the reduction of the oxidized methionine residues in proteins with thioredoxin as the reductant (Kryukov et al. 2002). There exist additional MsrBs (MsrB2 and MsrB3) and MsrA but without selenocysteines. In humans, MsrB1 in addition to the selenocysteine residue, one Zn^{2+} ion is bound per 12 kDa molecules via four cysteine residues (Kim and Gladyshey 2004). It exhibits the highest specific activity among three principal types of the MsrBs in humans.

15 kDa Selenoprotein (Sep 15)

Mammalian Sep 15 of 15 kDa size is localized in the endoplasmic reticulum where it binds the UDP-glucose: glycoprotein glucosyltransferase, a protein folding sensor and also having redox function (Korotkov et al. 2001). Sep 15 having the thioredoxin-like fold mediates anticancer influence of the dietary selenium (Ferguson et al. 2006). In humans also Sep 15 is localized in the ER and binds to protein folding sensor abbreviated as HUGT, UDP-glucose:glycoprotein glucosyl transferase 1 (Korotkov et al. 2001). It is mainly expressed in the prostate, testes, brain, kidney, and liver and also at low levels in the skeletal muscle, mammary gland, and trachea. HUGT-soluble ER enzymes function to correctly fold the glycoproteins in ER or transfer to the degradation pathways (Arnold et al. 2000).

Selenoprotein M/O (SelM/O)

SelM is a distant homolog of the Sep15 and also has a thioredoxin-like fold and a predicted redox motif (Ferguson et al. 2006). In humans, its SECIS element is unusual, as cytosines replace the invariant adenosines at the apical loop (Korotkov et al. 2002). Its mRNA is expressed at the highest level in brain and lowest in the liver/spleen. The SelM (122 residue in size) is localized (and retained) in the endoplasmic reticulum, since the first 23 residues contain an ER-signal sequence. The SelM sequence contains a CXXU motif indicative for a redox-active protein. The SelO in human is similar to the SelM, and its SECIS element is also unusually like M (Kryukov et al. 2003). It is a large protein of 669 residues, and its C-terminal Cys-XX-Sec motif may be indicative for a redox-dependent activity. Only vertebrate homologs of the SelO have Sec, which is located in the C-terminal penultimate position.

Selenoprotein P (SelP)

It is the only multiple Sec-containing selenoprotein (10 s in humans) (Burk and Hill 2005) and is the major plasma selenoprotein, synthesized in the liver, and delivers significant amount of the selenium to certain other organs and tissues (Hill et al. 2003). However, the brain synthesizes its own pool of SelP. Gene of the SelP plasma glycoprotein (Burk and Hill 1999) is transcribed in many tissues, yet the majority of the plasma SelP is secreted by the liver and presumably enters the target cells via a receptor-mediated mechanism. Two selenocysteines form a selenenylsulfide bridge with the cysteine (Ma et al. 2005). SelP is an established marker for the nutritional selenium status (Schweizer et al. 2005). The primary function of the SelP is for the storage and transport of the selenium. SelP-knockout experiments decreased the Se-plasma levels by 80–90 % and selenium tissue concentrations, and selenoenzyme activities dropped markedly in the brain, kidney, and testis (Hill et al. 2003).

Selenoprotein W/V (SelW/V)

SelW is the smallest mammalian selenoprotein (Vendeland et al. 1995), initially purified from the rat muscle (Vendeland et al. 1993) and later also demonstrated in most other tissues. Only trace amounts of the SelW are found in the liver, thyroid, pancreas, eye, and pituitary gland (Whanger 2002). In humans, the SelW is a small protein (9.5 kDa). The origin of the "W" is from the selenium deficiency-related white muscle disease in lambs where this protein was absent (Whanger 2002). SelW levels in the fetal heart and muscle correspond well to the selenium status in human fetuses (Whanger 2002). Since it is found associated with the glutathione, it may have a potential function in the redox metabolism (Jeong et al. 2002). Another protein, Sel V was identified using an in silico approach and shows homology to the SelW. SelV expression seems to be limited to the seminiferous tubules of the testis (Kryukov et al. 2003), and the CGLU motif in its sequence suggests its redox-related function.

Selenoprotein T (SelT)

It is a small selenoprotein with an N-terminal redox motif (Kryukov et al. 1999). In humans, its sequence contains a Cys-X-X-Sec motif, similar to the active site in thioredoxins and glutaredoxins, which suggests that the SelT has redox properties.

Selenoprotein H (SelH)

It is a small selenoprotein with a predicted redox motif. In humans, the SelH is a globular protein and comprises 122 residues, selenocysteine being the 44th (Kryukov et al. 2003). The genomic sequence is expressed in the numerous tissues. The CXXU motif suggests a redox function with the selenocysteine possibly forming a selenenyl sulfide bridge with Cys-40.

Selenoprotein K (SelK)

This small selenoprotein contains a single transmembrane helix in the N-terminal sequence that targets this protein to the plasma membrane (Kryukov et al. 2003). SelK homologs can be detected in many eukaryotes. In human, the SelK is a membrane protein (Kryukov et al. 2003).

Selenoprotein S (SelS)

Like SelK, the Sel S also has Sec in the C-terminal sequence and a single transmembrane region at the N-terminus (Kryukov et al. 2003). It plays a role in the retrotranslocation of the misfolded proteins from the ER of the mammalian cells to the cytosol for further degradation (Ye et al. 2004). The SelS was implicated in inflammation and immune response (Curran et al. 2005). In humans, the SelS was first predicted as a selenoprotein in silico. Computational secondary structure analysis indicates a single transmembrane helix (as well as many putative phosphorylation and glycosylation sites), and like SelK, it is a plasma and ER-membrane protein (Kryukov et al. 2003). SelS expression is inversely correlated to the plasma glucose concentration (as well as insulin and triacylglycerol). At least in rats, it is transcribed in almost all the tissues, but glucose levels only affect hepatic expression of the SelS in vivo (Walder et al. 2002).

Selenoprotein N (SelN)

It is one of the first selenoprotein discovered through bioinformatics approach (Lescure et al. 1999) and implicated in the role of selenium in muscle function (Moghadaszadeh et al. 2001). In human, the SelN was also identified using a computation approach as two splice isoforms. Both transcripts are detected in the skeletal muscle, brain, lung, and placenta, but the isoform 2 is always more abundant. SelN is retained within the ER (Petit et al. 2003) and seems to be a ubiquitously expressed glycoprotein, particularly during the fetal development, but also at lower levels in the adults.

Selenium and Selenoprotein Functions

Selenium is of fundamental importance to the human health and is an essential component of the several major metabolic pathways, including antioxidant defense systems, thyroid hormone metabolism, and immune function. The selenoproteins identified serve quite diverse functions. Glutathione peroxidases (GPx), thioredoxin reductases (TrxR), and thyroid hormone deiodinases (DIO) are well-characterized selenoproteins involved in the redox regulation of the intracellular signaling, redox homeostasis, and thyroid hormone metabolism. Antioxidant role of the selenium is primarily mediated by the expression of selenoprotein, glutathione peroxidase (GPx), which can reduce H_2O_2 and phospholipid hydroperoxides (Fig. 7.4). This decreases the propagation of free radicals and reduces hydroperoxides intermediates (Spallholz et al. 1990). Another member of the glutathione peroxidase family, phospholipid glutathione peroxidase (PHGPx) diminishes the phospholipids and cholesterol esters associated with the lipoproteins, therefore, reducing the accumulation of the oxidized low-density lipoproteins (Sattler et al. 1994). Selenoprotein thioredoxin reductase plays an important role as the first line of defense against free radicals in human keratinocytes and melanocytes. Glutathione peroxidase also causes attenuation of the prooxidant-induced oxidation of NADPH, NADH, lipids, and proteins in various tissues (Lei and Cheng 2005) and reduces the platelet aggregation, thus minimizing the risk of cardiovascular disorders (Neve 1996). Selenium supplementation provide well efficient antioxidant role against the oxidative stress developed during ischemia and reperfusion in the cardiac surgery (Guo et al. 2012). Also dietary selenium has been useful in improving the human health, especially in an associated heart disease and bone disorders in selenium deficiency (Weeks et al. 2012).

The Keshan disease occurs upon selenium deficiency along with the infection by coxsackie B virus and was prevented by selenium supplementation. GPx1 knockout studies showed that GPx1 is closely associated with the protection against this virus infection (Moghadaszadeh and Beggs 2006). Also, selenium and zinc as antioxidants above the daily recommended levels were found providing protection against oxidative stress associated with the viral hepatitis and AIDS (Stehbens 2004). The inclusion of selenium as an adjuvant in the management of HIV seropositive patients has been beneficial to prevent the additional damage caused by the free radicals (Ogunro et al. 2006). Adriamycin, an anticancer drug, causes free radical mediated cardiotoxicity. Selenium supplementation in such cases causes an increase in the total antioxidant activity through glutathione concentration and GPx and catalase activities, leading to a decreased generation of reactive oxygen metabolites and preventing adriamycin-induced cardiotoxicity (Danesi et al. 2006). In yet another study, selenium supplementation to infertile men produced a significant decrease in lipid peroxidation and improvements in sperm motility (Keskes-ammar et al. 2003) indicating its protective and beneficial effects on semen quality.

Selenium supplementation at the dietary excess levels, however, has been reported to cause the generation of free radicals by itself. This is one of the mechanisms accounting for the anticarcinogenic and growth modulatory effects of the selenium. These effects are largely dependent on

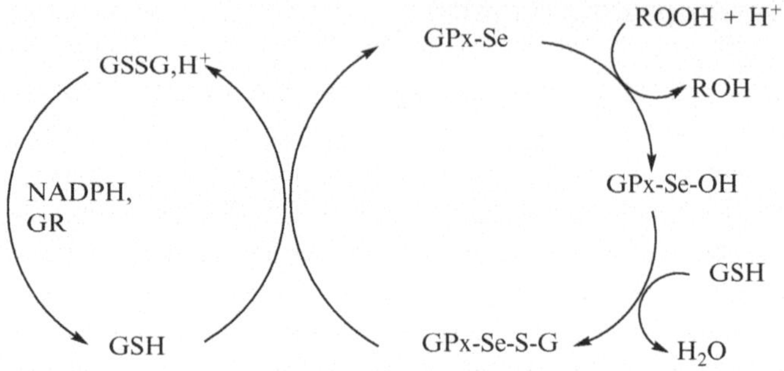

Fig. 7.4 Redox cycling of glutathione peroxidase

the GSH concentration and oxygen supply in the target cells. Sodium selenite inhibits growth and induces apoptosis in the NB4 cells, which correlates with the increased production of ROS in these cells and decreased levels of intracellular reduced glutathione (Li et al. 2003). In a major protective function, the selenium has been proposed to prevent the malignant transformation of cells by acting as a "redox switch" in the activation–inactivation of cellular growth factors and other functional proteins through the catalysis of the oxidation–reduction reactions of critical sulfhydryl (-SH) groups or disulfide (-S-S-) linkages (Schrauzer 1992). Diphenyl diselenide used in the synthesis of a variety of pharmacologically active organic selenium compounds also possesses prooxidant properties (Moreira-Rosa et al. 2005).

Further, while examining the interrelation of the known selenoproteins, mammalian selenoproteins were grouped as per location of the Sec in the protein (Kryukov et al. 2003): Sec very close to C-terminus (TrxRs, S, R, O, I, and K selenoproteins) and Sec close to N-terminus (GPxs, DIOs, H, M, N, T, V, W, and Sep15 selenoproteins). Most selenoproteins have the thioredoxin-fold structure, and some selenoproteins contain a CXXU motif, corresponding to the thioredoxin active-site CXXC motif (Archmann et al. 2007). Based on these structural characteristics, most selenoproteins have been shown to be involved in the redox-related reactions. The transcription of several selenoproteins such as TrxR1 and GPx2 is regulated by the redox-sensitive transcription factor Nrf2/Keap1 system (Banning et al. 2005). The basics of this system are explained in Chap. 5.

It is apparent that the selenoproteins for which the functions are known are mainly redox proteins. In these proteins, Sec is the catalytic residue that is employed because of its strong nucleophilicity and low pK_a (Kumar et al. 1992), and Sec reversibly changes its redox state during catalysis. Incorporation of Sec into selenoproteins uses a unique mechanism that involves decoding of the UGA codon (Squires and Berry 2008). This process requires multiple features such as the selenocysteine insertion sequence (SECIS) element in the 3′-untranslated region of the selenoprotein mRNA and several protein factors including the SECIS-binding protein 2 (SBP2). Translation of the selenoproteins depends on the integrity of the SECIS element–SBP2 interaction (Papp et al. 2007). Seeher et al. (2012) have well reviewed the role of these interactions during selenium level changes and further relation in the development of various pathological conditions in humans.

More specific roles of the selenoproteins have been revealed by the investigations using gene knockout techniques and by the mutant selenoproteins. Mutations in the selenoprotein N cause rigid muscular dystrophy and the classical phenotype of the multiminicore disease (Ferreiro et al. 2002). Expression of selenoprotein S was altered with change in the promoter sequence which influences the production of inflammatory cytokines such as TNF-α, interleukin-6, and interleukin-1β (Curran et al. 2005). Deletion of TrxR1, TrxR2, and GPx4 genes causes embryonic death in mice, hence their involvement in the embryogenesis. GPx1 knockout mice are more sensitive to the paraquat- and H_2O_2-induced oxidative stress. Three deiodinases deiodinate differently (Fig. 7.5). DIO2 knockout mice have

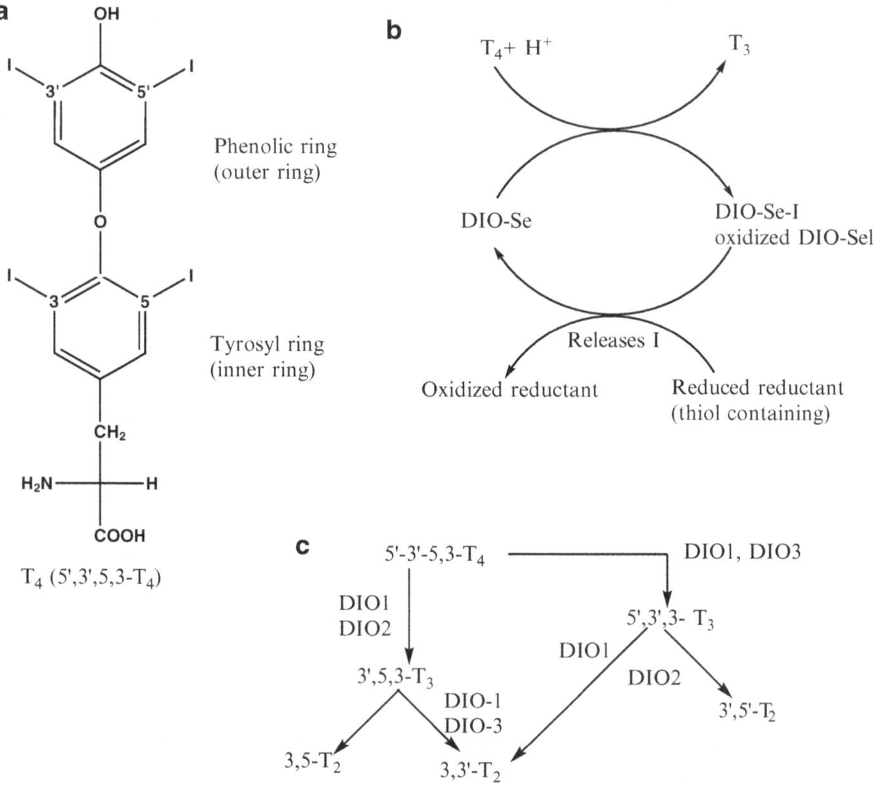

Fig. 7.5 Deiodinase redox activity and products

impaired auditory function and thermogenesis as well as mild brain function defects and temporary growth retardation (Moghadaszadeh and Beggs 2006). DIO1 knockout mice have abnormal excretion patterns of the thyroid hormone metabolites, including iodide. The DIO3 knockout model exhibits reduced viability, significant growth retardation, impaired fertility, and hypothyroid symptoms with significantly reduced T_3 and increased T_4 levels (Moghadaszadeh and Beggs 2006).

The biological activities of selenium as a nutrient, a cancer preventive agent, or even a toxicant are highly dependent on the dose and chemical form of the selenium (Ip et al. 1991). Bladder cancer risk in the humans has earlier been shown to be reduced with the increased circulating selenium; however less is known regarding involvement of selenoproteins. In an analytical study (Reszka 2012), high mRNA expression of various selenoproteins, namely, TrxR1, GPx1, Sep15, SelT, and SepW1 in human urinary epithelium, has been shown. However, bladder tumor showed increased selenium, GPx, and TrxR activity whereas the circulating selenium and GPx was decreased in these cancer patients. Involvement of the selenoproteins expression in the urinary epithelium has been speculated in the bladder cancer.

Epidemiological Evidences in Its Effectiveness in Pathologies

A number of epidemiological studies in the past have addressed the physiological and pathological effects of different selenium concentrations in different life forms. In several epidemiologic studies using different selenium status (low, moderate, and high selenium population) in prostate cancer, high concentration of selenium was found protective (Helzlsouer et al. 2000; Nomura et al. 2000; van den Brandt et al. 2003) and low plasma selenium levels were associated with increases in other cancers and human diseases (Rayman 2000). On the bases of the Nutritional Prevention of Cancer (NPC) trial of selenium on prostate cancer, selenium was included in the Selenium and Vitamin E Cancer Prevention Trial (SELECT) on a large number of men to test the role of supplementation with selenium and/or vitamin E in the prevention of the prostate cancer (Klein et al. 2003). Molecular and cellular bases for the published observations of selenium preventive activity in the prostate were studied and provided clarity for the clinical prostate cancer prevention trials of selenium, such as the SELECT. Selenium potentially affects cancer development through its effects on the oxidative stress, DNA repair, inflammation, apoptosis, proliferation, carcinogen metabolism, and immune function (Rayman 2000; Seo et al. 2002; Meuillet et al. 2004). Selenium effects can be indirect (through selenoproteins) and/or direct (though selenium metabolites) (Tapiero et al. 2003). The SELECT data and the informations from the ongoing mechanistic studies were expected to advance the understandings of the selenium in prostate carcinogenesis and many other diseases. The continued epidemiologic data on selenium from Li et al. (2004) supported its tremendous potential as a prostate cancer preventive agent. Klein (2004) also reviewed that a large body of epidemiological evidence, including observational, case–control, cohort, and randomized controlled clinical trials, support that selenium may prevent prostate cancer in humans. Molecular data demonstrate that selenium prevents clonal expansion of nascent tumors by causing cell-cycle arrest, promoting apoptosis, and modulating p53-dependent DNA repair mechanisms. Also epidemiological studies in the humans have shown relation between the chronic oral exposures to selenium and an increased incidence of death due to the neoplasms (Field and Youngson 2002; Tong et al. 2003). However, as per recent report (detailed by Nicastro and Dunn 2013) in SELECT studies, surprisingly, it was found that neither selenium nor vitamin E reduced the incidence of prostate cancer after 7 years of trial and vitamin E was associated with a 17 % increased risk of prostate cancer compared to placebo. Potential explanations for these negative findings include the agent formulation and dose, the characteristics of the cohort, and the study design. It is suggested that only specific subpopulations may be benefited from the selenium supplementation; future studies should consider the baseline selenium status of the participants, age of the cohort, and genotype of specific selenoproteins, in order to determine the activity of selenium in cancer prevention.

Selenium is important for the brain as well. Selenium deficiency besides affecting the turnover rate of some neurotransmitters also causes depression and other negative mood states such as anxiety, hostility, and confusion (Finley and Penland 1998). Further, genetic variations (e.g., SNPs) of the selenoproteins have been identified (Moscow et al. 1994; Kumaraswamy et al. 2000; Al-Taie et al. 2002; Villette et al. 2002), and these may respond differently to the selenium supplementation and suggest potential association with the pharmacogenetic differences in selenium's preventive effects (Hu et al. 2001). Epidemiologic studies, including two studies of the variant allele for the cellular antioxidant Gpx1 which was associated with increased risk for both lung (Ratnasinghe et al. 2000) and breast cancers (Hu and Diamond 2003) and a study of a GCG repeat polymorphism in Gpx1 which was not associated with prostate cancer (Kote-Jarai et al. 2002), have linked genetic variation to the disease. Development of selenoprotein transgenic and knockout models in mice will help to clarify the role of selenium and selenoproteins in cancer risk and prevention (Kumaraswamy et al. 2003). These studies become more important after the negative outcome of the SELECT studies.

Further, selenium and carotenoids play an important role in the antioxidant defenses and in the redox regulation involved in inflammation. An epidemiological study was carried with the hypothesis that low selenium and carotenoids predict mortality in older women living in the community (Ray et al. 2006). Women who were enrolled in the Women's Health and Aging Studies I and II in USA had serum selenium and carotenoids measured at the baseline and were followed for mortality over the 60 months. Higher serum selenium and higher serum total carotenoids were associated with a lower risk of mortality. It was concluded that the women living in community who have higher serum selenium and carotenoids are at a lower risk of death.

As mentioned earlier, large-scale clinical trials with selenium supplementation against prostate cancer are under way (Papp et al. 2007), and also many selenoproteins are involved in the antioxidant reaction in the protection of normal cells against oxidative stress. Further, when a normal cell turns into a tumor cell, selenoproteins in the tumor switch their role to protect the malignant phenotype. TrxR and Trx have been found to be overexpressed in many aggressive tumors. Moreover, the tumor cell may require enough activity of the Trx system for ribonucleotide reductase to keep up a constant DNA synthesis. Thus, the Sec-containing mammalian TrxRs have emerged as new targets for anticancer drug development (Arner and Holmgren 2006).

With the controversial results of selenium on the risk of diabetes, prospective studies were carried regarding the relationship between baseline plasma selenium concentration and occurrence of the dysglycemia (impaired fasting glucose or type 2 diabetes) in an elderly French cohort (Akbaraly et al. 2010). Epidemiology of Vascular Aging (EVA) study ($n = 1,389$, 59–71 years) is a 9-year longitudinal study. Risk of dysglycemia was significantly lower in men with plasma selenium in the highest level compared to those in the lowest, but no significant relationship was observed in women. This perspective study suggests a sex-specific protective effect of higher selenium status at the baseline on later occurrence of dysglycemia.

The relation of excess selenium exposure and human motor disease, amyotrophic lateral sclerosis (ALS) (Vinceti et al. 2010), has been suggested on the basis of the two epidemiologic investigations which found an increased risk of ALS associated with the residence in a seleniferous area or with consumption of drinking water unusually with high levels of inorganic hexavalent selenium. Further, critical illness with the systemic inflammatory response syndrome, SIRS (Hardy et al. 2012, is characterized by selenium depletion with high morbidity and mortality. Selenium supplementation for the critical ill can improve clinical outcome by reducing illness severity, infectious complications, and decreasing mortality in the intensive care unit (ICU). However, development of better biomarker to ascertain optimum selenium requirements for the individual patients is needed for the improvement in clinical practice guidelines.

Recently, importance of the selenium in reproductive health is highlighted (Mistry et al. 2012). Numerous reports implicate selenium deficiency in several reproductive and obstetric complications including male and female infertility, miscarriage, preeclampsia, fetal growth restriction, preterm labor, gestational diabetes, and obstetric cholestasis. However, inadequate information is available from small intervention studies to inform public health strategies. Therefore, larger intervention trials are required to get a beneficial role of the selenium supplementation in disorders of reproductive health.

Finally, new experiments need to be planned in light of these previous reports, namely, individual selenium status, type and dose of the selenium to be supplemented, and specially the knowledge of variants of the selenoprotein genes. The rapidly evolving field of selenium and selenoprotein biology promises to identify novel molecular targets for preventing or delaying the various cancer, cardiovascular, neurogenerative, and other diseases, in which selenium appears to play an important role.

References

Akbaraly TN, Arnaud J, Rayman MP, Hininger-Favier I, Roussel AM, Berr C, Fontbonne A (2010) Plasma selenium and risk of dysglycemia in an elderly French population: results from the prospective epidemiology of vascular aging study. Nutr Metab 7:21–27

Al-Taie OH, Seufert J, Mork H, Treis H, Mentrup B, Thalheimer A, Starostik P, Abel J, Scheurien M, Kohrle J, Jakob F (2002) A complex DNA-repeat structure within the Selenoprotein P promoter contains a functionally relevant polymorphism and is genetically unstable under conditions of mismatch repair deficiency. Eur J Hum Genet 10:499–504

Amer ES, Holmgren A (2000) Physiological functions of thioredoxin and thioredoxin reductase. Eur J Biochem 267:6102–6109

Anestal K, Arner ES (2003) Rapid induction of cell death by selenium-compromised thioredoxin reductase 1 but not by the fully active enzyme containing selenocysteine. J Biol Chem 278:15966–15972

Archmann FL, Fomenko DE, Soragni A, Gladyshev VN, Dikiy A (2007) Solution structure of selenoprotein W and NMR analysis of its interaction with 14-3-3-proteins. J Biol Chem 282:37036–37044

Arner ES, Holmgren A (2006) Thioredoxin system in cancer. Semin Cancer Biol 16:420–426

Arnold SM, Fessler LI, Fessler JH, Kaufman RJ (2000) Two homologues encoding human UDP-glucose glycoprotein glucosyl transferase differ in mRNA expression and enzymatic activity. Biochemistry 39: 2149–2163

Banning A, Deubel S, Kluth D, Zhou Z, Brigelius-Flohe R (2005) The GI-GPx gene in a target for Nrf2. Mol Cell Biol 25:4914–4923

Baum MK, Campa A, Miguez-Burbano MJ, Burbano X, ShorPosner G (2001) Role of selenium in HIV/AIDS (ch 20). In: Hatfield DL (ed) Selenium: its molecular biology and role in human health. Kluwer Academic, Norwell, pp 247–255

Behne D, Hilmert H, Scheid S, Gessner H, Elger W (1988) Evidence for specific selenium target tissues and new biologically important selenoproteins. Biochim Biophys Acta 966:12–21

Birringer M, Pilawa S, Flohe L (2002) Trends in selenium biochemistry. Nat Prod Rep 19:693–718

Bock A (2001) Selenium metabolism in bacteria. In: Hatfield DL (ed) Selenium: its molecular biology and role in human health. Kluwer Academic, Norwell, pp 7–22, 1

Bopp BA, Sonders RC, Kesterson JW (1982) Metabolic fate of selected selenium compounds in laboratory animals and man. Drug Metab Rev 13:271–318

Brigelius-Floh R (1999) Tissue-specific functions of individual glutathione peroxidases. Free Radic Biol Med 27:951–965

Brown KM, Arthur JR (2001) Selenium, selenoproteins and human health: a review. Public Health Nutr 4:593–599

Buettner C, Harney JW, Larsen PR (1998) The 3′-untranslated region of human type 2 iodothyronine deiodinase mRNA contains a functional selenocysteine insertion sequence element. J Biol Chem 273:33374–33378

Burk RF, Hill KE (1999) Orphan selenoproteins. Bioessays 21:231–237

Burk RF, Hill KE (2005) Selenoprotein P an extracellular protein with unique physical characteristics and a role in selenium homeostasis. Annu Rev Nutr 25:215–235

Butler JA, Whanger PD, Kaneps AJ, Patton NM (1990) Metabolism of selenite and selenomethionine in Rhesus monkey. J Nutr 120:751–759

Carlson BA, Martin-Romero FJ, Kumaraswamy E, Moustafa ME, Zhi H, Hatfield DL, Lee BJ (2001) Mammalian selenocysteine tRNA p. In: Hatfield DL (ed) Selenium: its molecular biology and role in human health. Kluwer Academic, Norwell, pp 23–32

Castellano S, Morozova N, Morey M, Berry MJ, Serras F, Corominas M, Guigo R (2001) In silico identification of novel selenoproteins in the drosophila melanogaster genome. EMBO Rep 2:697–702

Chu FF, Esworthy RS, Chu PG, Longmate JA, Huycke MM, Wilezynski S et al (2004) Bacteria-induced intestinal cancer in mice with disrupted Gpx1 and Gpx2 genes. Cancer Res 64:962–968

Combs GF Jr, Combs SB (1984) The nutrition biochemistry of selenium. Annu Rev Nutr 4:257–280

Combs GR, Liu L (2001) Selenium as a cancer preventive. In: Hatfield DL (ed) Selenium: its molecular biology and role in human health. Kluwer Academic, Norwell, pp 205–217, ch 17

Conrad M, Jakupoglu C, Moreno SG, Lippl S, Banjac A, Schneider M, Beck H, Hatzopoulos AK, Just U, Sinoutatz F, Schneider M, Beck H, Hatzo-poulos AK, Just U, Sinowatz F, Schmahl W, Chien KR, Wurst W, Bomkamm GW, Brielmeier M (2004) Essential role for mitochondrial thioredoxin reductase in hematopoiesis, heart development and heart function. Mol Cell Biol 24:9414–9423

Copeland PE, Stepanik VA, Driscoll DM (2001) Insight into mammalian selenocysteine insertion: domain structure and ribosome binding properties of Sec insertion sequence binding protein 2. Mol Cell Biol 21:1491–1498

Coppinger RJ, Diamond AM (2001) Selenium deficiency and human disease (ch 18). In: Hatfield DL (ed) Selenium: its molecular biology and role in human health. Kluwer Academic, Norwell, pp 219–233

Curran JE, Jowett JB, Elliott KS, Gao Y, Gluschenko K, Wang J, Abel Azim DM, Cai G, Mahaney MC, Comuzzie AG, Dyer TD, Walder KR, Zimmer P, MacCluer JW, Collier GR, Kissebah AH, Blangero J (2005) Nat Genet 37:1234–1241

Danesi F, Malaguti M, Nunzio MD, Maranesi M, Biagi PL, Bordoni A (2006) Counteraction of adriamycin-induced oxidative damage in rat heart by selenium dietary supplementation. J Agric Food Chem 54:1203–1208

Diskin CJ, Tomasso CL, Alper JC, Glaser ML, Fliegel SE (1979) Long term selenium exposure. Arch Intern Med 139:824–826

Escobar-Morreale HF, Obregon MJ, Escobar del Rey F, Morreale de Escobar G (1999) Tissue-specific patterns of changes in 3,5,3′-triodo-L-thyronine concentrations in thyroidectomized rats infused with increasing doses of the hormone. Which are the regulatory mechanisms? Biochimie 81:453–462

Fagegaltier D, Hubert N, Yamada K, Mizutani T, Carbon P, Krol A (2000) Characterization of mSelB, a novel mammalian elongation factor for selenoprotein translation. EMBO J 19:4796–4805

Ferguson AD, Labunskyy VM, Fomenko DE, Arac D, Chelliah Y, Amezcua CA, Rizo J, Gladyshev VN,

Deisenhofer J (2006) NMR-structures of the selenoproteins Sep15 and SelM reveal redox activity of a new thioredoxin-like family. J Biol Chem 281: 3536–3543

Ferreiro A, Quijano-Roy S, Pichereau C, Moghadaszadeh B, Goemans N, Bonnemann C, Jungbluth H, Straub V, Villanova M, Leroy JP, Romero NB, Martin JJ, Muntoni F, Voit T, Estournet B, Richard P, Fardeau M, Gulcheney P (2002) Mutations of the selenoprotein N gene, which is implicated in rigid spine muscular dystrophy, cause the classical phenotype of multiminicore disease: reassessing the nosology of early-onset myopathies. Am J Hum Genet 71:739–749

Field JK, Youngson JH (2002) The Liverpool Lung Project: a molecular epidemiological study of early lung cancer detection. Eur Respir J 20:464–479

Finley J, Penland J (1998) Adequacy or deprivation of dietary selenium in healthy men: clinical and psychological findings. J Trace Elem Exp Med 11:11–27

Floh L, Foresta C, Garolla A, Maiorino M, Roveri A, Ursini F (2002) Metamorphosis of the selenoprotein PHGPx during spermatogenesis. Ann NY Acad Sci 973:287–288

Foresta C, Floh L, Garolla A, Roveri A, Ursini F, Maiorino M (2002) Male fertility is linked to the selenoprotein phospholipid hydroperoxide glutathione peroxidase. Biol Reprod 67:967–971

Fu Y, Cheng WH, Porres JM, Ross DA, Lei XG (1999) Knockout of cellular glutathione peroxidase gene mice susceptible to diquat-induced oxidative stress. Free Radic Biol Med 27:605–611

Fu LH, Wang XF, Eyal Y, She YM, Donald LJ, Standing KG, Ben-Hayyim G (2002) A selenoprotein in the plant kingdom. Mass spectrometry confirms that an opal codon (UGA) encodes selenocysteine in Chlamydomonas reinhardtii glutathione peroxidase. J Biol Chem 277:25983–25991

Ganther HE (1979) Metabolism of hydrogen selenide and methylated selenides. Adv Nutr Res 2:107–128

Gladyshev VN (2006) Selenoproteins and selenoproteomes (Chapter 9). In: Hatfield DL, Berry MJ, Gladyshev VN (eds) Selenium: its molecular biology and role in human health, 2nd edn. Springer, New York, pp 99–114

Glass RS, Singh WP, Jung W, Veres Z, Scholz TD, Stadtman TC (1993) Monoselenophosphate: synthesis, characterization and identity with prokaryotic biological selenium donor, compound SePX. Biochemistry 32:12555–12559

Gromer S, Eubel JK, Lee BL, Jacob J (2005) Human selenoproteins at a glance. Cell Mol Life Sci 62: 2414–2437

Guimaracs MJ, Petrson D, Vicarari A, Cocks BG, Copeland NG, Gilbert DJ, Jenkins NA, Ferrick DA, Kastelein RA, Bazan JR, Zlotnik A (1996) Identification of a novel selD homolog from eukaryotes, bacteria and archaea: is there an autoregulatory mechanism in selenocysteine metabolism? Proc Natl Acad Sci U S A 67:1200–1206

Guo F, Monsefi N, Moritz A, Beiras-Femandez A (2012) Selenium and cardiovascular surgery: an overview. Curr Drug Saf 7:321–327

Hammond PB, Beliles RP (1980) Metals. In: Doull J, Klaassen CD, Amdur MO (eds) Casarett and Doull's toxicology, the basic science of poisons, 2nd edn. Macmillan, New York, pp 409–467

Hardy G, Hardy I, Manzanares W (2012) Selenium supplementation in the critically ill. Nutr Clin Pract 27:21–33

Hatfield DL, Portugal FH (1970) Seryl-tRNA in mammalian tissues. Chromatographic differences in brain and liver and a specific response to the codon, UGA. Proc Natl Acad Sci U S A 67:1200–1206

Helzlsouer KJ, Huang HY, Alberg AJ, Hoffman S, Burke A, Norkus EP, Morris J, Comstock GW (2000) Association between alpha-tocopherol, gamma-tocopherol, selenium, and subsequent prostate cancer. J Natl Cancer Inst 92:2018–2023

Hill KE, zhou J, McMahan WJ, Motley AK, Atkins JF, Gesteland RF, Burk RF (2003) Deletion of selenoprotein P alters distribution of selenium in the mouse. J Biol Chem 278:13640–13646

Hu YJ, Diamond AM (2003) Role of glutathione peroxidase 1 in breast cancer loss of heterozygosity and allelic differences in the response to selenium. Cancer Res 63:3347–3351

Hu YJ, Korotkov KV, Mehta R, Hatfield DL, Rotimi CN, Luke A, Prewitt TE, Cooper RS, Stock W, Vokes EE, Dolan ME, Glalyshav VN, Diamond AM (2001) Distribution and functional consequences of nucleotide polymorphisms in the 3′-untranslated region of the human Sep15 gene. Cancer Res 61:2307–2310

Ip C, Hayes C, Budnick RM, Ganther HE (1991) Chemical form of selenium, critical metabolites and cancer prevention. Cancer Res 51:595–600

Jakupoglu C, Pizemeck GK, Schneider M, Morlno SG, Mayr N, Hatzopoulos AK, deAnqelis MH, Wurat W, Bomkamm GW, Brielmeir M, Conrad M (2005) Cytoplasmic thioredoxin reductase is essential for embryogenesis but dispensable for cardiac development. Mol Cell Biol 25:1980–1988

Jenkins KJ, Hidiroglou M (1972) Comparative metabolism of ^{75}Se-selenite, ^{75}Se-selenate and ^{75}Se-selenomethionine in bovine erythrocytes. Can J Physiol Pharmacol 50:927–935

Jeong D, Kim TS, Chung YW, Lee BJ, Kim IY (2002) Selenoprotein W is a glutathione-dependent antioxidant in vivo. FEBS Lett 517:225–228

Jiang S, Robberecht H, Vanden Berge D (1983) Elimination of selenium compounds by mice through formation of different volatile selenides. Experientia 39:293–294

Kelly GS (2000) Peripheral metabolism of thyroid hormones: a review. Altern Med Rev 5:306–333

Keskes-Ammar L, Feki-Chakroun N, Rebai T, Sahnoun Z, Ghozzi H, Hammami S, Zghal K, Fki H, Damak J, Bahloul A (2003) Sperm oxidative stress and the effect of an oral vitamin E and selenium supplement on semen quality in infertile men. Arch Androl 49: 83–94

Kim HY, Gladyshey VN (2004) Characterization of mouse endoplasmic reticulum methionine-R-sulphoxide reductase. Biochem Biophys Res Commun 320:1277–1283

Klein EA (2004) Selenium: epidemiology and basic science. J Urol 17:S50–S53

Klein EA, Lippman SM, Thompson IM, Goodman PJ, Albanes D, Taylor PR, Coltman C (2003) The selenium and vitamin E cancer prevention trial. World J Urol 21:21–27

Korotkov KV, Kumaraswamy E, Zhou Y, Hatfield DL, Gladyshev VN (2001) Association between the 15-kDa selenoprotein and UDP-glucose glycoprotein glucosyltransferase in the endoplasmic reticulum of mammalian cells. J Biol Chem 276:15330–15336

Korotkov KV, Novoselov SV, Hattield DL, Gladyshev VN (2002) Mammalian selenoprotein in which selenocysteine (Sec) incorporation is supported by a new form of Sec insertion sequence element. Mol Cell Biol 22:1402–1411

Kote-Jarai Z, Durocher F, Edwards SM, Hamoudi R, Jackson RA, Ardern-Jones A, Murkin A, Dearnaley DP, Kirby R, Houlston R, Easton DF, Eeles R, CRC/BPG UK Familial Prostate Cancer Collaborators (2002) Association between the GCG polymorphism of the selenium dependent GPX1 gene and the risk of young onset prostate cancer. Prostate Cancer Prostatic Dis 5:189–192

Kryukov GV, Kryukov VM, Gladyshev VN (1999) New mammalian selenocysteine-containing proteins identified with an algorithm that searches for selenocysteine insertion sequence element. J Biol Chem 274: 33888–33897

Kryukov GV, Kumar RA, Koe A, Gladyshey YN (2002) Selenoprotein R is a Zinc-containing stereo-specific methionine sulfoxide reductase. Proc Natl Acad Sci U S A 99:4245–4250

Kryukov GV, Castellano S, Novoselov SV, Lobanov AV, Zehtab O, Guigo R, Gladyshev VN (2003) Characterization of mammalian selenoproteomes. Science 300:1439–1443

Kumar S, Bjornstedt M, Holmgren A (1992) Selenite is a substrate for calf thymus thioredoxin reductase and thioredoxin and elicits a large non-stoichiometric oxidation of NADPH in the presence of oxygen. Eur J Biochem 207:435–439

Kumaraswamy E, Malykh A, Korotkov KV, Kozyavkin S, Hu Y, Kwon SY, Moustafa ME, Carlson BA, Berry MJ, Lee BJ, Hatffield DL, Diamond AM, Gladyshey YN (2000) Structure-expression relationships of the 15-kDa selenoprotein gene. Possible role of the protein in cancer etiology. J Biol Chem 275: 35540–35547

Kumaraswamy E, Carlson BA, Morgan F, Miyoshi K, Robinson GW, Su D et al (2003) Selective removal of the selenocysteine tRNA [Ser]Sec gene (Trsp) in mouse mammary epithelium. Mol Cell Biol 23:1477–1488

Lee BJ, Worland PJ, Davis JN, Stadtman TC, Hartfield DL (1989) Identification of a selenocysteyl-tRNA^{Ser} in mammalian cells which recognizes the nonsense codon, UGA. J Biol Chem 264:9724–9727

Lei XG, Cheng WH (2005) New roles for an old selenoenzyme: evidence from glutathione peroxidase-1 null and overexpressing mice. J Nutr 135:2295–2298

Lescure A, Gautheret D, Carbon P, Krol A (1999) Novel selenoproteins identified in silico and in vivo by using a conserved RNA structural motif. J Biol Chem 274:38147–38154

Li J, Zuo L, Shen T, Xu CM, Zhang ZN (2003) Induction of apoptosis by sodium selenite in human acute promyelocytic leukemia NB4 cells: involvement of oxidative stress and mitochondria. J Trace Elem Med Biol 17:19–26

Li H, Stampfer MJ, Giovannucei EL, Morris JS, Willett WC, Gaziano JM, Ma J (2004) A perspective study of plasma selenium levels and prostate cancer risk. J Natl Cancer Inst 96:696–703

Low SC, Berry MJ (1996) Knowing when not to stop selenocysteine incorporation in eukaryotes. Trends Biochem Sci 21:203–208

Ma S, Hill KE, Burk RF, Caprioli RM (2005) Mass spectrometric determination of selenenylsulfide linkages in rat selenoprotein P. J Mass Spectrom 40:400–404

McKenzie RC, Arthur JR, Beckett GJ (2002) Selenium and the regulation of cell signaling growth and survival: molecular and mechanistic aspects. Antioxid Redox Signal 4:339–351

Meuillet E, Stratton S, Cherukuri DP, Goulet AC, Kagey J, Porterfield B, Nelson M (2004) Chemoprevention of prostate cancer with selenium: an update on current clinical trials and preclinical findings. J Cell Biochem 91:443–458

Miranda-Vizuete A, Damdimopoulos AE, Spyrou G (2000) The mitochondrial thioredoxin system. Antioxid Redox Signal 2:801–810

Mistry HD, Broughton PF, Redman CW, Poston L (2012) Selenium in reproductive health. Am J Obstet Gynecol 206:21–30

Moghadaszadeh B, Beggs AH (2006) Selenoproteins and their impact on human health through diverse physiological pathways. Physiology (Bethesda) 21:307–315

Moghadaszadeh B, Petit N, Jaillard C, Brockington M, Quijano Roy S, Merlini C, Romero N, Estournet B, Desquerre I, Cheiqne D, Muntoni F, Topaloglu H, Guicheney P (2001) Mutations in SEPN1 cause congenital muscular dystrophy with spinal rigidity and respective respiratory syndrome. Nat Genet 29:17–18

Moreira-Rosa R, De Oliveira RB, Saffi J, Braga AL, Rooesler R, Dal-Pizzol F, Fonseca-Moreira JC, Brendel M, Pegas-Henriques JA (2005) Prooxidant action of diphenyl diselenide in the yeast Saccharomyces cerevisiae exposed to ROS-generating conditions. Life Sci 77:2398–2411

Moscow JA, Schmidt L, Ingram DT, Gnarra J, Johnson B, Cowan KH (1994) Loss of heterozygosity of the human cytosolic glutathione peroxidase I gene in lung cancer. Carcinogenesis 15:2769–2773

Nakamuro K, Sayato Y, Ose Y (1977) Studies on selenium-related compounds VI. Biosynthesis of dimethyl selenide in rat liver after oral administration of sodium selenate. Toxicol Appl Pharmacol 39:521–529

Neve J (1996) Selenium as a risk factor for cardiovascular diseases. J Cardiovasc Risk 3:42–47

Ng L, Goodyear RJ, Woods CA, Schhneider MJ, Diamond E, Richardson GP, Kelley MW, Germain DL, Galton VA, Forrest D (2004) Hearing loss and retarded cochlear development in mice lacking type 2 iodothyronine deiodinase. Proc Natl Acad Sci U S A 101:3474–3479

Nicastro HL, Dunn BK (2013) Selenium and prostate cancer prevention: insights from the selenium and vitamin E cancer prevention trial (SELECT). Nutrients 5:1122–1148

Nomura AM, Lee J, Stemmermann GN, Jr Combs GF (2000) Serum selenium and subsequent risk of prostate cancer. Cancer Epidemiol Biomarkers Prev 9:883–887

Ogunro PS, Oqungbamigbe TO, Elemie PO, Egbewale BE, Adewole TA (2006) Plasma selenium concentration and glutathione peroxidase activity in HIV-1/AIDS infected patients: a correlation with the disease progression. Niger Postgrad Med J 13:1–5

Papp LV, Lu J, Holmgren A, Khanna KK (2007) From selenium to selenoproteins: synthesis, identity and their role in human health. Antioxid Redox Signal 9:775–806

Petit N, Leseure A, Rederstoriff M, Krol A, Moghadaszadeh B, Wewer UM, Guicheney P (2003) Selenoprotein N: an endoplasmic reticulum glycoprotein with an early developmental expression pattern. Hum Mol Genet 12:1045–1053

Ratnasinghe D, Tangrea JA, Andersen MR, Barrett MJ, Virtamo J, Taylor PR, Albanes D (2000) Glutathione peroxidase codon 198 polymorphism variant increases lung cancer risk. Cancer Res 60:6381–6383

Ray AL, Semba RD, Walston J, Ferrucci L, Cappola AR, Ricks MO, Xue Q, Fried LP (2006) Low serum selenium and total carotenoids predict mortality among older women living in the community: the women's health and aging studies. J Nutr 136:172–176

Rayman MP (2000) The importance of selenium to human health. Lancet 356:233–241

Reszka E (2012) Selenoproteins in bladder cancer. Clin Chim Acta 413:847–854

Rotruck JT, Pope AL, Ganther HE, Swanson AB, Hafeman DG, Hoekstra WG (1973) Selenium: biochemical role as a component of glutathione peroxidase. Science 179:588–590

Salvatore D, Low SC, Berry M, Maia AL, Hamey JW, Croteau W, StGermain DL, Larsen PR (1995) Type 3 iodothyronine deiodinase cloning, in vitro expression and functional analysis of the cloning: in vitro expression and functional analysis of the placental selenoenzyme. J Clin Invest 96:2421–2430

Sandalova T, zhong L, Lindqvist Y, Holmgren A, Schneider G (2001) Three-dimensional structure of a mammalian thioredoxin reductase: implications for mechanism and evaluation of a selenocysteine-dependent enzyme. Proc Natl Acad Sci U S A 98:9533–9538

Sarto C, Frutiger S, Cappellano F, Sanchez JC, Doro G, Catanzaro F, Hughes GJ, Hochstrasser DF, Mocarelli P (1999) Modified expression of plasma glutathione peroxidase and manganese superoxide dismutase in human renal cell carcinoma. Electrophoresis 20: 3458–3466

Sattler W, Maiorino M, Stocker R (1994) Reduction of HDL and LDL associated cholesterylester and phospholipids hydroperoxides by phospholipid glutathione peroxidase and ebselen. Arch Biochem Biophys 309:214–221

Schrauzer GN (1992) Selenium: mechanistic aspects of anticarcinogenic action. Biol Trace Elem Res 33: 51–62

Schwarz K, Foltz CM (1957) Selenium as an integral part of factor3 against dietary necrotic liver degeneration. J Am Chem Soc 79:3292–3293

Schweizer U, streckfuss F, Pelt P, Carlson BA, Hatfield DL, Kohrle J, Schomburg L (2005) Hepatically derived selenoprotein P is a key factor for kidney but not for brain selenium supply. Biochem J 386: 221–226

Seeher S, Mahdi Y, Schweizer U (2012) Post-transcriptional control of selenoprotein biosynthesis. Curr Protein Pept Sci 13:337–346

Seo YR, Kelley MR, Smith ML (2002) Selenomethionine regulation of p53 by a ref1-dependent redox mechanism. Proc Natl Acad Sci U S A 99:14548–14553

Shen H, Yang C, Liu J, Ong C (2000) Dual role of glutathione in selenite-induced oxidative stress and apoptosis in human hepatoma cells. Free Radic Biol Med 28:1115–1124

Soderberg A, Sahaf B, Rosen A (2000) Thioredoxin reductase, a redox-active selenoprotein is secreted by normal and neoplastic cells: presence in human plasma. Cancer Res 60:2281–2289

Spallholz JE, Boylan LM, Larsen HS (1990) Advances in understanding selenium's role in the immune system. Ann NY Acad Sci 587:123–139

Squires JE, Berry MJ (2008) Eukaryotic selenoprotein synthesis: mechanistic insight incorporating new factors and new functions for old factors. IUBMB Life 60:232–235

Stehbens WE (2004) Oxidative stress in viral hepatitis and AIDS. Exp Mol Pathol 77:121–132

Su D, Novoselov SV, Sun QA, Moustafa ME, Zhou Y, Oko R, Hatfield DL, Gladyshev VN (2005) Mammalian selenoprotein thioredoxin-glutathione reductase: roles in disulfide bond formation and sperm maturation. J Biol Chem 280:26491

Sun QA, Gladyshev VN (2002) Redox regulation of cell signaling by thioredoxin reductases. Methods Enzymol 347:451–461

Sun QA, Kirnarsky L, Sherman S, Gladyshev VN (2001) Selenoprotein oxidoreductase with specificity for thioredoxin and glutathione systems. Proc Natl Acad Sci U S A 98:3673–3678

Tapiero H, Townsend DM, Tew KD (2003) The antioxidant role of selenium and seleno-compounds. Biomed Pharmacother 57:134–144
Tham DM, Whitin JC, Kim KK, Zhu SX, Cohen HJ (1998) Expression of extracellular glutathione peroxidase in human and mouse gastrointestinal tract. Am J Physiol 275:G1463–G1471
Tong YJ, Teng WP, Jin Y, Li YS, Guan HX, Wang WB, Gao TS, Teng XC, Yang F, Shi XG, Chen W, Man N, Li Z, Guo XJ (2003) An epidemiological study on the relationship between selenium and thyroid function in areas with different iodine intake. Zhonghua Yi Xue Za Zhi 83:2036–2039
Urisini F, Heim S, Kiess M, Maiorino M, Roveri A, Wissing J, Flobe I (1999) Dual function of the selenoprotein in PHGPx during sperm maturation. Science 285:1393–1396
Van den Brandt PA, Zeegers MP, Bode P, Goldbohm RA (2003) Toenail selenium levels and the subsequent risk of prostate cancer: a prospective cohort study. Cancer Epidemiol Biomarkers Prev 12:866–871
Vendeland SC, Beilstein MA, Chen CL, Jensen ON, Barofsky E, Whanger PD (1993) Purification and properties of selenoprotein W from rat muscle. J Biol Chem 268:17103–17107
Vendeland SC, Beitstein MA, Yeh JY, Ream W, Whanger PD (1995) Rat skeletal muscle selenoprotein W: cDNA clone and RNA maturation by dietary selenium. Proc Natl Acad Sci U S A 92:8749–8753
Villette S, Kyle JA, Brown KM, Pickard K, Milne JS, Nicol F, Arthur JR, Hesketh JE (2002) A novel single nucleotide polymorphism in the 3′ untranslated region of human glutathione peroxidase 4 influences lipoxygenase metabolism. Blood Cells Mol Dis 29:174–178
Vinceti M, Bonvicini F, Bergomi M, Malagoli C (2010) Possible involvement of overexposure to environmental selenium in the etiology of amyotrophic lateral sclerosis: a short review. Ann Ist Super Sanita 46:279–283
Walder K, Kantham L, McMilan JS, Trevaskis J, Kerr L, DeSilva A, Stunderland T, Godde N, Gao Y, Bishara N, Windmill K, Tenne-Brown J, Augert G, Zimmet PZ, Collier GR (2002) Tanis: a link between type 2 diabetes and inflammation? Diabetes 51: 1859–1866
Weeks BS, Hanna MS, Cooperstein D (2012) Dietary selenium and selenoprotein function. Med Sci Monit 18:127–132
Weissman SH, Cuddihy RG, Medinsky MA (1983) Absorption, distribution and retention of inhaled selenious acid and selenium metal aerosols in beagle dogs. Toxicol Appl Pharmacol 67:331–337
Whanger PD (2002) Selenoprotein W. Methods Enzymol 347:179–187
Whitin JC, Tham DM, Bhamre S, Omt DB, Scanding JD, Tune BM, Salvatiessa O, Avissar N, Cohn HJ (1998) Plasma glutathione peroxidase and its relationship to renal proximal tubule function. Mol Genet Metab 65:238–245
Winger R, Bocher M, Floh L, Kolimus H, Brigelius-Floh R (1999) mRNA stability and selenocysteine insertion sequence efficiency rank gastrointestinal glutathione peroxidase high in the hierarchy of selenoproteins. Eur J Biochem 259:149–157
Ye Y, Shibata Y, Yun C, Ron D, Rpoport TA (2004) A membrane protein complex mediates retro-translation from the ER lumen into the cytosol. Nature 429:841–847

8 Future Perspective

ROS at the Helm of Pathogenesis: Targeting Oxidative Stress to Combat Diseases

Previous chapters have evidently addressed the fact that accumulation of ROS and RNS leading to oxidative stress is a common denominator in almost all pathological conditions ranging from cardiovascular diseases, autoimmune disorders, cancers, neurodegenerative diseases, metabolic disorders, and so on. Therefore, it is comforting to postulate that strategies targeted to reduce generation of ROS and RNS hold the potential in treatment of these disorders. In this direction, numerous efforts have been made, and there are plenty of studies that strongly project the beneficial effects of natural and chemical antioxidants in prevention and/or treatment of some of these diseases. Despite the great outcome and success in this research area, these antioxidant therapies are still not a part of regular treatment regimen in a clinical setup. To achieve these goals and draw a clinical relevance of impact of oxidative stress in clinical health, greater research impetus in the following areas is required.

Universal Clinical Markers to Monitor Oxidative Stress

ROS can oxidize all major biomolecules such as carbohydrates, lipids, proteins, and nucleic acids. The products of these oxidation reactions are routinely measured and considered as biomarkers of oxidative stress. We have discussed these biomarkers in detail in previous chapters. However, till date there are no sensitive, facile, and accurate assays to measure oxidative stress that can translate to requirement of specific antioxidant supplementation for individual's need. Also, specific subcellular compartments must be explored for oxidations to help designing the more specific and targeted redox modulations for therapeutic purposes. Furthermore, the fact that certain ROS or RNS are associated with specific pathologies warranties the need of using low-molecular-weight compounds or overexpression and knockdown technologies to modulate them particularly in context of diseases like cancers.

Therefore, efforts should be made to develop efficient ways to detect these markers of oxidative stress. In this direction, recently, the use of techniques such as gas chromatography with

M. Bansal and N. Kaushal, *Oxidative Stress Mechanisms and their Modulation*,
DOI 10.1007/978-81-322-2032-9_8, © Springer India 2014

mass spectrometry (GC-MS) is upcoming to monitor oxidative stress. The better versions of these techniques such as high performance liquid chromatography with electrochemical detection (HPLC-ECD) are being developed as a tool for the routine assessment of oxidative stress. It can be concluded that in the future there is a definite requirement of a detailed longitudinal study to evaluate the panels of oxidative biomarkers to be correlated to traditional clinical end points.

Mechanism-Based Therapies

Research over the decades has gained substantial insight into the mechanisms and consequences of oxidative stress. However, additional studies are required to further clarify how the molecular basis of oxidative stress-induced damage varies in different tissues. In this direction, additional analyses need to determine the role of various biochemical metabolites in the production of ROS and vice versa. Furthermore, in mechanistic studies, special emphasis must be laid down to decisively determine the molecular factors and event(s) that act as redox switch to induce oxidative stress and downstream signaling cascade.

It is well known that oxidative stress is known to potentiate the various pathologies via modification of redox-sensitive transcriptional factors and signaling molecules. Despite these understanding, the precise modification profile is thus far poorly defined. Therefore, further studies are warranted to study the specific modification to key proteins in order to open another avenue for understanding the role of ROS/RNS in these pathophysiological process and development of antioxidant therapies to combat these modifications. The development of these treatment regimes has to rely on the understanding of multiple extracellular and intracellular pathways involved in ROS and RNS production and destruction and of how they are integrated with the overall cellular signaling machinery.

Other questions that should be addressed in future research include the following:

- What are the redox sensors in the body that respond to the potentially harmful signal to disrupt the delicate balance between ROS at normal physiological levels and antioxidants?
- Recent trends in the field of nutrigenomics emphasize the need to know about the effects of specific types of nutritional factors and their safety levels on oxidative stress. These answers can be exploited to design more targeted therapies towards the mechanisms activated by that particular bioactive food component.
- The studies related to impact of oxidative stress and its association in various infections is scarce. More studies are warranted to study the effect of host redox state on the pathogens infectivity and virulence. Also, studying the pathogen antioxidants will open new vistas towards the novel therapeutic approach to target these pathogens exploiting the tool of ROS.

As basic information continues to emerge regarding the role of oxidative stress in disease development and the mechanisms underlying ROS-related cellular toxicity, these findings will lead to more rational antioxidant therapeutic approaches. Moreover, these findings could result in the development of more effective and selective new medications capable of blocking the actions of ROS.

Antioxidants from Lab to Clinic

Individualized Therapies

Numerous epidemiological studies have shown that susceptibility and extent of oxidative stress-mediated damage varies among individuals. These differences are mainly due to activation of different signaling pathways owing to the endogenous antioxidants, nutritional and environmental variations, and drug usage. Antioxidants being an integral part of the complex signaling network make it necessary to develop novel clinical antioxidant therapies based on individual needs. Till date, there is a lack of understanding of antioxidant therapies based on individual variations, which poses serious barriers to the introduction of antioxidant therapies into clinical medicine.

Another variable that adds to this challenge is establishing the decisive clinical biomarkers of inflammation. These issues can be addressed by:

- Performing a comprehensive analysis of individual's antioxidant status. This includes both the enzymatic as well as other prooxidant and antioxidants.
- Genetic variations and traits in certain populations should be monitored, and datasets should be analyzed to predict the susceptibility to oxidative stress and specific molecular markers of oxidative stress.
- As mentioned earlier in humans, ROS production and antioxidant status are affected by numerous nutritional, environmental, and drug influences; there is a need to develop and study better models to study the effects of ROS in response to these variables. This is required to extrapolate the findings humans more effectively.

Therefore, it can be concluded that there is a need to understand the molecular mechanisms by which physical activity, diet, drugs, chemicals, environment influences, and other etiological factors modulate oxidative stress and its pathophysiological outcomes. To achieve these goals, future studies using translational paradigms should be utilized. These novel therapeutic approaches that target oxidative stress may delay the onset or prevent disease progression.

The manufacturer's authorised representative in the EU is Springer Nature Customer Service Centre GmbH, Europaplatz 3, 69115 Heidelberg, Germany. If you have any concerns regarding our products, please contact ProductSafety@springernature.com

Printed and bound by CPI Group (UK) Ltd, Croydon, CR0 4YY
15/07/2026
02167650-0002